Below and overleaf: Eyes of estuarine crocodile *Crocodylus porosus*.

AUSTRALIA'S *dangerous* CREATURES

Text by David Underhill

PRINCIPAL EDITORIAL CONSULTANT
Struan K. Sutherland MD, BS, DSc, FRACP, FRCPA,
**Medical consultant, Commonwealth
Serum Laboratories**

Chief consultants

Joseph Baker OBE, PhD, FRACI
Director, Australian Institute of
Marine Science

Harold Cogger MSc, PhD
Deputy Director, Australian Museum

David Butcher BVSc, MRCVS
Director-designate, Royal Society
for the Prevention of Cruelty
to Animals (NSW)

Charles Kerr MB, BS, DPhil, FRACP, MFCM
Past Director, School of Public Health
and Tropical Medicine

The publishers also gratefully acknowledge the help and
advice given by the following:

Robert Endean, Associate Professor, Department of Zoology,
University of Queensland

Peter Fenner, Mackay

M.R. Gray, Curator of Arachnology, Australian Museum

Gustaaf Hallegraeff, CSIRO Marine Laboratories, Hobart

Merlin Howden, Associate Professor, School of Chemistry,
Macquarie University

Brian Kay, Queensland Institute of Medical Research

John Pearn, Head of Department of Child Health, Royal
Children's Hospital, Brisbane

Ronald V. Southcott, Past Chairman, Museum Board of South Australia

B.F. Stone, CSIRO Division of Tropical Animal Science

Edited and designed by Reader's Digest Services Pty Ltd

PROJECT EDITOR
ALISON PRESSLEY

ART EDITOR
ANITA SATTLER

RESEARCH EDITOR
VERE DODDS

Editorial assistant
Karen Wain

Art assistant
Caroline Goldsmith

Project co-ordinator
Robyn Hudson

PRODUCTION CONTROLLER SEAN SEMLER

FIRST EDITION
Published by Reader's Digest Services Pty Ltd (inc. in NSW),
26-32 Waterloo Street, Surry Hills, NSW 2010

National Library of Australia cataloguing-in-publication data:
Underhill, David, 1937-
AUSTRALIA'S DANGEROUS CREATURES, includes index. ISBN 0 86438 018 6
1. Dangerous animals — Australia — identification.
2. Dangerous plants — Australia — identification.
3. Dangerous fishes — Australia — identification.
I. Reader's Digest Services. II. Title.
591.6'5

AUSTRALIA'S *dangerous* CREATURES

READER'S DIGEST SYDNEY

INTRODUCTION: TO

No kind of creature poses a greater or more constant threat to human wellbeing than we ourselves do. Compared with the harm that people do one another, accidentally or maliciously, the impact of other species is seldom momentous. Yet it is deeply disquieting.

For animals to have power over us is an affront to our notions of environmental mastery. Injuries, envenomations and infections from supposedly inferior beings excite resentment and even hatred. Such emotions often mask a primitive fear – our inheritance from ancient man's struggle to survive in a world inhabited by many more dangerous creatures than exist today.

Australia's Dangerous Creatures, for all the array of terrors that it appears to present, is not intended to frighten people. On the contrary, in exploring the factual nature of risks and how they can be avoided, it seeks to offer security. In resolving doubts and demolishing fallacies and misconceptions, it aims to enhance everyone's enjoyment of the Australian countryside, from backyards to the bush.

This book will have truly done its work if, along with allaying many fears, it can help reduce the actual occurrence of injuries and illnesses related to animals and plants. Quite apart from the pain and distress suffered by the victims, these unfortunate encounters place a heavy burden on medical and paramedical services, and their total cost to the community as a whole is immense. For this reason 'danger' is taken in its broadest sense, to embrace not only those situations that threaten life or limb, but also all of the lesser misfortunes that can incapacitate victims and require treatment.

CONQUER FEAR

A further aim is to assist understanding of the animals themselves, and through that to promote their right to exist. In many cases of conflict, the human victims have a free choice of action – but most often the animals do not. With the exception of some particularly aggressive carriers of disease, our fellow creatures deserve not repugnance but respect. Only a resolve to live and let live can preserve the country's unusual biological richness.

Research for *Australia's Dangerous Creatures* probed every field of natural science. Nowhere can it claim to have the last word, because in no area is knowledge complete. Even while the text was being assembled, events and discoveries jolted scientific opinion. Unsuspected toxicity was found in crabs. A second species of paralysing tick emerged. Concern about spider bites swung from neurotoxic dangers to tissue destruction.

The author, untrained in the sciences, has drawn heavily on the life's work of numerous other people. Whether or not it has been possible to convey acknowledgments by name, he offers to each his gratitude and his admiration of such dedicated skills. Out in the field and in laboratories, the work of these people goes quietly on. Though of vital community concern it is largely unpublicised – and regrettably often hampered by lack of funding. Such research can only result in a safer and more enjoyable relationship with the natural world for all of us.

THE EDITORS

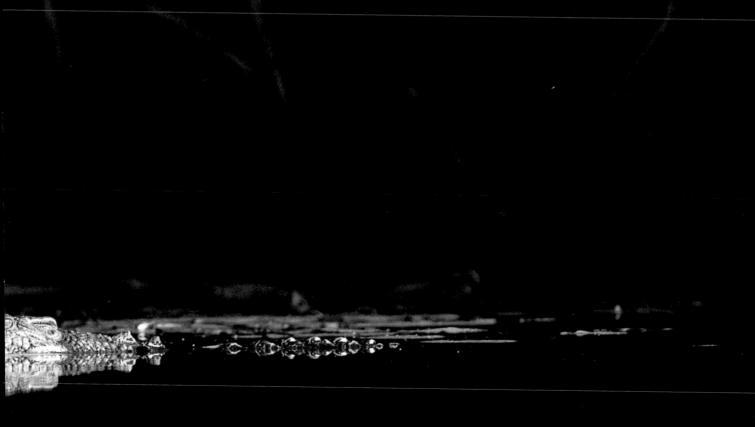

CONTENTS

Estuarine crocodile *Crocodylus porosus*.

Part 1
ANIMALS THAT CAN WOUND

Creatures ranging freely in their natural surroundings would sooner not know us. Usually it is our behaviour that sparks trouble.

'Boomer' kangaroos

Battering, grappling, biting and ripping, an enraged male displays fighting powers that are more than a match for an unarmed man.

Any marsupial bred in the wild is wary of humans, and will normally move away if we approach. But under unusual stress — most often after being harassed by dogs — a kangaroo may attack. The 'boomers', as hunters call mature males of the bigger species, are capable of overpowering and killing unarmed men.

The provocation can be innocent. Kangaroos coming in to feed in areas frequented by people may appear to be tame. But sometimes they are driven there in desperation, when drought afflicts other feeding grounds. A friendly approach — by children wanting to pat them, for example — can be seen as a threat to survival and met with violence.

In rare cases of attack without any provocation at all, naturalists suspect that the animals responsible were former pets that were turned loose when they grew too big to manage. They lack a wild kangaroo's natural fear of humans, and may even regard us as rivals for food or mates.

The total head and body length of a full-grown male of the tallest species, the red or plains kangaroo *Macropus rufus*, may be as great as 1.4 metres. In a resting attitude it does not stand much more than a metre high. But rearing on its long hind legs, with the solid prop of a massive muscular tail, it can tower more than 2 metres.

Grey kangaroos, wallaroos, euros and some wallabies may be nearly as tall, and many are more solidly built than the slender 'big red'. Weights exceeding 80 kg have been recorded. Adult females seldom reach half the size of their mates and weigh only about one-third as much. No females are known to attack. They do not even defend their young because they are always able to bear more, without mating again, through the marsupial phenomenon of delayed embryo implantation.

Fighting is a way of life

For the big males, however, combat is a regular exercise and a means of maintaining social order. In roaming mobs sometimes numbering scores, the leader or 'old man' enforces his domination by cuffing the others with his forepaws. In a manner much like boxing, younger males spar playfully among themselves

Propped on its tail, a male 'big red' slashes at the abdomen of a mating season rival.

AVOIDING TROUBLE WITH KANGAROOS

● Discourage children from trying to pat big kangaroos and wallabies, or even smaller ones if an adult male is nearby.

● Keep dogs under restraint where kangaroos are known to roam. If a dog is attacked, do not attempt to rescue it unless you are equipped to defend yourself.

— and fight fiercely when the time comes to win mates.

Kangaroos and wallabies make dexterous use of their forepaws in holding food and manipulating objects. In a serious fight they also use the forelegs for grappling, to pull an enemy within biting range — or worse still, to hold a victim helpless while a hind-leg toenail is brought into play. Many a hunting dog has been disembowelled in this way.

The mere barking of a dog may excite a 'boomer' to ferocity. The detestation that all kangaroos show for canines is presumably a response developed within the past 10 000 years, when the arrival of dingoes brought the first threat from carnivores fast and powerful enough to kill the bigger marsupials.

Kangaroos and wallabies are able swimmers and take readily to water. Pursued there by dogs, an angry male may go suddenly onto the offensive and succeed in drowning its tormentors. A tale, perhaps fanciful, is told of a kangaroo dealing with a wedgetailed eagle in the same way.

The shooter who claimed to have witnessed the battle, near Molong, NSW, in 1931, said the kangaroo was under repeated aerial attack as it fled for 200 metres. Reaching a dam, it plunged in up to its neck. Next time the eagle swooped it was caught in the kangaroo's clutching forepaws and held under water. Such resourcefulness in a 'boomer' is not surprising, but the eagle's behaviour is hard to explain.

A bite worse than a horse's

While the enlarged main toes on each hind foot are the most fearsome weapons of all kangaroos and wallabies, the bite of a big animal is also powerful. The sharp incisor teeth can do more damage than those of a stallion.

Country people in South Australia before World War I marvelled at a widely exhibited boot, its thick leather completely penetrated by the snapped-off tooth of a kangaroo. A farm boy at Penola had rammed his foot into the animal's mouth to restrain it after his dogs had pulled it down. His injuries put him on crutches for two months.

The only reliably documented killing of a human by a kangaroo in the wild occurred near Hillston, NSW, in 1936. The victim, a hunter who tried to rescue two of his dogs, died of head injuries. Wounds of varying severity have been inflicted in countless other cases in woodlands and grasslands throughout the mainland and Tasmania. Often the outcome could have been far worse, but the victims are aided by companions or they managed to grasp rocks to knock their assailants unconscious.☐

Bounding kangaroos can reach a speed of 55 km/h.

Red kangaroo
Macropus rufus.

Eastern grey kangaroo
Macropus giganteus.

Western grey kangaroo
Macropus fuliginosus.

Wallaroo or euro
Macropus robustus.

A rearing eastern grey dwarfs a man of average height.

Well fed, a western grey could weigh nearly 90 kg.

Dragged from a horse

When Flash the dog bought trouble, his owner collected it.

The morning was peaceful, though bitterly cold, when Lance Oliver rode out into the Victorian high country in July 1963. Oliver, 30, was a stockman on Barragunda station, near Mansfield. His task was to find any newborn lambs and take them to shelter.

On a ridge not far from the homestead, his dog Flash took off into the scrub and flushed out a mob of eastern grey kangaroos. Oliver sat astride his horse to watch the fun. But suddenly the 'old man' of the mob took up the attack and Flash made for safety — right under the horse's legs. The kangaroo paused and reached up. Grasping Oliver by the shoulders, it swept him from his horse and pinned him to the ground. The hind claws went into action on the stockman's chest, ripping through his thick jacket and sweater. Oliver wrestled desperately, getting behind the kangaroo and locking its forelegs in a 'full Nelson' hold. It pushed sideways and the combatants rolled off the ridge, 40 metres down into a creek. Grappling still and feeling his strength ebbing, Oliver feared he might drown. But the creek yielded a boulder that he managed to pound against the back of the roo's neck. The carcass of Oliver's attacker measured 2.35 metres from nose to tail — within about 25 cm of the record for the species.

Nearby at Eildon in April 1985, someone's excited dog provoked a terrifying attack on six-year-old Michael McClatchey. Fishing with his grandfather, all Michael heard was the barking. He turned and the kangaroo was on him, smashing at his face. Michael was slung to the ground 'like a rag doll', according to his grandfather. Though the little boy put up no resistance he was jumped on again and again. Dennis McClatchey, the grandfather, found a rock and beat the roo off. Michael was fortunate to escape with bruises and scratches.

When a grey kangaroo put 84-year-old Vincent Folland in hospital in 1983, he could not blame his dogs. They were trained to hunt — and they probably saved his life. After the three dogs bailed up the kangaroo on his nephew's farm near Williams, WA, Folland thought he saw a chance to 'dong' the animal from behind. He misjudged it. The kangaroo turned on him, bowled him over and jumped on him, striking with its weighty tail as well as its feet. Sticking to their task, the dogs worried at the kangaroo sufficiently for Folland — badly scratched and with many ribs broken — to be able to pull it down by the tail and throw his body across it. Then the dogs went for the throat, immobilising the kangaroo while Folland finished it off with a tree branch.□

Katie Schmidt: savaged for a friendly gesture.

Where the grass is greener

Thirst and hunger can override a wild roo's natural shyness.

Pastures and scrubland around Geraldton, WA, were parched after the hot summer months of 1985. But at Narngulu, on the city's southeastern outskirts, gardens in the grounds of an earthmoving company stood out like an oasis in a desert.

Though it was Easter, the regularly watered lawns sprouted lush, green grass — an inviting setting for a holiday barbecue put on for local families. Their noisy feasting, in the cool of the evening, went on late.

Apparently unperturbed, in near-darkness at about 8 p.m., a big western grey kangaroo

COLLISION COURSE

A bounding roo can wreck a car.

Through no fault of their own, kangaroos pose a special risk for night drivers on outback roads. In arid districts they frequently feed at the verges, on vegetation encouraged by the run-off of dew. Blinded by the headlights of an oncoming vehicle and panicked by the engine noise, a kangaroo may leap the wrong way.

The velocity of a big kangaroo in mid-bound could be 50 km/h. Meeting a fast-moving car, its impact is enough to put the vehicle out of action. Radiator damage alone could have disastrous consequences in inland regions. Unluckier still, the leaping animal may come through the windscreen.

Drivers whose vehicles are not protected by 'roo bars' are advised not to drive at night in the outback. If they must be on the road, they should keep to moderate speeds.

moved in and began to graze. A group of children walked over and Katie Schmidt, 12, reached out to pat the animal. Without warning it lunged and grabbed her by the hair. She was bitten on the head, neck and one ear.

Katie's father dived into the fray. Wayne Schmidt, 40 at the time, is a fit, strong boner at the Geraldton meatworks. But the kangaroo knocked him to the ground with a contemptuous swipe.

'As soon as I could get up he was onto me again, scratching and biting,' Schmidt recalls. 'He gave me no chance at all. But at least he let up on Katie.'

A quick-thinking farmer, armed with a fence post, dispatched the kangaroo. Katie's wounds healed well and the incident no longer worries her. Wayne Schmidt has scars on his arm that he expects to bear for life, and a heightened respect for the strength of kangaroos.

Saved by a 3 iron shot

Five years earlier, about 400 km to the southeast, it was the well-watered greens of a golf course that brought a 'rogue' roo into conflict with townspeople. Completing a round at Wyalkatchem, in the west's dry wheat belt, two women players were attacked.

Mrs Jean Ross and Mrs Heather Hutchinson had been amused when the animal followed them, peacefully enough, as they negotiated the 16th hole. But at the 17th tee it advanced on them, rearing to a height of 2 metres, and starting clawing at their heads.

The players were amply armed with their golf clubs. But with their first frantic swings they could not get in an effective hit. Then Mrs Ross dealt a shrewd blow to the back of the kangaroo's neck with her No. 3 iron, stunning the animal. As the women fled it recovered and bounded away, apparently unhurt.

Street pursuit

Early rising was the undoing of a train driver at Wycheproof, northern Victoria. Up before dawn to begin his shift in the drought-stricken winter of 1982, Daniel Pocock, 59, surprised a red kangaroo in his backyard. It was eating wheat left there for ducks.

Taking no notice, Pocock started his 400-metre walk to the railway station. The kangaroo jumped the fence and followed. After 300 metres it bounded up behind him and delivered a 'king hit', knocking him to the ground.

Fleeing, Pocock was chased and felled twice more, suffering deep scratches and gouges in his face, neck and arms. Railwaymen who came to his rescue were also attacked. But they succeeded in restraining the kangaroo, which was shot by a policeman.

A hammer came in handy

Bruno Schreiber was another early riser. In February 1965 at Gulargambone, northwestern NSW, where temperatures were passing the old Fahrenheit century mark day after day, it was the only way he could get through his work as a building contractor.

Schreiber, 36, was nailing a house roof together at six one morning when two workmates cried out from below. They were under attack from a kangaroo and appealed to him to come down quickly — and to bring his hammer with him.

The animal, matching him in height, turned on him as soon as he reached the ground and grasped him with its forepaws. It attempted to rip at his chest but he brought the hammer into play before any serious injury was suffered. Stunned by a blow to the head, the kangaroo lurched into the house and was killed there.

Regretting the action he was forced to take, Schreiber found no shortage of excuses for the kangaroo's behaviour. Dams and creeks in the district had been bone-dry for three months, he pointed out, and the heat had been enough to drive anything crazy. Worse still, yapping dogs had been on the run earlier that morning.☐

'The kangaroo knocked him to the ground with a contemptuous swipe.'

ONE TOE THAT CAN SLASH LIKE A SABRE
The special formation of a kangaroo's foot.

Kangaroos and wallabies have only four toes on each hind foot, formed for specialised purposes. The innermost two are tiny and fused together, with a split nail that is used as a fur-grooming comb. The outer toe is also small. In between is the great toe, with a long, strengthened nail and a pad reaching from the tip back to the heel of the foot.

The great toes, hooking into the ground, give the bounding marsupials anchorage and leverage for their leaps. Used in fighting, in combination with the immensely powerful muscularity of the hind legs, they can easily slash through the hides of other animals — or the clothing and flesh of human victims.

A kangaroo's great toe serves in bounding or fighting.

The kick of the cassowary

The peaceable giant of the rainforests attacks only in desperation. Then it hurls itself into the air, dealing slashing blows.

Tapering claws like daggers, up to 12 cm long, project from each of a cassowary's two inner toes. They are rarely used. Male birds clash occasionally, leaping into the air and kicking at each other with both feet. The battles are brief and most losers are injured only in their pride.

But when a cassowary cannot escape interference — if it is bailed up by dogs or foolhardy people — it may retaliate with overwhelming force. Wounds from the claws can be terrible. Even if no vital organ is damaged there is a likelihood of massive bleeding. Stiff quills on the shoulders, the flightless bird's vestiges of wings, can also inflict injury though they are not used deliberately in attack.

The biggest cassowaries are females. They weigh up to 60 kg and in an erect posture may reach a height of nearly 2 metres — not counting the blade-shaped, bony outcrop surmounting their heads. Called a casque, this growth is used by a running bird to ram its way through forest undergrowth.

Most flightless birds are drab in appearance, to aid concealment. But adult cassowaries are brightly coloured about the head and neck and carry vivid throat wattles. In the gloom under the closed canopies of rainforests, they can recognise their own kind. Mature birds have no enemies to hide from, other than humans and their dogs.

A shrinking kingdom

Australia's only species, the southern cassowary *Casuarius casuarius*, is fairly common in remnants of dense lowland rainforest north of Townsville, Qld. The same species is found in New Guinea, where two others also occur. One of these, a smaller type, is known from fossil records to have ranged into the Australian hinterland at least as far as central NSW when the climate was moister.

Cassowaries can eat insects, snails, fungi, leaves and sometimes the flesh of dead rats and birds. But they depend heavily on an ample supply of fallen fruits from tropical rainforest trees such as figs and quandongs. Anything that leads to a reduction of mature trees, whether it be climatic change, cyclone damage, logging or the felling of forests for development, threatens their existence.

Shortage of forest foods often brings cassowaries into conflict with fruit growers. The birds invade banana plantations and orchards, and commercial growers sometimes set dogs onto them. Tourist-oriented businesses, on the other hand, encourage the spectacular birds to visit. They are most often seen around Mission Beach, near Tully. Lured from their natural territories by handouts of food, many meet their deaths on busy roads.

Adult cassowaries lead solitary lives except at breeding time. In a forest, a nearby bird can escape the notice of walkers. But it may give away its presence by a low, rumbling noise — its expression of unease at an unfamiliar sight.

An empty threat

If approached quietly a cassowary will draw itself up to its maximum height, erect its feathers and hiss sharply. The display is meant to be intimidating, but it is not followed by any aggression. If the stranger does not move away, the bird will — sometimes conveying its irritation by a noisy stamping.

Cassowaries are fast runners, well able to evade pursuit through dense undergrowth. But their habit of first standing their ground to confront intruders can make them easy targets. Aborigines used to hunt them with dogs, usually finishing the birds off with spears.

A cornered or wounded cassowary is extremely dangerous, however. In New Guinea, where the birds have been more widely hunted, many people are said to have been killed when attempting to close in on their quarry. Australian and American troops serving in New Guinea during World War II were warned to leave the birds alone.

Unprovoked attack has been documented only once, in the early days of Australian administration of Papua. A cassowary terrorised a village where it had been raised in captivity and then allowed to 'go bush'.

Under repeated siege, the villagers were ordered by their chief not to go out alone. But one man went to tend his crops and disappeared. His mother went to search for him. Both were found dead. A magisterial inquiry confirmed that their injuries had been inflicted by a cassowary.□

Guarding eggs and rearing the chicks is the task of the male cassowary.

AVOIDING TROUBLE WITH CASSOWARIES

● Never pursue a cassowary or attempt to corner it.

● Keep dogs under restraint. (In those parts of the cassowary's range that are declared as national parks, dogs are forbidden.)

Turning the tables

When a hunter became the quarry, there was no time to flee.

Two farm boys near Mossman, far northern Queensland, were hotly competitive hunters in 1926. When one shot a feral pig his younger brother, 16, decided to outdo him by bagging a cassowary. He took his dogs out to feeding grounds in forested country nearby.

They soon found a big bird and the yelping pack took up pursuit. Suddenly the cassowary stopped and went onto the attack. The dogs took flight but the boy could not turn quickly enough. One slashing kick ripped open his neck. He managed to retreat about 400 metres towards his home before he collapsed and died. According to his sister, recalling the tragedy 60 years later, he had staggered and crawled 'until he ran out of blood'.□

'He managed to retreat about 400 metres before he collapsed and died.'

The elusive emu

Though rarely known for aggression, our national bird has the physique for it — and a skittish, unpredictable nature.

Generally a little taller than the cassowary though not as heavy, the emu is similarly capable of inflicting grave injuries with its claws. But it is a nomadic bird, conditioned to arid country and fluctuations in food and water supplies. So, unlike the cassowary, it does not defend a feeding territory.

Unless they are held captive, emus are most likely to avoid interference — running at speeds up to 50 km/h over short distances — rather than attempt to resist it. They are, however, inquisitive, not readily frightened, and sometimes inclined to be scatterbrained in their confrontations with people.

An alarming incident occurred at Canberra's Tidbinbilla nature reserve in midwinter 1975. Scott Woods, a toddler aged only 21 months and far too small to present any threat to an emu, was attacked by one of a free-ranging flock of about 40 birds. Hearing him screaming, his mother found him flat on his back with the emu standing on him.

Scott's father, who beat off the bird with a tree branch, said it had been using a ripping action that he likened to that of a hen scratching for worms. The little boy needed hospital attention for a 7 cm gash in his scalp, and also suffered facial scratches and bruising.

The Canberra attack would be inexplicable, but for one fact. Little Scott was playing with a ball. The emu was more than likely a male, and it is possible that it thought the child was interfering with an egg.

Above: A startled emu bolts for cover. The birds are not easily frightened, but near outback roads they can be panicked by motor vehicles. Plunging blindly into fences, many are killed or pitifully injured.

Below: The male parent takes charge of chicks for a year or more. Females in flocks often stand at a distance, on sentry duty while the other birds feed. They can reach a great age — over 40 years has been recorded in captivity.

Males care for young

Male birds take charge of eggs as soon as they are laid, rarely leaving them during the eight weeks they take to incubate. They lose 8 kg or more in weight during this period of near-starvation. When the chicks hatch the father undertakes their constant care, brooding them under his feathers at night when they are small and leading them around by day. The term of guardianship may be as long as 18 months where food is scarce.

Because of their parental duties, male emus are available to breed only every second season. But well-nourished females may mate twice in one season. They are prolific birds, usually laying clutches of about ten eggs but sometimes as many as 20. Emu populations decline severely in conditions of prolonged and widespread drought but recover quickly in better times, in much the same manner as kangaroos.

The emu *Dromaius novaehollandiae*, peculiar to Australia, is probably descended from an early type of cassowary that adapted to life without the shelter and food of rainforests. Now the natural range of emus is everywhere but in forests and absolute deserts. They have been driven away from densely settled areas in the south and became extinct in fairly recent times on Kangaroo Island, SA, and in Tasmania, though introduced birds flourish in some Tasmanian reserves.

Elsewhere emus are generally abundant. They eat herbs, fruit, flowers, seeds and insects — sometimes earning the gratitude of farmers during crop-damaging grasshopper plagues. When more natural foods are hard to find, the birds may graze on pastures and raid crops. In the southeast many agriculturalists are happy to see them feeding among sheep and cattle. But the birds are far from popular with wheatgrowers, especially in Western Australia where they are blamed for heavy crop losses and the destruction of fences.□

Hanging on — the Aussie ostrich

Leftovers from a century-old fashion fad are tenderly protected.

The world's biggest bird, and reputedly the most dangerously pugnacious towards people, is the African ostrich. Heights of nearly 2.5 metres and weights of more than 150 kg have been claimed for some giant individuals. Stories of disembowelling kicks and skull-crushing stampings are rife in South Africa's Cape Province, where ostriches are farmed. The birds are also said to be vigorous defenders of their nests in the wild.

Unknown to many Australians is the fact that ostriches also range in comparative freedom here, on an out-of-the-way cattle station north of Port Augusta, SA. Drought in the early 1980s reduced their numbers almost to vanishing point, from about 200 to 20. Carefully nurtured now, the flock is recovering.

The birds are descendants of a population of 700 that towards the end of last century stocked the biggest ostrich farm in the world. Birds or their eggs were first imported to Australia in the 1870s, to meet a demand for tail plumes to decorate the ornate hats of fashionable women.

Farms were set up in many parts of the country — including one in suburban Sydney, on sandy heights near South Head. Around the

Ostrich plumes: once the height of fashion.

turn of the century, the annual cut of plumes from one ostrich was worth up to ten times as much money as the seasonal wool clip from a sheep. And ostriches live for many decades, regrowing their feathers as readily as sheep grow wool.

But prices collapsed when fashions changed after World War I, and most flocks disappeared. The population near Port Augusta is preserved with the help of national parks and wildlife service officers. They make no pretence that the flock is of natural importance, but see it as a link with the colonial era that is well worth keeping.

Chicks are collected after they hatch and raised in small enclosures to protect them from predators — particularly foxes. As they grow they are gradually given more space until they have hectares to roam in. Adult birds show no resentment at their chicks being taken. In contradiction of the South African experience, they are good-natured and easily managed. Disclosure of their exact location is frowned on — not for public safety, but to ensure the well-being of the birds.□

WHEN THE ARMY WAGED WAR FOR WHEAT
Emu hordes made a mockery of machine guns.

Wheat farmers in Western Australia were struggling for survival in 1932, at the depth of the Great Depression. To add to their woes from the collapse of world markets, hundreds of thousands of drought-starved emus were raiding their crops. The birds had been declared vermin throughout the state since 1922, but single-shot culling by bounty hunters was too slow for the growers. They called for machine guns.

On the personal authority of the federal minister of defence, the army obliged with two Lewis gun teams. Each gun could fire 500 rounds a minute. With costs shared between the state and a wheatgrowers' group, the party trekked to the Lake Campion district, northeast of Merredin.

Instead of masses of emus milling about against a rabbit fence, ready to be mown down, the soldiers saw only scattered groups with alert sentinel females ready to raise the alarm at their approach. Two thousand rounds were expended and fewer than 30 birds were killed. Meanwhile farmers trying to herd a flock into range lost control of their truck and wrecked 12 metres of the rabbit fence.

The 'Great Emu War' became a public farce and a political embarrassment in Canberra as well as Perth. The machine gun detachment was recalled, but not before ambush tactics had been perfected. The emus' reprieve was shortlived.

The slaughter of tens of thousands of birds was achieved every year in the following decade, first by bounty hunters and then by farmers using state-supplied ammunition. Emus are now protected in the southwestern corner of the state, but not in the semi-arid wheatgrowing belt.

An ostrich flock in South Australia just before World War I.

Make way for a wombat

The 'weary Willy' of children's tales is a powerhouse of fighting energy, not afraid to tackle anything in its path.

Of all marsupials in the wild, wombats show the least fear of humans. Sometimes seen by day, though they are mainly nocturnal foragers, they amble about placidly. Some permit themselves to be touched, and they are readily tamed. But now and then a wombat gives a devastating display of ill temper. Little provocation is needed — just standing in its way can be enough.

In his book *Encounters With Australian Animals*, the naturalist Harry Frauca describes an attempt to take a close-up photograph of a wombat wandering in a farm paddock. Growling, it rushed at his legs, knocked him backwards and attacked again, sinking its teeth into the lower part of a leg.

Through a rubber boot, trousers and a thick woollen sock, a chunk of flesh the size of a 5 cent coin and more than 2 cm deep was bitten out. In spite of immediate first aid the wound became seriously infected and Frauca was incapacitated for more than a week. Terming the attack 'nothing out of the ordinary', he rated mature wombats as the most dangerous of vegetarian marsupials — worse than kangaroos, in other words.

Tree-cutting teeth

Wombats have developed powerful jaws and sharp, continuously growing teeth not because of their diet — which consists largely of grass — but to aid them in burrowing. They prefer to make their tunnel entrances at the bases of trees, and often need to bite through obstructing roots. Naturally enough, an animal capable of burrowing metres in an hour also has strong claws and is massively muscled. Big males, well over a metre long, can weigh more than 30 kg.

Dry conditions suit a hairy-nosed wombat.

INTRUDERS CAN BE CRUSHED
Force on a wombat is met with a muscular heave.

Wombats frequently have to cope with collapses of soil in their burrows. If they feel sudden pressure they immediately brace their legs, arch their backs and heave upwards. If the pressure happens to come from another animal invading a burrow, it will be crushed. Many a pet dog has failed to return from a foray into a wombat's burrow, or has been dug out with broken legs and ribs.

Much of our knowledge of the nature and extent of wombat burrows has been gained by adventurous children, small enough to wriggle into them. They run a considerable risk of suffocation if a tunnel collapses, or of injury if a mature wombat is encountered. An old story from Victoria was vouched for by no less an authority than the geologist and Antarctic explorer Sir Douglas Mawson. A boy was crushed to death at the very entrance of a burrow when the disturbed occupant rushed out underneath him.

A wombat's bite almost cost a nature-lover from Westleigh, Sydney, his manhood in June 1986. Picnicking with his family in Dharug National Park, beside the Hawkesbury River, he noticed the animal grazing nearby. It seemed unconcerned at his approach so he squatted down, legs apart, to take a closer look. The wombat charged, bowling him over. Flesh was torn from his thigh, high up on the inner side.

The common or hill wombat *Vombatus ursinus* inhabits open forests, but not usually rainforests, in the southeast. It abounds along the southern part of the Great Dividing Range, from central NSW into Victoria. The fur, coarse and stiff, varies in colour from buff to nearly black in different locations. The muzzle is bare.

Two species of hairy-nosed wombats — sometimes called plains or desert wombats — live in semi-arid country. They are grey, with soft, silky fur. The southern hairy-nosed wombat *Lasiorhinus latifrons* has its strongholds in scrublands west of the Murray River, reaching into the saltbush wilderness of the Nullarbor Plain, with some scattered communities in southwestern WA.

Endangered species

The northern hairy-nosed wombat *Lasiorhinus kreffti*, officially declared an endangered species, has disappeared from most of its eastern inland range. The only known population occupies

Common wombat
Vombatus ursinus.

Southern hairy-nosed
Lasiorhinus latifrons.

Northern hairy-nosed
Lasiorhinus kreffti.

aggression. But if a rash attempt is made to handle a possum or a koala, the claws can inflict severe injury.

The biggest and most abundant possum species, the brush-tailed *Trichosurus vulpecula*, has the size and strength of a domestic cat. In urban areas it often becomes a scavenger, and can cause a nuisance on roofs or in ceiling spaces. The removal of possums is a job best left to expert trappers, whose responsibility it is to take them to more suitable locations. All possums are legally protected.

The koala *Phascolarctus cinereus* avoids human habitations and spends most of its time high in eucalypt trees, where it usually escapes notice. But although koalas are night-time feeders they sometimes move from tree to tree during the day and can be seen on bushland roads. They may bury their long claws in the flesh of someone trying to pick them up — not from malice but in panic, clinging as they would to a tree trunk.

Mature males can reach a head-and-body length of 80 cm and weigh 10 kg. Even professional handlers find animals of such size difficult to manage, and cannot avoid a scratching. In 1982 hundreds of koalas from overpopulated French Island, in Melbourne's Westernport Bay, were moved to various parts of the Victorian mainland. Twelve wildlife service officers, assigned to capture them, check their health and tag them, all suffered their share of injuries.

Picking up a koala clumsily, or trying to pull one from a tree, may not only invite a scratching but injure the animal. While the limbs are powerfully muscled the body is comparatively delicate. Force applied to a koala, especially around its waist, can damage internal organs.□

100 or so burrows on a remote cattle station northwest of Clermont, central Qld. An area of 2600 hectares, incongruously named Epping Forest National Park, is fenced off to protect the community's habitat. Public entry is forbidden.

Wombats are the longest-living marsupials, sometimes surviving for more than 20 years. But they are not prolific breeders. Although females have two teats in their nursing pouches, they usually give birth to only one offspring each year. The young stay for up to six months in their pouches, which open at the back to prevent them from filling with soil when the mothers are digging.

Solitary in their habits except when they are breeding in winter, wombats dig individual burrows consisting of many tunnels. Those of the common wombat, each up to 20 metres long, diverge from a single entrance. The hairy-nosed types build interconnecting tunnels. Where the ground is too hard to dig — in rock-capped areas of Tasmania, for example — some wombats take advantage of hollow logs or natural caves to make nesting chambers.□

A common wombat skirts the well-concealed entrance to its burrow. Its sharp teeth cut easily through the shrub and tree roots it encounters in digging.

The anti-social climbers

Keep clear of the claws of possums and untamed koalas.

For all their endearing looks and sleepy de-meanour, our tree-dwelling marsupials are wild animals — not cuddly toys. Their strong, sharp claws are intended for climbing rather than

Above: A common brush-tailed possum Trichosurus vulpecula. *If a possum becomes a pest, enlist expert help to remove it.*

Left: In fright, a koala may cling to a person as it would to a tree.

Meat-eating marsupials

Quolls—often called native cats — are resourceful, bloodthirsty predators. Cornered, they fight ferociously for their liberty.

Fierce and fast, a quoll at bay is not afraid to leap at its challenger — even something as big as a man. Long canine teeth, developed for making quick kills through the necks of prey animals, can inflict serious wounds, as can the quoll's sharp tree-climbing claws.

The biggest quoll, *Dasyurus maculatus*, is more than a match for true cats and can fight off hunting dogs. Among early colonists impressed by its ferocity was the founding governor, Captain Arthur Phillip. He named the animal a spotted marten, likening it to the predatory mustelids of northern hemisphere pine forests.

Settlers called it a polecat or tiger cat. Modern zoologists prefer the term tiger quoll. Along with three other species of quoll or 'native cat', smaller but just as combative, it is closely related to the Tasmanian devil and the thylacine — the vanished Tasmanian tiger or wolf.

MARSUPIAL WOUNDS

Even minor bites and scratches from marsupials, including pets, carry a high risk of serious infection. Prompt medical attention should be sought for all such wounds.

Right: The tiger quoll is commonest in Tasmania. Its full distribution is mapped above.

An eastern quoll brings home its kill.

Raider of poultry flocks

Poultry owners persecuted the tiger quoll when it took to adding their birds and eggs to its natural diet. Its habitats shrank with the clearance of forests and woodlands for farming. But it survives in remaining eastern forests, sparsely from northern Queensland to Victoria and more commonly in Tasmania.

Tasmania is also the stronghold of the eastern quoll *Dasyurus viverrinus*, which rarely achieves a head-and-body length of more than 40 cm, compared with up to 65 cm for tiger quolls. A few eastern quolls may survive in remote parts of their former mainland range, from northern NSW to SA.

The western quoll or chuditch *Dasyurus geoffroii*, about the same size as the eastern quoll, now seems restricted to forests in south-western WA. The northern quoll or satanellus *Dasyurus hallucatus* ranges through well-wooded areas from northwestern WA to Queensland, as far south as Gympie. Less than 30 cm long, it may visit rural buildings.

Practising for the hunt

Baby quolls in litters of three or four start biting, wrestling and pouncing on one another as soon as their eyes open. Their pugnacious activities perfect the co-ordination and skills necessary for hunting and killing. At three months they bare their teeth and hiss defiance at any intruder in their den. Fighting becomes a constant part of their lives — between brothers and sisters, between maturing males, and often between males and females before mating.

Weaned at about four months, quolls are immediately capable of independent hunting, apparently without any parental training. They quickly become versatile, opportunistic predators, capable of killing anything up to their own size. The carrion of bigger animals is also eaten. Quolls are not exclusively carnivorous, often making a seasonal switch to soft fruits. But in the main their diet consists of small mammals, insects, reptiles, amphibians and birds and their eggs. Most feeding is at night. Agile climbers, quolls leap on roosting birds from higher branches, sweeping them to the ground and killing them in the fall.

Dens are made in hollow trees or logs, rock crevices and caves and sometimes under buildings. Eastern quolls in Tasmania dig burrows. When females are rearing their young they occupy only one den, with a grass-lined nesting chamber. At all other times, both sexes frequently change dens, travelling for up to a kilometre between resting places and temporarily sharing them.☐

Below: A diminutive northern quoll snarls defiance from its hollow-log den in a Queensland forest.

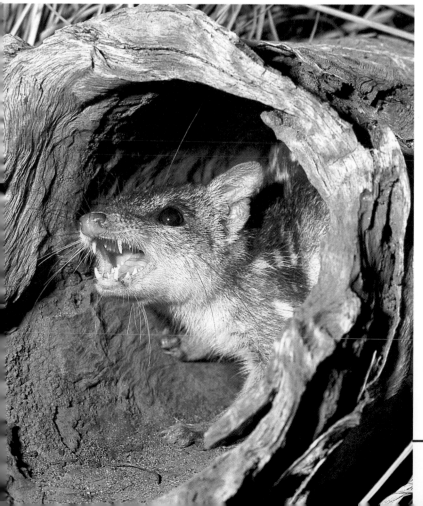

Ritual ferocity: the Tasmanian devil.

The devil got a bad deal

A noisy scavenger's undeserved reputation.

Unnerved by the night-time screams of Australia's only pitch-black marsupial, early settlers in Tasmania nicknamed it a devil — then manufactured a suitably hateful reputation for it.

This animal, the size of a stoutly built lapdog such as a corgi, was claimed by farmers to be a ruthless predator of livestock, seizing and pulling down cattle and sheep. And it was said to launch savage attacks on people trying to protect their stock. Modern knowledge indicates that neither claim could have been true.

The Tasmanian devil *Sarcophilus harrisii* is certainly a carnivore. But it is not a hunter like its relatives the quolls. It is the Australian equivalent of a hyaena — a scavenger almost entirely dependent on the carrion of dead animals. Restricted to a running pace of about 12 km/h, it is too slow to attack bigger animals, let alone strong enough to pull them down. Young domestic poultry can be subject to predation by devils, however. And bigger livestock, dying of other causes or immobilised for some reason, may be attacked. Farmers can obtain permits to shoot devils — normally protected — at lambing time.

In the main devils are a benefit to agriculture. By eating carrion lying in dense forests they remove inaccessible blowfly breeding sites and reduce the likelihood of sheep being dangerously infested. And if they eat dead sheep they help break the chain of such infections as sheep measles and hydatid cysts.

As for the possibility of attacks on people, the only recorded occurrences came about while devils were being captured or held in captivity. Devils are not normally aggressive.

In competition for food they scream at one another with a noise like circular saws and adopt threatening postures, but seldom fight. This ritual display makes them easy meat if they are confronted by a dog.

While the sharp teeth and bone-crushing jaws of a devil could easily amputate someone's fingers, there is no risk of its happening in the wild unless an animal is directly interfered with. Strictly nocturnal, devils are shy. They normally avoid open spaces.☐

NEW TIGERS FOR OLD

Thylacines — Tasmanian 'tigers' — were never abundant. Even spared a campaign of extermination by settlers, they were probably close to extinction.

Debate continues over the possible survival of a few, here and there in remote fastnesses.

Meanwhile, all over the arid heart of the mainland, an animal with more genuinely tigerish qualities has gained supremacy. Descended from domestic cats, it has bred by natural selection to become far bigger and stronger. Trained men dare not tangle with it.

TASMANIA'S HATED THYLACINE

Old-time tales of the savagery of the thylacine were undoubtedly overstated. It was not in farmers' interests to allow room for sympathy for an animal that they wanted exterminated, with the help of bounties of public money.

There is no doubt, however, that the doglike *Thylacinus cynocephalus*, rangy and more than a metre long, had the strength and stamina to run down kangaroos and wallabies and tear their throats out. Sheep stood no chance.

Stories of thylacines turning on people cannot be entirely discounted, though it is likely that the animals involved were cornered. From all accounts those kept in zoos were consistently ill-tempered towards their keepers.

Thousands of years ago the thylacine — commonly called a Tasmanian tiger or, more appropriately, a marsupial wolf — was widespread over continental Australia, though apparently not numerous. Its disappearance from the mainland is sometimes put down to competition from dingoes.

But the earlier arrival of Aboriginal hunters, using weapons and fire in their quest for the same prey, probably tipped the balance against thylacines. A species in decline, its way of life continually disrupted, could easily have been finished off by disease.

Extinction may also have been inevitable in Tasmania. During the period of extermination, which had its peak just before and after the turn of the century, it is unlikely that more than 5000 thylacines were killed. But that seems to have been virtually all there were. Most at home on grassy plains and in open woodlands, they did not have substantial reservoirs of breeding stock hidden in the untouched forests.

Hunted in pairs

Pitifully little scientific observation was made of the biggest carnivorous marsupial while there was the chance. Mated pairs are believed to have hunted together, mainly at night, with their young sometimes running with them. Though not particularly fast, and incapable of bounding like their prey, they pursued it to exhaustion. Blood-rich tissues from the throat, nose, liver and kidneys were eaten.

Some breeding may have taken place

A sheep farmer at Mawbanna, near the north-western tip of Tasmania, poses proudly with his kill and the dog that ran it down. Taken in 1930, this is the last known, undisputed photograph of a wild thylacine. On the coast a few kilometres away in 1966, zoologists examining a ship's boiler from an old wreck found what they declared to be traces of a female thylacine and her pups having used it as a lair. This district of Tasmania is the one most frequently in the news when subsequent sightings have been claimed.

throughout the year, but the peak of the mating season was midsummer. The young, born in litters of up to four, were nursed in the mother's pouch for about five months and stayed with the parents until the next breeding season. Dens were made in caves or under rock overhangs, in hollow trees or below dense shrubbery.

The last shooting of a positively identified thylacine occurred in 1930, and the last captive animal died in 1933 at Hobart Zoo. Sightings since then have been claimed on many occasions and in many localities, but material evidence of the species' survival is scanty. The Tasmanian National Parks and Wildlife Service gave credence to a 1982 sighting by one of its officers in the northwest. But a thorough search in the district proved fruitless.

Most authorities take the view that if thylacines had continued to breed, they would have enjoyed an abundance of food and should have shown a resurgence in numbers. Yet not even the remains of recently dead thylacines have been found since the 1930s. Even allowing the possibility that some still exist, the species is obviously so rare that it is doomed without special care and managed breeding.□

DESERT SUPERCATS NEED NO WATER

To rangers dedicated to conserving wildlife in the harsh conditions of Central Australia, the feral cat is public enemy No. 1. Far bigger and more ferocious than its domesticated ancestors, it is the scourge of all other animals in its territory. On the rare occasions that a cat is lured into a trap, there is no other course but to shoot it. The rangers know better than to get within striking distance of its teeth and claws.

The first cats to stray into the Australian bush may have come from Malay vessels visiting the far northern coast. More came on every European ship. But an explosion of the feral population occurred after the 1880s, when a misguided attempt was made to use cats to control rabbits. Thousands were released on inland grazing properties. Since then, the irresponsible dumping of unwanted house cats or kittens has contributed more every year.

Cats that survive in the wild become extraordinarily efficient and versatile hunters — the most successful land predators in this country. They kill mammals up to their own size, birds galore, lizards, snakes, frogs, fish and insects. And they are prolific. Well-fed females, breeding in spring and again in late summer, may produce more than ten offspring a year.

Mixing of the various domestic breeds brings about a reversion towards wildcat origins. Whatever the appearance of the tame forebears, after a few generations almost all feral cats are striped tabbies. Forest cats tend to be greyish while plains cats are usually ginger — the hues that are least conspicuous in their respective habitats. The variation occurs because unsuitably coloured kittens are more likely to fall victim to eagles, foxes and dingoes.

Feral cats occupy every type of habitat, from snow country to coastal heaths and arid wastes. Usually solitary, each patrols its own hunting territory at night, resting by day in a cave, a hollow log or an unoccupied burrow.

Survival of the fittest

Cats living in forests, where food supplies are most ample, grow fatter than their domestic counterparts but are otherwise not much bigger.

A feral cat makes its home high in a river red gum, finding ideal cover in a natural cleft.

This large ginger feral cat (above) was shot by a National Parks and Wildlife ranger after it had left a trail of destruction of native wildlife through a National Park. Desert cats are typically ginger after a few generations, to blend in with red soils; highland cats (left) are typically grey.

In arid conditions, however, the scarcity of food brings about a process of rigorous natural selection. The biggest, strongest cats take the largest territories and most of the available prey. And the most powerful toms seize all of the breeding opportunities. As a result each new generation tends to grow even bigger.

'Supercats' occasionally seen in the Simpson and Gibson Deserts grow to nearly a metre in head-and-body length. They stand about twice as tall as domestic cats and are many times heavier. Some monsters tip the scales at 15 kg — and most of that is muscle. Such animals are capable of pulling down well-grown wallabies.

Drought is no handicap to cats. Provided that they eat enough fresh-killed prey, they have no need of water. The blood and other body fluids of their victims provide moisture enough. Cats have been seen as far as 150 kilometres from any known source of water.

Feral cats are extremely wary of humans, and the chance of an unprovoked attack is remote. Risks arise from attempts to take cats alive from traps, or in accidental encounters when cats are cornered in their dens. Perth Museum used to display the stuffed pelt of a giant cat that was said to have injured two children.

Expensive campaigns of trapping and shooting have little impact on feral cat populations in the outback, and poisoned baits cannot be dropped without endangering other animals. The only hope of control seems to lie in the release of more cats — infected with feline enteritis to start deadly epidemics.□

The big goannas

Their awesome challenge is usually a bluff — but don't count on it. And take care you're not mistaken for a tree!

Among Australia's 20 species of goannas — lizards of the genus *Varanus* that are usually called monitors elsewhere — are three giants. In size and bulk they are exceeded only by their close relatives, the rare 'dragon' of the Indonesian island of Komodo.

No goanna or other lizard is inclined to be aggressive towards humans. Their first instinct is to take refuge. But if none is at hand they may 'freeze', flattening themselves to escape detection. This behaviour sometimes tempts curious people into potentially dangerous confrontations. No attempt should be made to pick up a big goanna — or to call its bluff if it suddenly stages a dramatic show of anger.

The biggest goanna is the perentie *Varanus giganteus*. It commonly exceeds 1.5 metres in length, more than half of which is taken up by its muscular tail. Like all reptiles it keeps growing throughout its life, and some specimens have measured more than 2.5 metres.

Ground-dwellers, perenties occupy an arid belt across Central Australia from the west coast to far western Queensland. They shelter in burrows or crevices in rock outcrops, and are rarely seen in their home territories. They are more conspicuous, waddling with a slow, swaying gait, when they roam over deserts and salt pans in search of food.

Kangaroo killer

Perenties forage for the carrion of dead animals and prey on other lizards, snakes, insects, low-nesting birds and their eggs, and mammals. Most mammalian prey is small and swallowed whole, but a perentie has been observed catching and killing a young kangaroo. It pinned the carcass with its forefeet, in the manner of a dog, and bit off chunks of flesh.

A tree-dweller of the east, the lace monitor *Varanus varius*, grows almost as big. The average length is about 1.5 metres but animals exceeding 2 metres have been found. The lace monitor is the heaviest goanna, weighing as much as 25 kg. Its diet is as diverse as that of the perentie, but it preys more frequently on nestling birds and eggs.

If disturbed while foraging on the ground, the lace monitor makes immediately for a tree. It keeps the trunk between itself and the intruder — so it is possible to walk all around a tree and not know the huge creature is on it, unless the scrabbling of claws is heard. Lace monitors are most numerous on the Great Dividing Range but are not restricted to forests. They may also be found on the coastal strip and in well-wooded country on the adjoining inland plains.

The third of the giant goannas is the sand monitor *Varanus gouldii*, also called the ground goanna or Gould's goanna. It is the most widespread species, adapted to grasslands and sandy deserts as well as to open forests and woodlands. If shelter in hollow logs or piles of leaf litter is not available, it digs a burrow or takes over the burrow of another animal. The species can exceed 1.5 metres in length.

WOUNDS FROM LIZARDS

Bites and scratches inflicted by goannas and other lizards carry an extreme risk of tetanus, or of septicaemia from other bacterial infections. Most lizards are carrion-eaters, with particles of rotten flesh often clinging to their teeth and claws.

Any wounds should be disinfected as quickly as possible and prompt medical attention should be sought — especially if the injured person has not had an antitetanus injection in recent years.

Note: The frequency of infections, in pioneering days before the nature of bacteria was understood, gave rise to a bush myth that goanna bites never healed. Many people believed that the reptiles must introduce some sort of tissue-destroying venom. In fact no Australian lizard is venomous.

The lace monitor or tree goanna is an exceptionally skilful climber, though the heaviest of the group. It negotiates slender trunks and delicate branches with ease in its search for the eggs and chicks of birds.

Threatening posture

All goannas, if forced to move at top speed, will run for short distances on their hind legs. Walking in scrub or long grass, they may pause and prop themselves up on their tails to get a better view of the terrain. And if they are cornered, they take this upright stance to show their readiness to fight.

The sudden rearing of any of the big goannas, perhaps to the height of a man's thighs, is a frightening experience. The animal sucks in air to puff up its body, hisses fiercely and thrusts its heavy, wedge-shaped head at its adversary. The forked tongue is held out, quivering.

Such a threat is generally regarded as a bluff. Most authorities doubt whether even a perentie would launch an attack on a human unless an attempt is made to grasp it. But few people stay around to find out. Aboriginal tribes who are traditional hunters of the perentie are said to regard it as highly dangerous.

A frightened goanna may also lash out with its tail, in the same manner as a crocodile. Adults have been knocked off their feet and badly bruised in encounters with perenties, and dogs' legs have been broken. Sometimes it is not the person interfering with a goanna who gets hurt, but a quiet bystander. In its instinctive urge to make for a tree, the panic-stricken reptile may mistake a stationary human for a place of refuge. The result will be an onslaught of desperate clawing. An Aboriginal woman near Alice Springs, badly scratched all over her body by a perentie, shrugged off the incident as a commonplace hazard among her people.

The lesser lizards

While the biggest goannas are most to be feared because of the size of their teeth and claws, they are far from being the only lizards to show ferocity in self-defence. Significant bites and scratches can be inflicted by many other *Varanus* species if they are interfered with. Even the common frill-necked, jew and blue-tongued lizards, from different groups and much smaller, can be formidable biters.□

A lace monitor scavenges scraps in eastern Victoria.

The power of pythons

*Though not venomous, their bites go deep.
And one northern type may grow
big enough to squeeze the life out of people.*

An amethystine python, shot in high country halfway across the base of Cape York Peninsula in 1948, is said to have measured 28 feet — more than 8.5 metres. That is a staggering length for any snake, placing the species among the biggest in the world. Even the notorious boa constrictor of Central and South America does not exceed about 6 metres.

A more usual length for the amethystine, scrub or rock python in Australia is about 3.5 metres. A python of such size, probably weighing more than 20 kg, is formidable enough. As well as exerting immense power in its constricting coils, it has a huge gape and long teeth that can inflict severe bites.

Pythons and their relatives the boas — including the biggest snake of all, the 10-metre anaconda of South America — do not crush prey animals. They simply squeeze the breath out of their victims to subdue them for eating, alive or dead.

Prey is usually gripped first by biting. Then coils are thrown around it and slowly tightened by muscular contraction. To achieve this the snakes do not, as is sometimes supposed, need any anchorage for their tails.

Mammals can be strangled by a coil around the neck, or gradually suffocated by constriction of the rib cage. With each exhalation of breath the pressure is slightly increased. Eventually the victim can no longer draw in enough air and loses consciousness. If the pressure is continued

Least colourful of the family, the olive python easily escapes notice.

AVOIDING TROUBLE WITH PYTHONS

Rules to cut the risk of venomous snakebites (pages 123-124) apply also to pythons.

In addition, bushwalkers should bear in mind that some pythons often bask on tree branches, where they are less able to detect the ground vibrations of people approaching. Special care should be taken if trees are used as handholds in negotiating steep slopes.

Prompt medical attention should be sought for python bites, even if the tissue damage appears to be slight. The chances of tetanus or other infection are high.

the heart stops. When victims cease to struggle they are swallowed. Scores of long, back-curving teeth are used not for chewing but to draw the snake's head around its prey. Extreme elasticity in the skin and the jaw connections of pythons allows them to engulf animals of improbable size, including kangaroos in Australia and tigers in India.

Children swallowed

Human victims have been claimed in many overseas reports. While all published accounts have been second-hand and few can be substantiated, there is no reason to doubt that some children at least have been eaten in India, Africa and tropical Asia.

In theory adults too could be overpowered — though it is hardly likely that they would be eaten — by the biggest pythons. A man taken by surprise, and perhaps stunned after being thrown to the ground, could find both his arms pinned to his sides. He would not be able to fight off further coils around his chest and neck.

Pythons do not hunt people. Like other snakes they do their utmost to stay out of our way. Most of the hair-raising incidents reported from overseas have occurred in places where huge pythons live alongside heavily populated urban areas.

That is not the case in Australia. The biggest pythons are rarely found in well-frequented districts. And in the bush the chance of encountering one of freakish, life-threatening size, in a situation that gives it no option but to attack, is far too slight to mar anyone's prospect of enjoying the outdoors.

Smaller pythons, incapable of immobilising people, still carry some risk and should not be interfered with. Although their bites are not venomous, the wounds that they cause are likely to be contaminated with bacteria. The needle-sharp teeth, arranged in double rows, are many times longer than the fangs of a venomous snake of comparable size. Wounds are deep and difficult to disinfect.

The amethystine python, the giant of Cape York Peninsula.

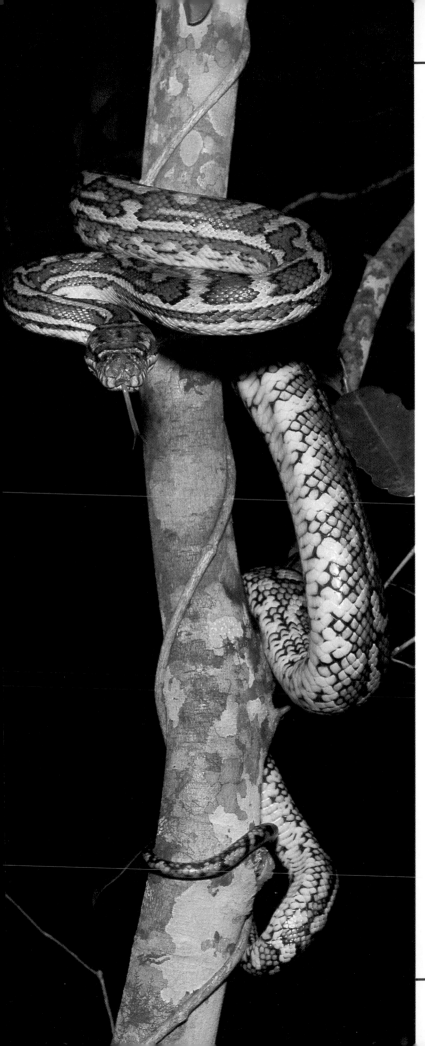

Venomous snakes can be rendered helpless if they are short enough to be lifted entirely off the ground by their tails and held at arm's length. But a python can twist upwards and bite at the point of restraint, or suddenly coil up its attacker's arm or leg. Professional handlers catch pythons by gripping them just behind their heads. But even the experts, if they are not quick enough in bagging bigger specimens, may find themselves in trouble with coils around their necks. Then they are forced to release their grasp, usually incurring a bite.

The big three

The amethystine python *Liasis amethystinus* seems to be restricted to the eastern half of Cape York Peninsula and some of the Torres Strait Islands. But its habitats are surprisingly varied. A ground-dweller that is most often seen basking on rocks, it has been found in rainforest gorges, dry, grassy woodlands, monsoonal vineforests and the scrub of coral cays.

Next biggest of Australia's ten species is the olive python *Liasis olivaceus*. It normally averages about 2.5 metres in length but specimens up to 4 metres have been found. Preferring rocky ridges in woodlands or vineforests, it has a range across the north from near Carnarvon, WA, to Cape York.

The large pythons that Australians are most familiar with are diamond and carpet snakes. These are classed as variations of one species, *Morelia spilotes*. Occupying every conceivable habitat from rainforests to deserts, they occur throughout the country except in Tasmania and parts of Victoria and South Australia. Two metres is a good size for them, though lengths of up to 4 metres have been recorded.

Diamond and carpet snakes are the only pythons likely to enter buildings, foraging at night for rats and mice. If touched when their retreat is cut off they may bite savagely. But they can be conditioned to handling. Before they were accorded legal protection, many were captured and sold to farmers and warehouse owners as rodent controllers.□

The diamond snake (above) and carpet snake (left) are variations of the same widespread python species.

Pigs that went wild

Hardened by generations of breeding in the outback, feral pigs are not only lean but also mean. Hunters are most at risk.

Domestic pigs have been escaping into the Australian bush ever since the founding landholders of the New South Wales colony were allotted their share of First Fleet stock. Others came in from the north in pioneering days, from Timor, New Guinea or the Pacific Islands, and were abandoned after short-lived attempts to establish outpost ports along the tropical coastline.

Pig breeding then was not the refined genetic science it is today. The animals were not far removed from their ancestors, European wild pigs. Their descendants, interbreeding and mixing the primitive agricultural strains, took on a look very like those forebears of the European forests — all head and shoulders with small

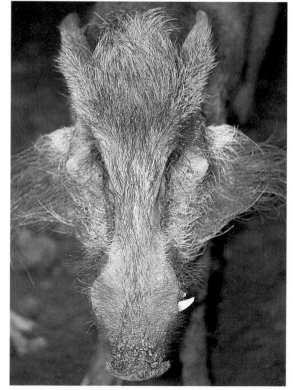

Feral pigs emerge from shelter towards sunset, making for water. After drinking copiously they wallow in mud, often fouling the water supplies that agricultural stock rely on. The rest of the night is spent foraging, resting, then foraging again around sunrise. From 18 months of age they are mainly solitary, joining groups only to breed. Mating can occur at any time of year, with sows bearing two litters a year — each of up to 10 piglets if feeding conditions are favourable.

Left: Sharp tusks grow out and up from the lower jaw.

rumps and hind legs. Narrow along their backs, and usually carrying a bristling mane, feral pigs of the old stock were called razorbacks or ridgebacks. They became notorious for their aggression if disturbed by settlers. Very few genuine razorbacks remain in settled areas. They have been shot out or have bred with escaped animals of more modern strains.

The tens of thousands of feral pigs roaming the outback now have forms that do not differ much from those of farmyard stock. But they are noticeably leaner, with narrower backs and more heavily developed shoulders and necks. Mature boars grow a horny plate of cartilage under the skin around their shoulders, to protect their ribs when they fight.

Colours vary but most feral pigs are black, reddish black or black and white. Their bristles are long and coarse and may form a mane from the head along the back. The tail is tufted.

Tusks keep on growing

A pair of lower canine teeth, triangular in section and exceptionally strong, curve out, up and back from the rear of the mouths of adults. These tusks, sharply pointed, continue to grow throughout a pig's life, which in the wild may be as long as 12 years. The older an animal gets, the more dangerous it becomes.

Weights vary markedly with the availability of food, as does the ability of females to breed. Among modern feral pigs, boars reach about 135 kg and sows 70 kg. The remaining razorbacks attain weights only half as great.

Major populations are spread from inland NSW through nearly all of Queensland and much of the Northern Territory to the Kimberleys in WA. The pigs must have regular access to water with undisturbed shelter nearby. They can eat almost anything, but show a strong preference for agricultural crops. By their treading and wallowing alone, they damage pastures and degrade water sources.

Also eaten are carrion and live animals including reptiles, small birds and wild mammals, and newborn lambs. In 1954, on the northwestern outskirts of Sydney, the 61-year-old caretaker of a poultry farm was savaged by pigs after he collapsed from a heart attack. He was found dead with part of his head and one hand gnawed away.

When pigs attack

Pigs — usually boars — may charge when people unwittingly interfere with a food supply, or when piglets are hidden in a nearby nest. But most attacks are provoked by hunters when they corner big animals. Shooting or roping feral boars is a dangerous pastime for amateurs, though some experts have made pig hunting their fulltime profession.

Feral pigmeat sales are prohibited in Australia because of disease risks. Millions of dollars a year are earned from frozen exports, however, mainly to West Germany. The strong-tasting meat, soaked in wine and heavily sauced, passes for the prized flesh of genuine wild pig.

Farmers and nature conservationists have reservations about commercial and sporting hunting because they tend to disperse pig populations rather than destroy them. Understandably, some hunters are not anxious to kill sows or piglets. Agricultural authorities prefer the institution of planned shooting programmes or the use of poisoned baits.□

Above: Waterfowl nesting grounds near Katherine, NT, are ruined by trampling pigs.

Below: Hunting by night near Moree, NSW, shooter Leo Schultz makes his living from selling feral pig carcasses to export butchers.

A hero at nine

A little boy took on a raging boar to save his father's life.

Farmer Alf Collins and Alf junior, with the medals and trophy that the Queen presented for bravery.

After completing his quota of correspondence school lessons, nine-year-old Alf Collins liked nothing better than helping his father. There was always work to be found on their beef-raising property near Marlborough, central Queensland. But one afternoon late in September 1984 was special. They were going to try to shoot pig. Alf Collins senior, 36, had noticed a big, black boar the previous day, rooting up spoil from a grading job. It was new to the district and he had no intention of letting it start damaging his dams and fouling water supplies. Driving his truck back to the same area, he had a rifle with him now.

Collins had a hunch that the boar would have been attracted to the carcass of a recently dead cow nearby. Leaving three younger daughters in the truck, he led Alf junior on foot through a patch of thick scrub. Sure enough, the carcass had been mutilated. But there was no other sign of the pig.

Then, with no more warning than a rustling in the scrub, the boar came charging at them. The boy dived for cover as it slammed into his father's legs, knocking him over. With tusks and teeth it started savaging the farmer's feet, slicing apart his left boot.

Collins had always thought that the attacks of feral pigs were short-lived. But the big tusker — he guessed its weight at more than 130 kg — kept at him. The muscle of his right calf was torn away and blood spurted.

As the scything tusks and snapping teeth tore at his knee and thigh, Collins tried to wrestle the pig off by grasping its head. More blood flowed from a gashed hand. Desperately frightened now, he appealed to the boy for help.

Young Alf, 28 kg, armed himself with a fallen branch and dashed into the fray. Belaboured with the stick, punched and shouted at, the pig was momentarily distracted but renewed its attack on the wounded man. Alf tried again, this time kicking the boar's rump. It turned and trotted back into the scrub, shaking its head.

The boy managed to get his fainting father back to the truck, retrieved the rifle and ammunition and then competently drove the 7-tonne vehicle home. After first aid by his wife to stem the bleeding, Collins was taken 130 km to Rockhampton Hospital. With a transfusion to replace lost blood, he was treated for dozens of wounds to his legs and hands. But one hand remains partly crippled.

The nine-year-old scooped the pool of national bravery awards — the Royal Humane Society's Clarke gold medal for the most oustanding action of the year, the Wilks trophy for courage shown by children under 13, and a Star of Courage medal for valour in perilous circumstances.

Alf had to wait until March 1986 to receive his awards. But it was worth it. They were presented by the Queen herself, at a televised ceremony in Canberra.

'I liked her,' the shy farm boy told interviewers. 'She's got lovely soft hands.' □

DISEASE WORRIES

It's not just what feral pigs have, but what they may get.

Feral pigs in various parts of the country can harbour diseases that are dangerous in food, including leptospirosis, brucellosis and sparganosis. Meat sales are prohibited here; exported frozen supplies must pass inspection first. Some populations also carry the virus of Australian encephalitis, helping to pass it between mosquitoes.

Agricultural authorities are especially worried by what could happen if foot-and-mouth disease or swine fever broke out here. Epidemics could be controlled among domesticated livestock, but if the germs spread to feral pigs they could be impossible to eradicate, with disastrous effects on farm export earnings.

A bull buffalo, living as long as 20 years, can reach a height of 1.8 metres at the shoulder and weigh up to 1200 kg. Females in the same conditions attain only about half the weight. Breeding extends from spring to late autumn but peaks in March. The gestation period is longer than a human's — up to 11 months.

Right: Trophy horns from big bulls are prized possessions in the Northern Territory.

To roam no more

*Buffalo herds head for domesticity
or destruction. Trophy hunters mourn but
ecologists rejoice at the last round-up.*

To romantics the 'Top End' will not be the same without its wild water buffaloes, spread in countless numbers across the coastal floodplains east of Darwin. Tour operators will miss the lucrative promotion of shooting safaris. After 1992, if a joint programme of the federal and Northern Territory governments goes to plan, there will be nothing to shoot.

Huge buffalo horns will remain a Territory symbol, with mounted pairs the pride of many homes and public places. But only the old ones will be hunting trophies. New ones will be the byproducts of abattoirs.

In the early 1980s the buffalo population was thought to be approaching 300 000. When the programme is completed the only herds left will

be behind fences. They will be farmed for large-scale meat and milk production — successfully domesticated and guaranteed free of disease. Others will be exported live. The rest, perhaps 200 000 or more, will have been destroyed.

It is a drastic measure, but one that is amply justified. Buffaloes provide the only reservoir in Australia for the bacteria of bovine tuberculosis. They are also widely infected with brucellosis — up to 25 per cent in some herds, and about three per cent on average.

Neither disease can be eradicated unless the animals are managed. Without eradication the tropical beef cattle industry is in jeopardy — and there can be no future for an expanded buffalo meat industry, with its promise of more Aboriginal employment and export earnings.

Ruination of wetlands

Furthermore, the buffaloes have been an eco-logical disaster. For more than 150 years in some districts, their habitual tracking and tram-pling have degraded a terrain that had never before suffered the impact of immensely heavy, hoofed animals. Their erosion of clay soils beside tidal rivers has led to the intrusion of salt water and the death of paperbark forests. Lotuses and other waterlilies die in muddied billabongs. Waterfowl habitats are destroyed, threatening rare bird species. Overbrowsed higher ground is denuded of eucalypt saplings and vines.

Some of the worst damage occurred, or was imminent, where the animals were given sanc-tuary from hunting. It threatened the integrity of the most precious of tropical wetland en-vironments, the World Heritage listed Kakadu National Park.

People have been at risk, too. From time to time drought-tormented beasts have invaded camps and townships, even causing trouble on the outskirts of Darwin. Bad-tempered bulls have charged walkers and motor vehicles. Near Pine Creek in 1973, one young bull took on a

Above: Family groups range in size from 50 to about 250 animals, sometimes associating in herds of up to 500. They follow habitual paths between drinking areas, mud wallows and grazing grounds. In favourable conditions they sleep at night in woodland camps and feed by day. In severe heat and drought the routine is reversed — the buffaloes graze at night and spend their day wallowing.

Right: Wetland bird habitats are degraded by constant trampling, and trees are killed by salt water intrusion.

moving goods train, derailing the 60-tonne locomotive.

But public safety was not a significant issue in the debate over the fate of the buffaloes. Such incidents are recalled by Territorians with more fondness than fear, because they enhance an image of frontier adventure. They certainly did the tourist trade no harm.

An animal that can stand as tall as a man and weigh more than a tonne, with horns nearly a metre long, is undoubtedly dangerous. But the water buffalo is seldom vindictive — unlike the cunningly vengeful Cape buffalo of South Africa, which big game hunters are said to fear more than any other creature. Nor do water buffaloes migrate in vast, thundering herds as the prairie 'buffaloes' of North America, more properly called bison, once did.

Like shooting cows

Old-time shooters who worked on horseback or foot to earn buffalo hides and trophy horns may have taken grave risks. But with the use of trucks, and then fast four-wheel-drive vehicles, the greatest hazard has been crashing into trees or termite mounds. Buffaloes are no harder to hunt and kill than cattle — for that is virtually what they are.

The water buffalo *Bubalus bubalis* is an Asiatic ox that has been domesticated for thousands of years as a draught animal, beast of burden and source of food. It is employed not only in most Asian countries but also in Central and South America and the Mediterranean region. Italian mozzarella cheese was originally made from buffalo milk.

Buffaloes survive on many plants that ordinary cattle breeds cannot eat — even the leaves of pandanus palms. They revel in monsoonal flooding, avoiding the worst of the summertime heat by wallowing in water. This habit, giving them a coating of mud, also protects them from many of the insect bites that debilitate cows.

Between 1824 and 1838, buffaloes were shipped from Timor and neighbouring islands to provide meat and milk for the three abortive northern settlements that preceded Darwin. One site was on Melville Island, the others to the east on Cobourg Peninsula. As each port was abandoned the animals were turned loose.

Another introduction, in 1857, was accidental. A barque was loaded in Burma and Malaya with buffaloes for slaughter in Victoria, to feed the sudden influx of goldrush diggers. A cyclone wrecked the ship in Cambridge Gulf, near the Northern Territory-Western Australian border, and the animals that survived the wreck swam to an uninhabited shore.

More fertile than European cattle, the untended buffaloes bred prodigiously. Beginning in about 1880, shooters killed tens of thousands on Melville Island alone, and many more on the mainland floodplains.

Easily tracked

Hunting was simple because the movements of buffaloes are highly predictable in normal seasons. Family groups of 50 or more favour the same home ranges year after year, moving from higher ridges in the wet season to swampy plains in the dry.

Each family territory has separate areas where the animals feed, drink, wallow, scratch themselves against trees, and sleep. Their habits are so ingrained that they wear tracks from one site to another.

Until the 1950s the only money to be made out of buffaloes was from selling their hides to the tanning industry. Shooting was often brutal, with no ammunition wasted on acts of humanity. One bullet was aimed to shatter a buffalo's spine. It would be left overnight, usually not dead, to await a skinner who would cut its throat. Carcasses were abandoned to rot or be eaten by scavengers.

Attitudes changed when a profitable demand for buffalo meat was discovered — for pet food here, and for processed use in sausages overseas. Hunters took to using mobile chilling plants, and then tranquilliser darts so that they could bring their quarry back alive.

With the advent of herd management and reliable disease control, buffalo meat gained approval for human consumption in Australia. Now farming corporations are concentrating on the production of prime steers, ensuring richer markets both here and abroad. The buffalo, though never again to be king of the northern plains, has a future of a sort.□

Time is running out for the wild buffaloes of Kakadu.

Saved by a toddler's cry

The maddened bull had a young mother at its mercy.

Late in 1961, the worst drought in three decades drove buffaloes into widespread conflict with Territorians. The town of Adelaide River was besieged. Out in Arnhem Land, mission station water supplies were raided by normally timid animals. Even at Nightcliff, in suburban Darwin, gardens were trampled and shrubs were stripped and uprooted.

About 15 kilometres from the city Mrs Sheila Jenkinson, 29, was driving with her two-year-old son Tony when her car hit and injured a wallaby. To put the animal out of its agony, she followed it into roadside scrub with a rifle.

From behind her, a lean bull buffalo charged. Mrs Jenkinson got off one shot, but has no idea whether it struck. The buffalo knocked her down and stood over her, its eyes just centimetres from hers. Her screams seemed to incite it to greater anger. It started to gore her, rolling her body over twice and kneeling astride her.

Then the little boy, still back in the car, gave a piercing call for his mother. Puzzled, the bull looked for the source of the noise. Mrs Jenkinson scrambled to her feet and ran. She escaped with a badly gashed knee, a broken rib, a skinned nose and a black eye.□

Going for the doctor

'Perfect picture' subject came for a close-up.

Alistair Marshall, a New Zealand doctor, was taking the trip of a lifetime in 1974, sailing the world in a 12-metre yacht. He and his colleague Dr Ralph Reeves had mapped out an itinerary of exotic places where they could put ashore and pick up relieving medical work. At Gove Hospital, in eastern Arnhem Land, they found more adventure than they bargained for.

Towards the end of their stay, emerging from their quarters for a pre-dawn swim in the hospital pool, they saw a huge buffalo on the front lawn. It started to walk away. The two men decided to go and fetch a camera, then follow the buffalo until the light was good enough to take photographs.

After about an hour the buffalo knelt down to wallow in a mud pool. Using a telephoto lens from about ten metres away, Marshall composed what he called the perfect picture, taking care to get in 'horns as long as billiard cues'. He had taken half a dozen shots when the buffalo surged out of the mud and charged. Marshall turned and fled but the animal, thought to have weighed well over a tonne, quickly caught him.

Marshall remembered no more. According to Reeves he beat the buffalo off by swinging the camera at its head — but not before a horn had knifed through his thigh, smashing the femur.

Flown to Darwin and then to Melbourne, Marshall was three months in hospital. The grand tour, with outer Indonesian islands as the next ports of call, was wrecked.□

Taking the bull by the horns

A deadly wrestling match in the kunai grass.

'The buffalo battered at him with its head and tried to gore him.'

For visitors to Kakadu National Park, Nourlangie Rock is a must. Galleries of intricate Aboriginal cave paintings and etchings date back many thousands of years. It takes a tough walk from the vehicle track to reach them — and in the 1960s, before the park was declared and buffaloes were brought under control, it could be unexpectedly dangerous.

Frank Muir, 68, and his son-in-law Harold Smith, 45, were on a safari holiday from Melbourne. Leaving their car, and prudently armed with a loaded .303, they started the climb to the rock through head-high kunai grass. A buffalo bull, fortunately not fully grown, suddenly appeared and lunged at Muir before he had time to raise the rifle.

Muir lay on the ground, the buffalo battering at him with its head and trying to gore him. Smith searched for the rifle, lost in the grass. But when he found it and saw a chance for a shot, the mechanism jammed.

His collarbone and ribs already broken, Muir grabbed for the buffalo's horns and twisted its head to one side. The animal sank onto its flank. Smith leapt in, reversing the rifle, and smashed the butt against its muzzle until it got up and backed away.

For three kilometres as the two men struggled to their car — Smith also received injuries — the buffalo followed at a distance. In the safety of the vehicle Muir got the rifle working and took a long-range shot, wounding the buffalo in a leg. It ran off into the kunai grass but hunters tracked and killed it.□

Rats: a gruesome greed

They will feed off anything we let them have — including our flesh as we sleep. Helpless babies and invalids can be maimed.

Startled by a human, a rat sits back on its haunches and erects its fur like a cat. This is not a posture of defiance but one of sheer terror. If a wild rat's escape route is cut off and it is approached more closely, it may leap towards its adversary — not in attack, but in an effort to get past. Finally, if a rat cannot get away and an attempt is made to touch it, the desperate animal will bite.

But the presumed fighting ability of a cornered rat is overrated. And the risk need never arise unless we bring it upon ourselves.

Rats are infinitely more dangerous when they are left to their own devices. They spread germs wherever they go in their bold explorations for food. And they try to eat virtually anything — even people who are asleep or unable to move.

Happily the incidence of severe ratbite injuries in Australia is low, reflecting generally good standards of sanitation and food hygiene. Any slackening of vigilance would foster a frightening upsurge not only of epidemic diseases but also of cases of maiming among infants and bedridden invalids.

Two worldwide species of rats, both probably of Asian origin, have adapted to living at close quarters with humans. Infesting ships and their cargoes, both kinds accompanied the earliest British colonists to Australia. They continued to arrive freely from foreign ports until 1900, when bubonic plague deaths prompted measures to halt their influx.

By then, local rat populations had been established in enormous numbers. The black rat *Rattus rattus* had spread throughout all coastal communities and agricultural districts. The brown rat *Rattus norvegicus* had made cities and towns its stronghold.

They outnumber people

The total of introduced rats now in Australia is anyone's guess, but it certainly exceeds the human population. In Sydney in the early 1970s, when underground blasting for the Eastern Suburbs railway brought a plague of rats up into the central business district, it was estimated that the city had more than seven million altogether.

When rats are seen creating a public nuisance they are almost invariably described as being 'as big as cats'. In fact black rats rarely reach 20 cm in head-and-body length. The maximum for a brown rat is about 25 cm according to zoologists, though pest controllers claim to have caught some up to 27 cm.

A brown rat Rattus norvegicus *raids a grain store. Rats eat anything that humans can, and many things that we can't — such as uncooked cereals. Young rats follow their elders to sources of food, so losses through an unchecked infestation can be heavy. And what is not eaten is often polluted by urine and droppings.*

While their common names suggest a reliable colour distinction, both species can be dark grey or brownish. The black rat has the bigger ears, about half as long as its head, and the longer tail of the two.

When both species live in the same area they choose different habitats. In a building the brown rat, naturally a burrower and swimmer, is likely to occupy the basement. The black rat,

The black rat Rattus rattus *is the more common of the two introduced species. It does better in hot climates and has spread from coastal towns into rural districts.*

Rats' teeth keep on growing throughout their lives. The chisel-like upper incisors automatically sharpen themselves as they are worn down. A rat's mouth is small, but with these honed teeth it can bite out large chunks of flesh.

Newborn brown rats, shielded by their mother, are blind and naked. Their fur will have grown by the time they are ten days old; their eyes will open after 15 days; they will be weaned at three weeks. If they are born in a colony, young males will be forced out to search for a new home. Females will stay to breed. And within about four months, these babies will themselves be parents.

more adept at climbing and jumping, is likely to nest on upper levels. This difference accounts for two of their many alternative names — sewer rat and roof rat.

A rat's mouth is small, but its bites chop out big chunks of flesh. The top front teeth are gnawing incisors, characteristic of rodents. They are shaped like inward-curving chisels, continuously growing and self-sharpening. Hard enamel at the front of the teeth is backed by softer dentine. As they grind over the bottom teeth the dentine is worn away, leaving honed edges of enamel.

Rapid multiplication

Both introduced species can live for up to three years, though the average life span is about six months. They are ready to breed at three or four months. Long-lived females can produce as many as seven litters in a year, each containing 10-20 young. If food and water supplies are unlimited, it is possible for one pair of rats to have more than 1000 descendants in a year.

Such rabbit-like fecundity, coupled with the ability of rats to gnaw through building materials and nest in inaccessible places, makes large-scale extermination impossible. Even in limited areas, control measures are useless unless they are sustained.

Pest destruction companies may use traps and fumigants to make a quick impact on major infestations. But better long-term results are achieved with baits. Chemicals that interfere with rats' blood-clotting or breeding capabilities are now replacing directly lethal poisons.

Baiting produces results more slowly because of an intriguing aspect of rats' behaviour. Searching for food, they readily enter new places and sample unfamiliar substances. But if anything is placed on one of the regular pathways that they use while foraging, they will not touch it for days.

This wariness is apparently part of the adaptation of rats to living among people. It is not a sign of particular intelligence, for the suspicion is undiscriminating — the object they avoid may be a heap of perfectly good food.

Rats do show a fair capacity for learning, however. While exploring they memorise complicated routes that allow the quickest unseen escape back to their nests. Given choices of food they favour those that are most nutritious. And if they survive a poisoning, they will not eat the same preparation again.□

AVOIDING TROUBLE WITH RATS

If a problem of rat infestation arises on your property, block their entry if you can, store food in metal containers and remove all edible waste materials. Seek advice from the health department of your local council if the rats persist. Baits may be available free of charge, but if a professional extermination service is needed you will have to foot the bill. If a serious problem arises on someone else's property in your neighbourhood and the owner seems unwilling to rectify it, inform the council. Local authorities are empowered to compel householders or companies to control rats, or else send in exterminators to have the job done at their expense.

ACTION AFTER A BITE Any bite from a rat is highly likely to be infected. Seek prompt medical attention even if a wound seems slight.

Death in a pram

The sick baby may have been mercifully unconscious.

After a seven-week-old girl died of ratbite wounds near Launceston, Tasmania, in 1969, her parents told a coroner that another little daughter had been bitten a month before. A health inspector described their home as providing better conditions for rat breeding than the local garbage tip.

The 31-year-old mother told the court she had left the baby asleep overnight in her pram. Next morning, while she was dressing another child, she heard a teenage daughter scream. 'When I got to the pram I saw the baby was dead and half its face was eaten away,' she said.

A pathologist gave evidence that skin and flesh on the baby's nose and the left side of her face had been eaten to the bone. Her left forearm, hands and fingers were similarly injured. A police constable, also describing the wounds, added that a storeroom next to the bedroom contained discarded mattresses, rags and waste paper.

The coroner noted medical evidence that the baby had been suffering from a form of pneumonia that could have rendered her unconscious. This illness may eventually have proved fatal. 'But I cannot escape the conclusion,' he said, 'that the injuries caused by the attack of rats ruled out any chance of survival.' No action against the parents was recommended by the coroner or proposed by the police.□

'Skin and flesh had been eaten to the bone.'

Plague rats of the outback

Graziers pay a price for good growing seasons.

Australia has more than 20 native rat species — not marsupials but more highly evolved placental mammals that reached here from Southeast Asia in relatively recent times. They are largely inoffensive, living off the seeds and foliage of native plants and avoiding human settlements, though one species has become a canefield pest.

Bush rats are not as fecund as their introduced cousins, but in seasons of unusually good plant growth their populations can explode. In the case of the long-haired rat *Rattus villosissimus* the phenomenon is rare — a decade may pass without major problems. But it can be devastating. Swarms move in millions over grasslands in northwestern NSW, western Qld and Central Australia. The worst-infested area on most occasions is the Barkly Tableland, straddling the Qld-NT border.

Because of their sudden appearance in hordes, eating grass down to the roots and destroying crops and fodder supplies, rats of this species are frequently called plague rats. It is an unfortunate name, leading some people to associate them with bubonic plague. In fact they eat their way across country too quickly to pose any significant risk of disease.

But the chance of physical injury cannot be ignored. In the last phase of infestation, as the rats denude the land of natural food, they try anything else. Stockmen and drovers sleeping out have woken to find rats gnawing at their bedrolls. Sleeping cattle have been lamed, their hooves and parts of their feet eaten.

The native long-haired rat Rattus villosissimus *was not discovered until the 1840s, and was presumed rare. A plague in 1869 was blamed on brown rats.*

Young children in infested areas are kept inside, behind screened doors and windows, until the rats have all gone. This may take months as waves of them migrate, each generation hungrier than the last. But the plagues are eventually self-limiting — starved rats cease to breed and start eating one another.□

Spotlight on the dingo

A dog from the dawn of civilisation, damned for attacking livestock, was winning favour when a fresh charge was laid — baby-killing.

British settlers found Australia's mainland wildlife ruled by dogs — the oldest pure breed in the world. Quickly the dingoes were condemned as enemies of pastoral agriculture. In retaliation for occasional raids on livestock, dingoes were trapped, shot or poisoned in hundreds of thousands.

Dingoes in the Great Dividing Range, their last eastern stronghold. Forest edges are the favoured habitat.

To help eliminate the breed from the farmed grasslands of the southeast, a fence longer than the Great Wall of China was built. The states declared 'wild dogs' noxious animals, making it obligatory for landholders to destroy all dingoes on their properties.

These unique dogs remain common only in the arid hinterland and in unpopulated regions of the far north. In the other strongholds of dingoes, ranging along the Eastern Highlands, the breed is disappearing through mixture with runaway domestic dogs.

Branded a coward

Throughout many decades of vilification and persecution, the dingo's possible menace to people was seldom mentioned. It was more fashionable to regard the animal as a cur, altogether too cowardly to attack humans.

That view was tenable because the contact between dingoes and settlers declined, and animals remaining in farmed areas became extremely wary. If people are tending to think differently today, it is because of a renewed familiarity — brought on mainly by the development of tourism in Central Australia.

During the pioneering era, reports of dingoes snatching children and attacking adults were not uncommon. Few contained much detail. But John Gale, a respected newspaper proprietor and historian, gave complete credence to one such tale.

In *Canberra: Its History and Legends*, published in 1927 when Gale was 98 years old, he told of the death of a girl at Mt Tennent in 1845. The girl, 12, hid in bush near her home to escape punishment from her mother. Her 'gnawed bones' were found next day.

Gale concluded his account by noting: 'It is well known that dingoes in numbers, and when pressed by hunger, would attack even an adult. And it was by no means an uncommon thing in the far-back days to come upon human remains that had been gnawed by dingoes.'

Attacked while saving a calf

Most other stories involved children of two years or younger. But Sir John Cleland recounted an attack by a single dog on an adult. A young farmer, hearing a cow bellowing at night, found it defending its newborn calf from a dingo. When the farmer stooped to pick up the calf the dingo sprang, knocked him over and 'badly damaged' him.

Virtually nothing is known of the experiences

of tribal Aborigines, who before they adopted European dogs lived at close quarters with semi-domesticated dingoes and competed with the wild ones for food. The initial coronial inquiry into the disappearance of Azaria Chamberlain (page 43) received second-hand evidence of the scars borne by a North Queensland Aborigine who was said to have been seized by the head and dragged from his parents' campsite as a boy.

Near Innaminka, SA, in the 1930s a young stockman is said to have been stalked and set upon by dingoes after he was thrown from his horse and broke a leg. But the inclination of dingoes to track people and vehicles was generally put down to curiosity and an appetite for scraps, rather than to any menacing intention. 'Dingoes never attack men.' That accepted view was stated twice by a leading zoologist, Dr A.J. Marshall of Monash University, Melbourne — even as he described an attempted attack on himself in 1965.

Writing in *Australian Natural History* magazine about a university expedition to an unpopulated district of the Kimberleys, in northern WA, Marshall told how he peppered a

Coats are typically ginger with some white underneath. Though colours vary, the pricked ears and white 'boots' are characteristic.

The world's longest fence — more than 8000 kilometres — was erected in the hope of excluding dingoes from the most prized grasslands. Its meandering course through SA, NSW and QLD is shown above.

snarling dingo with birdshot as it charged at him. He believed that this animal, never having seen humans, had mistaken him for an oversized wallaby.

Night of terror

When a policeman in far western Queensland was beset by a dingo pack in 1978, he knew that veteran bushmen would scoff if he told his story. And they did. But he remains convinced that the dogs were determined to bring him down, and would have killed and eaten him if they had succeeded.

Stranded at night on the edge of the Simpson Desert when his utility broke down, Constable Dale Searle faced a long, lonely walk to his base at Bedourie. But he soon had company: for more than 50 kilometres, he was followed and harassed by dingoes.

Throughout the night Searle, 26, wielded a tree branch as dogs rushed in to snap at his heels or tried to jump on him. He nursed a box of matches, lighting one occasionally to scare them off. But they stalked him relentlessly. Next morning when he reached a main road and waved down a tour bus they were still there, watching from just off the road.

At that time, late in the 1970s, public opinion had swung in favour of dingoes. Thorough field studies had shown that they rarely took domestic livestock unless drought made marsupial prey scarce. Then it was mostly the carrion of dead stock that they ate. Destructive raids, in which stock was killed or maimed apparently 'just for the hell of it', were often found to be the work of packs of renegade farm dogs or pets, sometimes hybridised with dingoes. Many graziers in semi-arid country, especially those running only cattle, began to welcome wild dingoes for the good they did in keeping down feral pigs, rabbits and goats.

Limited protection

In national parks that did not adjoin grazing properties, dingoes gained protection as native wildlife. Meanwhile, in towns, more and more

people openly flouted state laws against keeping dingoes as pets.

As conservationists lobbied to have the breed fully protected, and 'native dog' societies vigorously promoted the cause of private ownership, the popular perception of dingoes was turned topsy-turvy. Speedily, a despised animal became an object of admiration.

The media, even while still recording alarming mishaps with dingoes in captivity, gave prominence to items demonstrating the good nature and intelligence of the breed. Triumphs in dog obedience classes, where dingoes frequently outpointed dobermann pinschers and German shepherds, were headlined. Proposals were advanced to train dingoes for guard work or police and military use. Australia Post featured the breed in a stamp issue, and an army unit in Queensland adopted a dingo as its mascot.

The dingo deserved recognition as an important breed of dog. But the rash of publicity in its favour, particularly in the cities, obscured the fact that the dingoes of the outback were wild, predatory animals.

Misguided sentiment

People on holiday made the mistake of treating dingoes as if they were shy pets, coaxing them into camps and overcoming their normal wariness with offerings of food. In doing so they bought trouble not for themselves, but for other travellers who followed.

Breaking down natural barriers and converting powerful hunters into scavengers was bound to promote conflict. Around tourist sites, especially in Central Australia, dingoes became noticeably more audacious, entering tents at night and carrying off food, clothing and items of camping gear.

At the former Ayers Rock motel and camping ground, signs urging people not to encourage dingoes and warning of unfortunate consequences were up long before the Chamberlain incident. That case brought to light previous attacks on children by emboldened but untamed dingoes.

Removal of all Ayers Rock camping facilities to the Yulara tourist complex, outside the boundary of Uluru National Park and behind fences, has minimised the chances of further conflict in that area. But it can arise in any other place where dingoes and wild hybrid dogs have become over-familiar with holidaymakers.

In April 1986 at the popular Freshwater camp in Cooloola National Park, north of Noosa, Qld, a five-year-old girl gathering pipis on the beach was savaged by four dogs. She was ripped and punctured about the head, throat, arms, chest and legs. The pack leader, a German shepherd-dingo cross, was shot. Two of the other dogs appeared to be purebred dingoes.

Worthy of respect

It would be regrettable if dingoes were to

Below: Active night and day, dingoes hunt in short spells alternated with frequent periods of rest.

more advanced seafaring people. Sheep, goats and fairly similar dogs were taken to Timor at least 3500 years ago, perhaps from India.

Aborigines and dingoes

Knowledge of the relationships betweeen Aborigines and dingoes before they came under European influence is sketchy. Observations in one locality could not be assumed to apply in another. But it seems that most dogs were wild, and none were fully domesticated.

With the exception of captured or orphaned pups, which were sometimes suckled by women, camp dingoes were never given food. Instead they were a source of it. Live pups were carried by travelling parties as reserve supplies. An important role of the tamest dogs was as wintertime bed-warmers.

Some tribes are said to have used dingoes in hunting, exploiting their ability to track and bail up game. In other tribes they seem to have been regarded as a hindrance. It is significant that Aborigines forsook dingoes and acquired more tractable European hunting breeds as soon as they could. 'Dingo', incidentally, was a word that Sydney natives coined for these new breeds. They called their own type *warrigal*.

Scientists can measure variations in skull and tooth formation between dingoes and domestic dogs of similar size and build. To untrained people there are no major differences in appearance. An adult dingo's paws are slightly bigger and its ears are always pricked. In the wild, it droops its tail in the manner of wolves. And dingoes do not bark — they yelp and whine, and communicate over distances by howling.

Coats are short and smooth in hot regions, longer and fluffier in the south. The typical

become victims of another wave of revulsion and mass destruction, simply because some people misunderstand the nature of wild predators. Left alone to follow their own ways, they present very little danger. They treat humans with a healthy respect. And as rare links with the roots of human civilisation, they deserve the same from us.

The dingo belongs to the genus *Canis*, with wolves, jackals and all dogs. Most zoologists classify it as *Canis familiaris dingo*, a subspecies of the world's domestic dogs that is able to interbreed with them. But while all the others have been hybridised again and again, selectively or accidentally, dingoes in Australia were isolated for thousands of years.

Domestic dogs are generally thought to have descended from wolves. Dingoes bear resemblances to a primitive type of wild dog that was distributed in an equatorial band from North Africa across southern Asia to New Guinea, and domesticated in Southeast Asia at least 5000 years ago. The ancestor of this type may have been the Asiatic wolf.

The oldest convincing fossil records of dingoes, found in southern Australia, have been dated to not much more than 3000 years ago. Dingoes were presumably in the north for some time before then. But authorities agree that their arrival was a relatively recent event, certainly within the last 10 000 years, and that it must have been aided by humans.

An assumption that dingoes were brought as hunting companions by a late wave of Aboriginal immigrants has lost favour. More weight is now given to a theory that they were carried for trading or in an abortive settlement attempt by

Above: Family groups share prey, making the biggest kills they can manage together. But their meetings are brief — adults spend most of their time alone.

Below: A young pet dingo sits in on a recital of dingo 'Dreamtime' mythology by NT Aborigines.

A brindle dingo — unusual, but still purebred.

colour is a gingery yellow, lighter underneath, but a minority of dingoes are black and tan or almost white. Whatever the body colour, a purebred always has a white-tipped tail and nearly always white 'boots'.

One litter a year

Domestic dogs can breed twice a year, but dingo bitches come on heat for only a few days each autumn or winter. Making dens in rock shelters, hollow logs or the enlarged burrows of other animals, they bear litters of four to seven pups about two months later.

Pups can be fed on meat swallowed and regurgitated by the parents, but if possible the carcasses of prey animals are carried back to the den and shared. Later the pups are led out to the sites of kills, over a gradually widening range. Though weaned at four or five months, the young may be permitted to hunt with their parents throughout their first year.

In natural conditions dingoes seek the biggest possible prey and feed on it together, so relatively few victims are needed. Kangaroos, wallabies and wombats are the favoured prey when they are available. Lizards, smaller marsupials, rodents, birds and their eggs and even insects may also be taken.

Mr Robert Harden, a wildlife research officer who devoted eight years to the study of dingoes in the New England region of NSW, found them active both by night and by day, usually in periods of less than an hour interspersed by even shorter rests. They rarely moved in packs, but members of small groups met from time to time, co-operating to suit their needs and the size of available prey.□

Born in spring, dingo pups depend on their mothers for about four months.

A black winter at the Rock

Azaria Chamberlain's disappearance revealed a spate of attacks by dingoes.

The fate of 10-week-old Azaria Chamberlain, who vanished from a tent at the foot of Ayers Rock in August 1980, has become Australia's most controversial mystery. The awesome setting of a forbidding island-mountain in the stark, flat 'Dead Heart', and Lindy Chamberlain's claim that her baby had been carried off by a dingo, ensured that every chapter in the judicial saga that followed would be attended by unprecedented publicity.

After all the twists and turns of the case, during which the reputations of law officers and forensic scientists were made and unmade and Mrs Chamberlain spent 2½ years in prison convicted of the child's murder, a complete explanation of what happened may never surface — at least not to everyone's satisfaction.

What did emerge early in the case, and remains unshaken, was evidence that a dingo *could* have taken Azaria. Until then very few Australians had any inkling that such a danger existed. But as witnesses came forward at a succession of hearings, it became clear that all had been far from well between tourists and the fearless dingoes at Ayers Rock in the winter of 1980. In retrospect it may seem that a serious injury, if not a tragedy, was inevitable.

Dragged from a car

Among the last cases to come to light — not until Lindy Chamberlain's unsuccessful appeal in 1984 — was that of Amanda Cranwell. She was three years old when she visited the Rock with her parents late in June 1980, less than two months before Azaria's disappearance.

Amanda was left on the front seat of the family car as her parents, dairy farmers from Victoria's Latrobe Valley, set up their caravan. Hearing a slight noise, Mr Cranwell went to investigate and found his daughter on the ground with a dingo standing over her. Her head and neck were bleeding. The dingo believed responsible was understood to have been shot. A month later, the trial jury was told, three-year-old Paul Cormack was attacked, receiving a deeply gashed ear. His father, also from Victoria, said he found the dingo standing astride the boy, who was on his back. The animal had been seen at the campsite two or three times previously. It showed no fear when he ran to rescue his son, merely plodding away.

The same jury was told of two incidents just one day before Azaria's disappearance. A

Azaria's parents, Lindy and Michael Chamberlain.

student, sitting outside her tent writing in a diary, saw a dingo nearby but took no notice. Without warning it seized her by an elbow. That evening, according to a nursing sister from NSW, a dingo knocked her son to the ground while he was having his meal. She chased the animal off before any injury was suffered.

The theoretical ability of a dingo to seize a baby and carry it a long distance was attested to by a man at the forefront of campaigning for protection of the breed. Mr Leslie Harris, president of the Australian Dingo Foundation, told the court that he had watched a dingo kill a wallaby weighing 10-13 kg, pick it up by the back and carry it for more than a kilometre with only its tail touching the ground.

Referring to a dingo's technique with smaller mammals, Harris said it would usually seize the whole head in its jaws and crush it. 'Usually it will accompany this with a sharp shake which is calculated to snap the neck of the animal,' he added. Harris told the court that in his opinion a dingo could assess a baby as available prey if its head were visible, and seize it.

At the Morling inquiry, opened in 1986 to review all the judicial proceedings, evidence was heard of seven dingo attacks on children in the three months before Azaria's disappearance. And tests on dolls and models appeared to contradict an earlier opinion — from British zoologists — that a dingo's jaws could not open wide enough for it to grasp a baby's skull. □

Dogs on the rampage

Runaways form big packs, exciting one another's blood-lust. Survivors in the wild often interbreed with dingoes.

Renegade dogs — runaway pets or farm animals — wreak worse havoc than dingoes among domestic livestock. Banding together in big packs, they develop a shared enjoyment of killing that can turn into a frenzy. Often they indulge in senseless slaughter and maiming, far in excess of their feeding needs.

Such dogs are not known to attack humans by choice. But on the fringes of cities and towns, rampaging packs sometimes harass pets. Size does not daunt them; they may terrorise anything from cats and ducks to deer and ponies. People trying to protect their pets — or children simply playing with them — could become victims of indiscriminate savagery.

In 1983 at a school near Penrith, in Sydney's outer west, a dog pack raided sheep that were kept there for a husbandry course. Only one was killed but many others were maimed. Police, concerned for the safety of children, found the marauders and cornered one dog. It escaped after attacking an officer. The school hired armed guards to protect its pupils.

A few weeks later, a warning came from Queensland that it was only a matter of time before feral dogs claimed human victims. Officers of the Lands Department stock route unit reported being menaced by packs that were becoming increasingly aggressive. Among the dogs that they had managed to trap and destroy was a wolfhound two metres long, easily capable of pulling down fully grown cattle.

Australian cattle dog.

Kelpies.

HELPFUL HYBRID

Uncontrolled breeding between domestic dogs and dingoes spells nothing but trouble. But one deliberate crossbreed, with some adaptation, has proved highly useful in managing beef cattle herds.

Early runholders sought a type of dog more able to cope with heat and drought than British breeds. Among the experiments was a crossing of dingoes with blue-mottled, smooth-haired Scotch collies called merles. With the addition of dalmation and kelpie genes, the Australian cattle dog or blue heeler was developed.

The kelpie itself, which derived from black-and-tan Scotch collies with a mixture of some other breed, is also believed by many people to have dingo genes.

Below: The result of a feral dog attack on livestock near Armidale, NSW. The photograph was taken by the district Pastures Protection Board's noxious animals inspector. This sheep survived, but attacks by packs of renegade dogs often turn into frenzied massacres of livestock. Worries are mounting that a human will be next.

Breeding in the bush

Not all domestic dog breeds are equipped to fend for themselves in the wild, let alone protect and provide for their young. Unlike cats, which become more vigorous by natural selection, feral dogs often mate unsuitably. The progeny of mismatched parents may lack the physical qualities needed for efficient hunting and survival against predators.

Dogs are never such capable and versatile hunters as cats. That is why they victimise livestock — it cannot escape. The main impact of feral dogs on native wildlife is not so much destruction as disruption. Their barking alone drives many animals, especially birds, away from habitats that may be hard to replace.

In remote areas, successful feral dogs eventually come into contact with dingoes. Dogs of unsuitable type and size are likely to be killed. Others sooner or later mate with the dingoes.

Natural hybridising was once thought to be rare. But recent studies indicate that more than three-quarters of supposed dingoes in the Eastern Highlands have some blood of domestic breeds. And it is probably the hybrids that cause the greatest trouble to farmers.

Crossbreeding combines the superior hunting ability of dingoes with the greater fertility of domestic dogs. The dingo hybrid is capable of breeding twice a year instead of once, and bitches may bear more pups in a litter. Family groups become larger and the tendency to hunt in packs more common, heightening the potential danger to people.□

Insect surprises

Fortunately, in view of their numbers, few insects can cause significant injury. But their defenses are often startling.

Apart from some well-known venomous species and disease carriers, remarkably few insects in Australia are capable of hurting us. Trouble can arise from unexpected quarters, however. Many bugs and beetles secrete blistering chemicals. Some can squirt them over a distance. It is unwise to squash strange insects with your fingers, or to interfere with them while your eyes or mouth are near. Most of these chemicals are purely defensive, to dissuade predators. A few are sex attractants.

Rarely will insect chemicals be used against you if the creatures are left alone. If they crawl on you, however, it is very hard to resist the impulse to brush them off — and that may be enough to cause trouble.

Probably the worst blistering insect in this country is the rove beetle *Paederus cruenticollis*. Ant-like, but mostly orange in colour and with a characteristic cocked tail reminiscent of an earwig, it is common in soils that remain damp in summer. Squashed on the skin, it can cause painful, long-lasting 'whiplash' lesions.

Insects of many kinds will defend themselves by attempting to bite with their jaws or by using claws and pincers. The nips may be painful, but seldom break the skin.

Certain more formidable species, including various assassin bugs that are common in gardens, have piercing equipment that they normally use to impale prey. These may be able to inflict a stab wound.

Whether from a bite or a puncture, tissue damage caused by an insect will be slight. Care must be taken, however, to make sure that any wound stays free of infection.□

Lethocerus insulanus, the fish-killer or giant waterbug of tropical pools and streams, grows to 7 cm and spears frogs and fish with a strong, blood-sucking rostrum.

Right: The bronze orange bug Musgraveia sulciventris, a citrus pest, sprays a strong-smelling chemical.

Assassin bugs such as the bee-killer Pristhesanchus — here feeding on a fly — can inflict stab wounds.

ANTS: A HAZARD FOR HOSPITALS
Their only threat is to helpless people.

Australia has no ants of the 'driver' type so feared in South America, migrating in millions and consuming any living thing in their path. People can be hurt, however, if swarms of ants are attracted to their bodily secretions and they have no means of moving or defending themselves. Even quite small species of omnivorous ants, working together, can gnaw away tender flesh.

In the bush, ants have contributed to the deaths of people immobilised by injury, exhaustion or drunkenness. A baby boy, abandoned by his mother near Thornton, NSW, in 1929, suffered ghastly mutilations.

But in none of these cases is it likely that ants were the killers — the principal cause of death would have been exposure.

Ant infestations are of greater practical concern in buildings where people may be immobilised. Hospitals in particular must be on their guard. Many patients are not only infirm, but also helpless through the effects of anaesthetics or other drugs. And they may have lesions or surgical wounds that give voracious ants an easy starting point for an attack.

Familiar black house ants and meat ants can be troublesome in this regard. The Argentine ant, because of its sheer numbers and the speed with which it invades buildings, causes greater problems. Worst of all is another imported species, *Monomorium pharaonis* — so specialised that it is known as the hospital ant.

The dive-bomb brigade

Unprovoked aggression by magpies can make springtime a misery. Their swooping attacks are unnerving at best; at worst, they inflict painful wounds.

Spring in Australia is marked, and for many people marred, by magpie terrorism. For up to three months, birds that normally enjoy cordial relations with humans exhibit some of the boldest aggression known in the avian world. Their onslaughts, sometimes in big gangs, can make outdoor activities unnerving for adults and a frightening ordeal for children.

Nearly all attacks are bluff. But some birds are pugnacious enough to carry them through. Since the usual approach is a swoop from above and behind the target, the likeliest injury is a scalp furrowed by the sharp bill, or a gash in the back of the neck. Frontal and side attacks do occur, however, raising a risk of eye damage.

Cyclists can face the gravest consequences. Diving magpies do not need to strike them — only alarm them enough to make them lose control. At Currumbin, Qld, close to the spot where Karen Hatchett (whose story is told under the heading 'An eye ripped open') was attacked, another girl tumbled from her bike into speeding traffic and was lucky not to be hit.

The Australian magpie *Gymnorhina tibicen* is a ground-feeder, living mainly off insects and worms. But where possible it nests in tall trees, at heights of up to 20 metres. People who are dive-bombed are often nowhere near nests.

Left: The pied butcherbird Cracticus nigrogularis *takes its name from a habit of wedging whole prey — mice, small reptiles or the nestlings of other birds — in tree forks or impaling it on twigs. Then it dissects the kill with a powerful, hooked bill.*

Below: The Australian magpie is a butcherbird that took to ground feeding, on insects or scraps of human food. Back patterns vary across the country from mostly white through grey to mostly black, but the different 'races' form only one species, Gymnorhina tibicen. *The European magpie is a type of crow, and not related.*

Many other birds are vigorous defenders of their nests. But the aggressive behaviour of adult magpies, while they are incubating eggs and rearing their young, is something more. Joined in often by non-breeding adults, it seems to be an attempt to forestall any disturbance of the feeding area — which is soon to become a nursery.

When chicks are fledged, but before they are accomplished fliers or foragers, they follow their parents to the ground. Hopping and flapping about and pleading for tidbits, they are easy marks if predators or other enemies have been allowed to enter the feeding territory.

Young may join the fray

But protection of the fledglings is no explanation for the behaviour of some groups late in the season, when the young hone their new-found aerial skills by taking part with the adults in attacks on people. Nor does it explain why some birds mount their patrols over roads and sealed pathways where they could have no wish to feed themselves or train their offspring.

Certain magpies, it seems, have an unusually determined dislike of humans. It may stem from some previous persecution. Magpies are long-lived, often exceeding 15 years, and have good memories. In the days when there were clear and consistent distinctions between the dress of boys, girls and adults, it was noted that magpies picked almost exclusively on boys. And boys, of course, commonly compete to throw sticks and stones at birds' nests, or climb trees to take eggs. Nowadays, with men, women and children frequently wearing similar garb, anyone can take the brunt of what in some cases may be acts of revenge.

Magpies are highly social birds. The strongest band together in groups of up to about 12 in the east, and sometimes 20 or more in the west. They find a territory big enough to meet their year-round needs and endeavour to hold it throughout their lives.

All of the adults help to defend the territory,

occasionally exhibit similarly threatening behaviour and have been responsible for some injuries. Their hooked bills can be more damaging. But they live in pairs rather than groups and as perch-and-pounce hunters, roving widely while the young wait in the nest, are less angered by intrusions at ground level.

The biggest, most powerful and most fearless members of this family of birds are the currawongs. Fortunately, though they may throng home gardens and urban parks in winter, they do not remain there to breed. Instead they retreat to remote forest homelands, where there is little likelihood of their coming into conflict with people. □

but usually only one dominant male breeds, mating with several females. Only those to which he brings food and gives special protection while they are incubating eggs are likely to produce viable hatchlings. Should the dominant male die, his leading rival takes over the breeding role. For this reason, destroying an aggressive bird can make the problem worse. If the new breeding male fertilises unmated females, the season of attacks will be extended.

In any case magpies, their nests and their eggs have the full protection of the law. Police and wildlife officers are empowered to get rid of birds known to present a serious threat to human safety, but rarely are any destroyed.

The preferred method of control is by trapping and wing pruning. If certain feathers are plucked — not clipped — on only one wing, the bird is not impaired in its normal flights in search of food. But it cannot accomplish the aerobatics needed for a swooping attack.

Butcherbirds, close relatives of the magpie,

In spring at Raymond Terrace infants' school in the Hunter Valley, NSW, children are not allowed to leave their classrooms unless they don 'magpie hats' — old ice-cream containers. Magpies nesting in a century-old pine tree are beloved pets at other times, but in the breeding season children are often attacked. The ground keeper, working in the danger zone all the time, wears a motorcyclist's crash helmet. The practice of head protection has been widely adopted in schools.

AVOIDING TROUBLE WITH MAGPIES

●Stay away from areas where magpies are swooping if you can. Otherwise wear a hat of strong material or carry an umbrella or a stick.
●Try not to duck your head if a magpie threatens. Look straight up at it and keep on walking. Magpies rarely complete an attack if they are being watched — they normally wait to swoop from the rear.
●Some wildlife authorities seriously recommend wearing sunglasses back-to-front, offering a big extra pair of 'eyes' to dissuade the birds.
●Warn children who ride bicycles to dismount and walk if they see swooping magpies ahead.
●If magpies nesting in your garden or on a neighbouring property cause trouble, you can reduce their aggression by leaving out generous quantities of meat scraps during the breeding season. It will probably turn them into year-round beggars, but their visits at other times are always friendly.
●Should a bird persistently strike at people, rather than merely threaten them, notify the police.

WOUNDS If the skin is broken in an attack by a magpie or any other bird, obtain medical treatment to avoid infection.

An eye ripped open

Karen hitched a ride home — only glad she wasn't bleeding.

Magpie breeding was in full swing on Queensland's Gold Coast in September 1983 when 16-year-old Karen Hatchett, of Tugun, cycled off to see a friend at Currumbin. Head down and pedalling for all she was worth, her only warning was a dark shape suddenly flapping beside her. Then she felt a stinging pain in her left eye.

Fluid streamed down Karen's cheek as she dismounted. She was relieved to find that it was not blood — just what seemed to be tears. With the eye quickly closing, she was driven home by a passing motorist.

The family doctor called and was horrified at what he found. The eyeball was torn across from the outer corner and had collapsed. The cornea as well as the iris was severed.

Microsurgeons at Southport Hospital faced the problem of re-forming the eye before they could repair it. For the first time they tried a technique just introduced from South Africa. Instead of using an air pump to inflate the ball, they inserted a soluble bag of fluid.

Karen, newly apprenticed to a hairdresser, spent an anxious week waiting to know whether the eye would work at all. When the bandages came off her sight was partially restored, and it has improved gradually to about 80 per cent of full vision. The only blind spot is in the corner. Scarring, once 'like a lightning flash', has been reduced to a fine thread that only she and her family know is there.

Karen's life has changed, however. Her home, close to the famous Currumbin bird sanctuary, receives daily visits from its flocks of lorikeets. In happier times it was her joy to feed them. Now she is frightened to go near any bird. And she has never ridden her bike again. □

Magpie victim Karen Hatchett, who saw 'just a dark shape flapping' before she was struck in the eye.

Defending the nest

Too close an approach to eggs or chicks invites a wrathful response. And once roused to attack, some birds can't stop.

Birds in their relationships with humans seek only peace. But in defence of their nests or their young, many species show a formidable antagonism. Some of these, once their fury is roused, know no restraint. An intruder may be given no chance to retreat.

The powerful owl Ninox strenua *can kill full-grown possums.*

Birds of prey are capable of inflicting the greatest damage, ripping with their hind talons as well as striking with their hooked, flesh-tearing beaks. It has to be said, though, that the only people in danger are those who go looking for trouble, by actually disturbing nests in search of eggs or chicks. Eagles, hawks, owls and the like are otherwise highly tolerant of human activity.

People who used to indulge in bird's-nesting — generally illegal now — found that their fiercest and most relentless opponent in the wild was the peregrine falcon. The Australian hobby or little falcon was ranked not far behind. Australia's biggest hawk-owl, the possum-killing powerful owl, is another species to be treated with the greatest respect.

Strangely our biggest birds of prey, the wedge-tailed eagles and sea-eagles, are said to make no effort to protect their eggs or young from human interference. Yet wedge-tailed eagles in captivity sometimes display extreme aggression. It would be unwise to invade the nests of either species in the belief that they never attack.

Killed by angry swans

The most dangerously aggressive birds, because they occasionally attack people who are not even aware that nests are nearby, are swans. A

male swan, or cob, perceiving some threat while eggs are being incubated or cygnets are being reared, is usually content to chase the intruder away. But sufficently enraged, it may launch a leaping onslaught.

The swan uses neither beak nor claw. Its weapons are the knobbly joints of its half-folded wings — the equivalent of our elbows. Blows from them can break the arms of adults and the legs of children, or knock any person unconscious. Once a buffeting starts it goes on mercilessly. Using what seems to be a tactical sense, some swans overseas have driven their victims into the water and continued to beat at them until they drowned.

White or mute swans seen in Australian parks, the descendants of ornamental imports, are semi-domesticated and nearly always good-natured. The affability of native black swans —

many of which are truly wild though they also may breed in parks — should not be taken for granted. And the touchy season of breeding may come at any time of year.

Geese, too, are capable of buffeting assaults. Cape Barren geese, as heavy as swans and nearly as big, can be especially pugnacious. An angry gander could batter a child to the ground with ease. Fortunately these geese have their breeding grounds on rocky islets, and are placid at other times when they visit the mainland.

Safety in numbers

Sea birds such as gulls, and even huge albatrosses, are similarly harmless in their seasons of coastal occupation. But boating parties, going ashore at the southern islands where most of them breed in big colonies, should take care not to disturb nesting grounds. These birds mount mass attacks, dive-bombing and pecking, to resist intruders.

The sea bird species with the worst reputation for ferocity in defence is the piratical southern skua, which terrorises other birds and forces them to drop food. Its subantarctic breeding grounds include Macquarie Island. □

The peregrine falcon Falco peregrinus *(above) and the Australian hobby or little falcon* Falco longipennis *(below) are the most feared birds of prey.*

Jets and birds don't mix

Airport authorities were forced to clean up their acts.

Airfields, typically built on swampy wastelands, used to be meccas for Australian waterfowl. Geese, herons, ducks and so on flocked to them in tens of thousands. In many centres, airfields were adjoined by municipal garbage dumps as part of swamp reclamation. This attracted even more birds — the scavengers such as seagulls, kites and pigeons.

In the 1960s all that had to change. Birds had created no great problem for propeller-driven aircraft. But with the advent of jet airliners the landing grounds became mass killing fields — not only for birds but potentially for passengers.

Jets acting like giant vacuum cleaners could suck in huge numbers of birds and be forced out of operation. The danger of 'bird-strike' was tragically illustrated in the United States in 1960, when an Electra turbo-prop flew into a cloud of starlings at Boston airport. All engines failed and the plane crashed in a river, killing every one of the 61 people aboard.

All over the world, laborious efforts were made to shoot, poison or scare off airport bird flocks. The efforts were largely useless. As long as food and covering vegetation still beckoned, more birds turned up.

Major drainage and reclamation works have been necessary at most Australian coastal airports, and garbage dumping practices have had to be revised. Some dumping is still possible, provided that the refuse is tipped by night and completely bulldozed over before morning. Rivers and creeks near airports are often strung with wires to prevent waterfowl alighting. □

Wedge-tailed eagle Aquila audax.

A MYTHICAL MENACE

Eagles lack the strength to abduct people.

Tales abound of eagles carrying off children to their eyries, especially in undeveloped countries where folklore holds more sway than verifiable fact. Scientists are disinclined to believe any such story. Though eagles sometimes kill live prey and take it home for their young, they are unable to carry more than about half their own weight. Some overseas tests indicate that the biggest eagles may manage about 6 kg over significant distances.

It is just possible that eagles have occasionally mistaken small humans for prey, and when they swooped have entangled their talons in clothing. There is one fairly convincing account from Norway of an undersized four-year-old girl being carried off in 1932. She was dropped unharmed about two kilometres away.

A world record wingspan of about 2.7 metres is claimed for the Australian wedge-tailed eagle *Aquila audax*, but 2 metres is a more usual span. Though the third or fourth biggest eagle in the world, it is unlikely to carry off prey larger than a rabbit. Bigger kills are eaten on the ground. Once persecuted because it was believed to prey on livestock, the wedge-tailed eagle is now known to attack only ailing or dead lambs. In helping to control feral animals such as rabbits and cats, the eagle is of great benefit.

An Airport Safety Officer fires a cartridge from a flare gun into the air. This will explode over a flock of birds and scare them from the runway area.

Estuarine crocodile Crocodylus porosus.

FROM THE AGE OF DINOSAURS

Crocodylus porosus, *king of Kakadu National Park, NT.*

Well over 100 million years ago, swamps all over the world were ruled by crocodilian reptiles remarkably like those we know today. They co-existed with dinosaurs, preying on smaller ones. And when a climatic crisis killed off the dinosaurs, the crocodilians survived.

Evolution does not always dictate the alteration or overthrow of animal types. If a creature has utter mastery of its environment and enjoys supremacy in competing for food, it need not change. Crocodiles, alligators and the like, though their portion of the world shrank, held unchallenged dominion over equatorial wetlands and waterways for 65 million years after dinosaurs had gone.

Not even humans could seriously threaten them, until high-powered rifles came to hand. Then in every accessible habitat a slaughter began. No sense of admiration, such as we accord lions, tempered men's fear-ridden hatred, and later their greed. Crocodiles were reviled as monsters — hideous and somehow evil. Quickly the most ancient and majestic survivors of the reptile epoch faced extinction.

In Australia respect for the grandeur of the world's biggest crocodiles, and appreciation of their ecological role, came almost too late. For years after legal protection was imposed, the estuarine species was seldom seen. But in the 1980s it was clearly back — with what seemed like a vengeance. The renewal of a tragic conflict, with people among the victims now, raises agonising questions about our ability to share our country with some of its most extraordinary natural inhabitants.

SALTWATER LABEL GIVES A FALSE ASSURANCE

Because northern Australia has some small, inoffensive crocodiles restricted to fresh or brackish water, the others — the potential killers — are often called saltwater crocodiles. That distinction is misleading. It can lull people into taking grave risks in unsafe places.

The huge *Crocodylus porosus* does favour salty estuaries and tidal waterways. And maturing males in search of breeding grounds sometimes put to sea. Thus Australia shares the species with New Guinea, Indonesia and continental Asia, from Malaysia to India. Ocean-going crocodiles have turned up in places as isolated as Fiji. But the estuarine species does just as well in fresh water. It can be found hundreds of kilometres inland, inhabiting rivers, swamps and floodplain billabongs, sometimes moving considerable distances overland between them.

Its natural range is more or less continuous from the Kimberleys in WA, across the 'Top End' and the Gulf of Carpentaria and down almost the whole of the Queensland coast. It occupies the big islands near Darwin, along with those in the Gulf and some in Torres Strait.

The southern breeding limit is around Tin Can Inlet, opposite the foot of Fraser Island. Rivers feeding the inlet share their catchment area with the Noosa River. It should come as no surprise if some day a crocodile appears down-river, near the hub of Sunshine Coast tourism.

Crocodylus johnstoni, the freshwater species, has a similar range. Narrow-jawed, it does not exceed 3 metres in length. Wounded or cornered 'freshies' can inflict severe injuries. But normally they show no aggression. In the Kimberleys their dry-season waterholes are sometimes shared by many Aboriginal children

The estuarine crocodile can be hard to spot even when in plain view. But its tracks over mudflats (below) are unmistakeable.

at play. Estuarine crocodiles, however, are undeniably dangerous. Australia probably has some males longer than 7 metres and weighing more than a tonne. Such giants would be survivors of the hunting era, extremely wary of any human contact. But bolder specimens in the 4-5 metre range are becoming common.

Most victims drown

A crocodile that big can cripple people with a lashing tail and easily drag them under water. Victims are helpless against the spinning action that crocodiles use to tear flesh from large prey. Most are thought to die quickly from drowning.

By no means all attacks are made by hungry 'man-eaters'. Sometimes a lashing with the tail is incidental to a crocodile's instinctive retreat to the water, after it has been surprised while basking on a riverbank. Assaults on small boats may start because big males mistake them for challengers to their domination of a territory.

In any case, in spite of their sometimes enormous bulk, crocodiles are not the voracious feeders that we may imagine. Typically reptilian, they do not use up calories to maintain a constant body temperature. They absorb the sun's heat by day and avoid the chill of night by staying in the water. And they are economical in their movements, spending most of their time inert. Crocodiles can exist on about ten per cent of the nourishment that mammals of the same size would need.

On firm ground, raising its bulk on stiffened legs, a crocodile can be a surprisingly fast runner. Slithering over mud, it takes advantage of webbed hind feet to push itself along. And it is well adapted for aquatic life.

Eyes and nostrils are positioned so that the crocodile can float or swim with only these showing. The legs fold back along the body, with all propulsion and steering coming from

Mountain streams feed the Daintree River, Qld, with ever-fresh water. But so-called 'salties' thrive there, and this is a typical habitat.

Snouts of the freshwater crocodile (above) and estuarine species (below).

TELLING THE SPECIES APART

Aggressive estuarine crocodiles grow far bigger than their timid freshwater cousins. And the two are easily told apart by their snouts. The freshwater species' is long, narrow and smooth, while that of the estuarine crocodile is broad, bumpy and relatively short.

Colours are similar — usually grey or brown above and whitish below. But some freshwater specimens vary towards an olive shade and some estuarine crocodiles are almost black on top.

Freshwater females dig nesting holes in sandbanks, unlike the sometimes huge mounds of vegetation built high on riverbanks by their estuarine counterparts.

the paddling of the flattened tail. Valves seal the nostrils if the animal completely submerges, and another flap shuts the windpipe off from the gullet, enabling it to feed under water.

The myth of rotting prey

A frightened crocodile can stay under for about an hour before it is in danger of drowning. But it may keep out of sight for much longer if it has a lair under a riverbank, where air is trapped by a rising tide. Food that the crocodile cannot eat at once may also be taken to such a hiding place.

Crocodiles have been regarded with particular disgust because of a long misunderstanding of their feeding habits. They do not store victims under water and wait for them to rot. It is true that they cannot chew their food, any more than snakes or lizards can. But prey is eaten fresh — gulped down whole if possible. Bigger kills are torn apart in the grip of sharp-pointed, interlocking teeth, with a wrenching force supplied by the rotation of the crocodile's body. Others may share the feast, taking turns in an orderly fashion. If parts of victims are found decomposing in lairs it is because there was more than enough for all.

Feeding mainly by night, crocodiles are carnivorous from the time they are born. But their diet depends on their size. Hatchlings start on insects, tiny fish, crustaceans and frogs. As they grow they seize bigger fish, waterfowl and other reptiles. Mature crocodiles prefer mammals — or to cannibalise their own juniors.

Like all reptiles, crocodiles go on growing for as long as they live. Their teeth are perpetually replaced and their armour-plated hides get tougher. They become ever more skilful and powerful predators. And nothing else in their natural world preys on them.

For the males, breeding is the prerogative of the old — those that hold territories. Males attaining breeding age, unless they can find places of their own, are killed off by veterans perhaps twice their size and many times their weight. This apparently wasteful system of seniority makes crocodile populations — in the absence of human interference — stable and secure. The young are forced to roam, probing into every swamp and creek in the hope of finding an unoccupied niche. If one of the invulnerable elders dies of old age, a strong claimant to his kingdom soon arrives.

How many — and how big?

Surveys at selected NT sites in the mid-1980s indicated a total estuarine crocodile population of scores of thousands and an astonishing increase of about eight per cent a year.

Such a recovery rate reflects the drastic depletion by shooting and trapping before the 1970s. At what point the numbers will level out naturally, and whether they can be reconciled to the demands of public safety, no one can say.

Nor does anyone readily know what age, or more to the point what size, *Crocodylus porosus* can achieve. Few authorities doubt that the species is capable of outliving humans. The greatest length claimed is 9.6 metres, for an animal shot long ago near Mackay, Qld. That record is poorly authenticated.

For future generations to see living crocodiles approaching such a size would be a towering achievement of conservation. □

SQUEAKS MEAN IT'S TIME TO BE BORN

Eggs of the estuarine crocodile, averaging about 50 to the clutch, are laid on a riverbank in a scratched-up pile of rotting plant matter and mud or sand. The mother covers them with more of the same material to provide a constant internal heat during about three months of incubation. If the sun should overheat the nest, urine is sprayed on it.

The mother spends much of her time guarding the mound, supposedly to ward off predators that dig for eggs. But flooding is a far greater threat, for unlike the freshwater crocodile this species lays in the wet season. Researchers have seen every nest along a flooded river system destroyed.

Scientists for many years could not fathom how the young inspired one another to hatch together, or how the mother could know the precise time to open the nest and remove their suffocating covering, often 25 cm thick. Work on African crocodiles has revealed that the young set up a collective squeaking that she can detect through the shells and the mound.

Hatchlings from the one nest are of the same sex, evidently determined by the incubation temperature. Relatively few eggs hatch, and even fewer young survive their first year. They come out snapping, ready to feed or fight, and they are dispersed into the water fairly quickly. But predation by fish, waterfowl and older crocodiles is heavy. Little over 20 cm long when they hatch, survivors gain about 45 cm a year during their juvenile phase.

Males reach sexual maturity at about eight years, females sooner. Both sexes grow more slowly after that. □

A mother-to-be guards her huge riverbank nesting ground (above). When she senses that the young are ready she will scrape away the top so that the hatchlings emerge into the open air (left). The rotting of the nest material generates a steady internal heat to incubate the eggs.

The traffic and the terror

Resurgence of the estuarine crocodile — hunted to the brink of extinction — makes it once more a target of hatred and fear.

To northern Aborigines, having to share hunting and fishing grounds with crocodiles was a fact of life — and by no means undesirable. What danger may have existed was offset by the advantage of an ever-reliable and easily located supplementary source of food. Even little children could be given the task of collecting eggs, most of which would otherwise be destroyed in floods.

Some crocodiles were speared or trapped for their meat, but only to satisfy immediate needs. Big specimens were widely held sacred — they were believed to harbour the spirits of ancestors. As is often the case, ancient lore had a conservationist thrust. It is the oldest, biggest male crocodiles that sire new generations.

Cattlemen trying to establish tropical herds were first to declare war on crocodiles. Startling tales were told of the reptiles seizing steers, dogs and even a prize Suffolk draught horse, imported at great expense and said to weigh a tonne. Drovers made detours of hundreds of kilometres to avoid moving stock through rivers.

Graziers ordered the destruction of every nest that could be found. They shot crocodiles and left them to rot without even bothering to skin them. Some offered good money to itinerant hunters and trappers. The era of professional shooters dawned.

Soon they were joined, in wildernesses where crocodiles were in no conflict with agriculture, by wealthy amateur 'sportsmen' and adventurers seeking trophies. Huge specimens of *Crocodylus porosus*, already hard to find in Asia, were an easy target in Australia.

Lashing its tail to provide thrust, a crocodile can leap right out of deep water. Occasionally one lands in a boat.

Left: A potential killer in surf off Nhulunbuy, where a spotting tower is manned for the protection of bathers.

Shooting for skins

After World War II an insatiable international demand for skins to make handbags and shoes put high values on Australian crocodiles — the bigger the better. And the far north did not lack for restless returned servicemen, skilled enough in marksmanship to get in head shots without damaging the skins. Freshwater crocodiles were spared this first onslaught, because bony lumps under the belly scales made their skins less pliable and presented difficulties in tanning.

Everywhere within safe reach of settled areas, the estuarine crocodiles were shot out by the early 1960s. In most of Queensland there were hardly any to be seen. The species was not yet endangered, however. In the Northern Territory considerable numbers remained naturally protected by the remoteness of their habitats.

Then reliable light aircraft and outboard-motor boats became available. The slaughter resumed, now reaching far into untamed regions whose only populations were sparse scatterings of Aborigines. Often they were enlisted in hunting, receiving pittances for their efforts.

Before long the kills again declined. Expensive expeditions brought little reward for marksmen. But other hunters started laying hooked baits by the hundred in rivers that were supposedly shot out. Rare survivors, if they did not fall victim to this method, drowned in the fish nets that were set for another lucrative specialty of the far north — barramundi. Along many rivers, extermination of what the hunters called 'salties' was complete.

The turn of the 'freshies'
Later in the 1960s, with the trade coming to a standstill, tanneries devised techniques for processing the skins of freshwater crocodiles. Once a price was offered, a quick and easy massacre of the 'freshies' began. Few bullets were necessary — dragnets and hooks were sufficient to clean out most waterholes and streams.

No crocodile of either species was too small to bring in some return. Trapped specimens under 3 ft (91.5 cm) in length were slitted, gutted and stuffed as tourist souvenirs. At the end of the 1960s a 75 cm 'stuffy' could sell for about $20.

Nature conservationists by then were united in protest. Much of their argument related to the upsetting of delicate ecological balances. But in the southern centres of political power a more emotional chord was struck. Many politicians and voters learned for the first time of the ruthless thoroughness of the extermination, and were appalled. Legal protection of crocodiles was first enacted in 1971 in Western Australia, where crocodiles had never been abundant. Federal authorities imposed similar restrictions a year later in the Northern Terri-

Above: Freshwater crocodiles, harmless to humans, were also at one stage killed for their skins.

Below: Two photographs of Arnhem Land Aborigines showing off a good-sized crocodile skin. The pliable belly skin was the part prized for shoe and handbag manufacture, so, as the photographs clearly show, crocodiles were slit along the back to keep the valuable belly skin intact.

tory. The Queensland government, responding to tropical rural opinion, held out.

One of the first acts of the Whitlam government at the end of 1972 was to use its overriding Commonwealth powers to prohibit the export of crocodile products. But a trade in souvenir 'stuffies' and the mounted heads of bigger crocodiles continued within Queensland. With that market beckoning, hunting went on illicitly in the Territory and the west until 1974, when Queensland at last fell into line.

When the trouble began
Throughout a quarter-century of mass killing and during the conservation debate, the question of public safety from crocodiles was seldom put with much force. Aborigines were said to have been taken from time to time, but authorities knew of no reliable record of fatalities — certainly none involving Europeans.

Ironically the first fatal attack of modern times happened in Queensland in 1975, just a year after that state had granted protection to crocodiles. An Aborigine was taken near the bauxite mining town of Weipa, on the Gulf of Carpentaria coast. Sizeable crocodiles were such a rare sight in Australia then that it was regarded as a freak occurrence — perhaps the work of an animal from Papua New Guinea.

In 1979 aggression by crocodiles was so unusual that the antics of 'Sweetheart', the inhabitant of a lagoon near Darwin, were regarded with amusement. He was credited with more than 20 attacks on fishing boats, damaging their hulls or outboard motors. Wildlife officers attempted to relocate him for his own safety — they feared that a resentful fisherman might shoot him. Drowned when he was struck by a floating log while groggy with a tranquilliser, 'Sweetheart' measured 5.5 metres and tipped the scales at 780 kg.

All amusement ceased later that year when on the opposite side of the Gulf near Nhulunbuy on Gove Peninsula a Melbourne

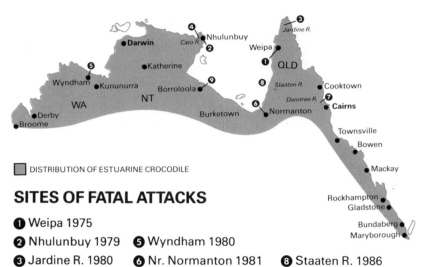

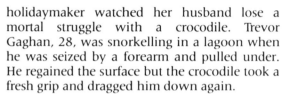

DISTRIBUTION OF ESTUARINE CROCODILE

SITES OF FATAL ATTACKS

1 Weipa 1975

2 Nhulunbuy 1979   **5** Wyndham 1980

3 Jardine R. 1980   **6** Nr. Normanton 1981   **8** Staaten R. 1986

4 Cato R. 1980   **7** Daintree R. 1985   **9** Borroloola 1986

Signs warning against swimming appear wherever visitors have access to the known habitats of dangerous crocodiles. But they have to be replaced frequently. Some tourists, with an irresponsibility that borders on the criminal, take them as souvenirs.

holidaymaker watched her husband lose a mortal struggle with a crocodile. Trevor Gaghan, 28, was snorkelling in a lagoon when he was seized by a forearm and pulled under. He regained the surface but the crocodile took a fresh grip and dragged him down again.

Gaghan died of drowning. His body was found intact, though extensively bitten, on a creek bank about a kilometre away. Rangers tracked the crocodile across mudflats and harpooned it. The killer was a mere 3.5 metres long. Nowadays crocodiles of greater size are common enough for Nhulunbuy authorities to have installed a croc-spotting tower at the town beach to warn swimmers.

Three in a year
Worries mounted in 1980 when three people were taken by crocodiles at widely scattered locations. Telecom maintenance men saw a 62-year-old Aboriginal helper seized as he swam in the Jardine River, near the tip of Cape York Peninsula. At the Cato River — again near Nhulunbuy — police recovered the torso of a 30-year-old Aboriginal woman who was taken while fishing. And at Wyndham, WA, two crocodiles were found with the mutilated body of a young Perth truckdriver who went swimming at 1 a.m. near the town abattoirs.

The following year a 43-year-old Gulf Country station hand was taken during an early morning swim in Walkers Creek, 30 km from Normanton. A female companion heard his desperate screams in the dark but could do nothing to help him. On Cobourg Peninsula and at Channel Point, south of Darwin, two other men escaped with injuries.

In 1983 a police officer stationed at Bamaga, Cape York, was tossed into the air by a crocodile that attacked him on land beside Cowal Creek.

He was uninjured but his dog was taken. And at Byno Harbour, southwest of Darwin, a crocodile only 2 metres long made an alarming assault on a fishing boat, leaping in and snapping at the arms of a young airman.

Lone ordeal in Kakadu
The most graphic account of a crocodile attack was given in February 1985 by Val Plumwood, a lecturer in environmental studies at Macquarie University, Sydney. At the height of the wet season in Kakadu National Park, NT, she made a lone canoeing expedition down the East Alligator River. Most of the park was closed by heavy flooding.

Exploring a paperbark swamp near Cahills Crossing, Ms Plumwood saw what she first took to be a floating log. Edging in for a closer look she realised it was a crocodile but could not back off in time. The crocodile began a battering assault on the canoe, pursuing as she paddled away. Fearing that she would be thrown into the water, Ms Plumwood made for the edge of the swamp and reached up to grasp some overhanging branches.

She was seized by the left thigh and pulled into the water — luckily little more than a metre deep. Although the crocodile spun her around, her face came back to the surface. At last she grasped a mangrove stem and held tight, not attempting to struggle.

Thwarted in its 'death roll' manoeuvres, the crocodile decided to change its grip. Ms Plumwood broke away, swam behind a paperbark and started to climb it. The crocodile lunged from the water and seized her again. The same routine followed: she was spun in the water but managed to cling to a mangrove.

Sliding towards destruction
This time when the crocodile tried to change its grip she flung herself onto a steep, muddy bank. At first she was unable to climb it, slipping back again and again towards the waiting crocodile. Eventually she found that if she spread her fingers and dug them fully into the mud, she could haul herself up.

Blood pumped from her torn thighs and buttocks. Tearing off strips of her clothing to make tourniquets, she began an agonised struggle through sodden bushland to reach a park rangers' station. Four hours later, the rangers came upon her. Alarmed because she had not returned, they had started a search.

It took all night, and relays of boats, vehicles and a helicopter, to extricate Ms Plumwood from the flooded park. After emergency surgery at Darwin Hospital and further repairs in Sydney, she made a good recovery.

Before the year was out another woman had a less fortunate encounter. At Barrett Creek, a tributary of the Daintree River in far northern Queensland, a Christmas party ended tragically when a local shopkeeper, 43-year-old Beryl Wruck, decided to cool off in the water. A crocodile warning sign stood just metres away.

Standing only knee-deep, Mrs Wruck — the mother of three children — disappeared in a swirl of water. Fingernails, toenails and what were assumed to be her bones were found three weeks later in a trapped 5.6 metre crocodile.

A new bloodbath starts
The incident provided a trigger that many north Queenslanders seemed to have been waiting for. Even before the remains were found, shooters and trappers were killing every crocodile they could find on the Daintree. Along the northeast coast, wherever big crocodiles had been seen, other hunters followed suit.

Queensland's protection laws had already been relaxed to allow people to use their own judgement on crocodiles, and destroy any that were considered 'immediately or potentially' dangerous to human life. Now police made it clear that they would allow the most liberal interpretation of this provision.

Wildlife officers sought time to remove potentially troublesome animals to breeding farms or fauna parks. But within the first few months of 1986, according to the estimates of some authorities, about 10 per cent of all known crocodiles on the east coast were killed.

The slaughter was given added impetus in February 1986 when Katie McQuarrie, a young deckhand on a fishing trawler, was killed while in the Staaten River, on the Gulf of Carpentaria coast. The boat's owner shot the crocodile but it sank with its victim. Her limbless body and the dead reptile were found next day.

With public antipathy to crocodiles mounting, Northern Territory wildlife officers kept their fingers crossed. But their turn came when in September Lee McLeod, a Queensland visitor to the fishing town of Borroloola, went to sleep on the bank of the McArthur River after a drinking session. In the morning only his shirt could be seen. His torso was found in the stomach of a 4.5 metre crocodile two days later.

If a child should die
In the circumstances of that attack there was no great sympathy for the victim. Territorians are proud of their wild crocodiles, as they were of their water buffaloes. They are acutely aware of the reptiles' value to tourism.

But if ever a victim is taken in complete innocence, while doing nothing to invite trouble — a small child sitting on a beach, for example — conservationists fear that there will be no way of preventing a blood-letting on the same scale as Queensland's. □

THE TUG-OF-WAR THAT FOILED A MAN-EATER
A schoolgirl waded in to drag her friend from death.

Twelve-year-old Peta-Lynn Mann grew up among Northern Territory wildlife. At Channel Point on the Timor Sea coast, a tough 180 km drive southwest from Darwin, her parents ran a safari business. Hunters had a choice of feral pigs or water buffaloes, while photographers revelled in a diversity of birds. But the star attractions were crocodiles, abounding in nearby paperbark swamps.

In April 1981, just back from boarding school, Peta-Lynn was treated to a homecoming tour by her parents' partner, 23-year-old Hilton Graham. Taking a four-wheel-drive truck, the two travelled 20 km south to the swamps, where an airboat was kept.

Their exploration went smoothly until almost nightfall, when the boat accidentally grounded. As Graham climbed out to push the craft clear, his pistol slipped from its holster. He was kneeling in shallow water, groping for the weapon, when a crocodile attacked.

Graham raised an arm to defend himself. A snapping bite broke it in two places. He got to his feet before the reptile, nearly four metres long, lunged again. It seized his right thigh and began pulling him into deeper water.

Throwing out his uninjured arm, Graham called for Peta-Lynn to anchor him. Fearlessly the youngster waded into the water, braced her legs and pulled on Graham's arm like a tug-of-war contestant hauling on a rope. The crocodile spun in the water, sweeping both man and girl off their feet and dragging them under. But Peta-Lynn regained her footing and kept on pulling.

She brought Graham's head to the surface and began to drag him and his attacker towards dry land. With a few paces to go the crocodile released its grip. But as Graham staggered away it surged out of the water and tore at his buttocks. Peta-Lynn wrenched him towards the bank and they scrambled up.

Leaving her weakened companion propped against a tree some 50 metres from the swamp, Peta-Lynn ran to find the truck. Graham had taught her to drive it when she was eight. She brought the vehicle to the swamp, got him aboard and drove home to Channel Point, her passenger semi-conscious.

No one was at the safari base but radio contact was made with a homestead 70 km away. After Peta-Lynn had covered Graham's wounds with antiseptic powder and wrapped him in a sheet, they set off for Darwin. Along the track they were met by Graham's fiancee in a faster vehicle. Within six hours of the attack the injured man was under hospital treatment. Though scarred, he recovered his full fitness.

In October 1982, when the Queen began an Australian tour in Darwin, her first official duty was to present the Royal Humane Society's gold medal for bravery and the Australian government's Star of Courage to Peta-Lynn Mann.

The Queen presents Peta-Lynn with her bravery awards.

Farmed crocodiles, receiving their meat ration, catch it in mid-air.

Keeping the peace

Potential killers, taken out of harm's way, become breeding stock for a humanely managed revival of the skin export trade.

Wildlife conservators in the Northern Territory perform a dual role, doing all they can to foster the recovery of the crocodile population while ensuring that public safety is not compromised. Officers of the NT Conservation Commission stand ready to act swiftly if estuarine crocodiles appear near settled areas. Warning signs are erected and the potential trouble-makers are captured and removed.

For the biggest crocodiles, rope snares are rigged at the waterline and baited with meat. A log is poised to fall if the snare is violently disturbed, closing and anchoring it. Smaller crocodiles are lured into cylindrical traps of wire

mesh, in or out of the water. Nooses are slipped over their jaws for handling and transportation.

Traps are laid permanently on the inner reaches of Darwin Harbour. These waters became so infested early in the 1980s that nightly 'spot-a-croc' trips in lashed-together canoes were a tourist attraction. Elsewhere the removal teams respond to specific emergencies.

Homing instinct

Attempts to give crocodiles new habitats in the wild, at remote sites where they cannot cause worries, have not succeeded. The animals have a strong homing instinct and the ability to make long journeys overland as well as through river systems and coastal waters. Some taken more than 100 km from their breeding grounds have found their way back within weeks.

All captured crocodiles are now taken to farms. The Territory has three, one of which, on the Stuart Highway about 20 km south of Darwin, is open to the public. Queensland has two farms. The oldest-established, at the Edward River Aboriginal mission on the Gulf of Carpentaria coast, has thousands of crocodiles.

Protected from cannibalism and the jealous savagery of older breeding males, young crocodiles develop on the farms in far greater numbers than would ever occur in natural situations. With these reserve stocks the survival of the estuarine species should be assured.

The farms also offer a way of reviving the lucrative skin export trade without again

Queensland wildlife officers tie a potential menace to a ladder before relocating it by tractor.

AVOIDING TROUBLE WITH CROCODILES

● Refrain from swimming, riverbank angling and small-boat fishing where warning signs are erected. They are placed where crocodiles of dangerous size are known to live.
● Elsewhere within the range of the estuarine crocodile (see map, page 58), seek local advice before taking risks.
● Do not interfere with nesting mounds.
● If you come upon a basking crocodile, retreat quietly. Head inland rather than towards the water. On no account place yourself between a crocodile and water.

Left: NT Conservation Commission researchers locate a flood-prone nest in Arnhem Land, just at the onset of the wet season. Below: The nest is uncovered and the eggs are taken away for experimental incubation in laboratory conditions. Scientists hope to learn how to predetermine the sex of hatchlings and to ensure the best rates of growth.

Until 1985, under an international convention, only crocodiles originating in captivity could be commercially exploited. Now the estuarine species has been placed in a less endangered category that permits controlled trade in the products of wild populations.

Cheating the floods

In Australia this relaxation does not mean any resumption of shooting or trapping in the wild. But it allows the collection of eggs from flood-threatened nests, and the farming of the hatchlings for later skin production as well as for breeding purposes.

Researchers are experimenting to determine the most favourable conditions for hatching the eggs on farms. They need a better understanding of the relationship between egg incubation times and temperatures and the sex and growth rates of the hatchlings. Once the programme is in full swing, the NT Conservation Commission intends to take about 4000 eggs a year from the wild — eggs that in most cases would otherwise never get a chance to hatch.□

The chances are that few of these farm-bred juveniles would have hatched in the wild, let alone survived.

depleting wild populations. The excess numbers of juveniles can be profitably culled at three or four years of age. Like buffalo farming, that promises a boost to Aboriginal employment and the general tropical economy.

When turtles snap

The seafaring giants are mild-mannered until they are picked up. Some smaller river dwellers have the worst tempers of all.

Turtles are toothless. But their jaws are equipped with tough, sharp-edged plates. Careless interference, with a hand within reach of the darting head of one of these reptiles, invites a snapping bite as damaging as that of a big bird.

Huge ocean-going turtles, frequenting warmer parts of the coast, are surprisingly placid. In the far north, where Aborigines and Torres Strait Islanders are allowed to hunt these otherwise protected animals, children sometimes ride on their backs.

Each summer at Mon Repos Beach near Bundaberg, Qld, one of the principal mainland nesting grounds, hundreds of turtles are handled. They are tagged, and some are weighed, as part of a state wildlife service research programme. Most belong to the world's biggest species, the carnivorous loggerhead turtle *Caretta caretta gigas*, with a shell length of up to 1.5 metres. Some flatback turtles, also carnivorous, and vegetarian green turtles also visit the beach to lay eggs.

Colin Limpus, head of the programme, knows of no instance of unprovoked aggression

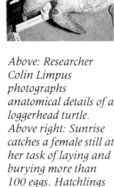

Above: Researcher Colin Limpus photographs anatomical details of a loggerhead turtle. Above right: Sunrise catches a female still at her task of laying and burying more than 100 eggs. Hatchlings emerge about two months later.

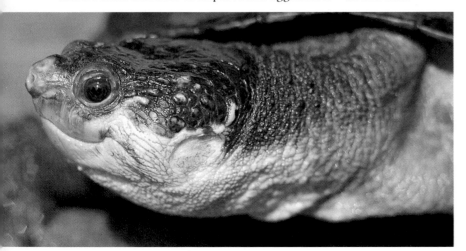

Above and left: Snapping tortoises of the Elseya *group are characterised by horny shields on top of their heads, warty skin over the temples, and two projecting tubercles, called barbels, under their chins. Their bites can inflict severe wounds.*

toward people — on land or in the water — by these species or by the more common hawksbill turtle (page 258), another carnivore. But he warns that any of them may attempt to bite if they are picked up. They can lacerate or crush fingers, and perhaps break them.

Limpus is aware of tales from overseas of swimmers having drowned with their hands or feet trapped under the shells of marine turtles. He is not inclined to believe any of them. Experienced since boyhood in handling turtles, he has never known anyone to be trapped in such a manner.

Inland snappers

Most freshwater turtles — commonly called tortoises in Australia — are capable of delivering painful bites if they are interfered with. Their jaw structures vary, as do their sizes, and few can cause significant damage. Members of one group, however, are so notoriously ill-tempered that they are known as snapping tortoises. If attempts are made to pick them up or restrain them they snap at anything they can see and bite whatever is within reach.

The most formidable of the group, the deep-shelled snapper *Elseya dentata*, inhabits the bigger rivers and overflow lakes of the north from the Kimberleys to the Gulf of Carpentaria and perhaps farther east. Feeding on molluscs, crustaceans and fish, it can reach a shell length of almost 40 cm. Older animals commonly develop oversized heads. Their bites can inflict severe wounds, with the damage increased by an extra cutting ridge in the upper jaw.

The saw-shelled snapper *Elseya latisternum* is said to be just as aggressive in its resentment of handling. But it is smaller, not often exceeding 20 cm in shell length. Ranging along the east coast from Cape York to northern NSW, the species is easily recognised by serrations in the rear edge of the upper part of its shell.□

King of the crabs

Tasmanians fish up the world's most massive species, with a pincer as big as a man's forearm. One blessing: it's slow-moving.

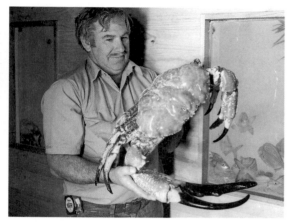

At depths of 90-150 metres in eastern Bass Strait and off eastern Tasmania — and apparently nowhere else — lives a crab that in body size dwarfs all others. The giant Tasmanian crab *Pseudocarcinus gigas* can measure as much as 45 cm across its carapace and may weigh up to 15 kg. One Japanese species has a claim to greater overall size, but it is a small-bodied, long-legged spider crab.

Smaller Tasmanian specimens turn up occasionally in crayfish pots. They are usually eaten. Full-grown crabs, accidentally trapped in shark nets or hooked in long-line commercial fishing, are rare finds. They are snapped up for public display in aquariums, the most ambitious of which is near the resort township of Bicheno, on the east coast. Numbers of the giants live there for years.

In the males, one crushing, cutting pincer-claw is about as long as the carapace is wide. The other is only about half as big, but still awesomely powerful. Females have two claws of moderate size.

Bottom-feeding scavengers by nature, the crabs are ponderous in their movements. That is just as well for fishermen trying to land them. Handlers approach cautiously from behind, planting a foot on the back of the carapace while the claws are lassoed and safely lashed.

There are some close calls, however. The broomstick handles of big dip-nets, used to lift the crabs out of the water, have been snapped like matchwood by resentful giants. The fishermen are in no doubt that an arm could be amputated just as easily.

Muddies and sandies

On a smaller scale, a similar danger is presented by many crabs inhabiting shallow waters on tropical and subtropical coasts. In a sense it is a

Left: Bill Bailey, manager of the Sea Life Centre at Bicheno, eastern Tasmania, shows off a local catch.

Below, centre: A marine crayfish or 'spiny lobster' has no claws. Injuries can be received from its spines or from the sharp edges of its shell segments. Shown here is the main commercially fished species of the southeast, the southern crayfish Jasus novaehollandiae.

Below, bottom: Handling the mangrove or mud crab Scylla serrata is best left to experts.

greater danger because the crabs are relatively common — and often accessible to children.

Mangrove or mud crabs — commonly called 'muddies' on the Queensland coast — and blue swimmer or manna crabs — 'sandies' — can measure about 20 cm across and are massively armed. These are the types most highly prized for eating. But people should not attempt to handle them without training in how to secure and tie them. Other types reaching much the same size include coral crabs, ghost crabs and red-spot crabs. Males of any of these are probably capable of crushing or cutting off fingers. Children should be warned as a matter of course not to play with any big crab.

Comparable injuries can be inflicted by the pincers of large freshwater crayfish. Often called yabbies, some of these river-bottom scavengers can grow to 30 cm long. Sea crayfishes — marketed as lobsters — have no claws. But they still require careful handling: the spines can cause painful punctures and fingers caught between segments of the shell can be deeply cut. Either type of wound is likely to be infected with marine micro-organisms.□

Octopuses: fact and fable

*Drowning in the clutches of evil-eyed monsters is a fantasy. The true risk —
if you go asking for it — is of suffocating.*

In the tiny but highly venomous blue-ringed *Hapalochlaena* (page 210), Australia has the world's most dangerous octopus. By any reasonable assessment — barring freakish mishaps that people go out of their way to create — it is probably the *only* dangerous octopus.

During World War II, when naval divers in the Allied forces were issued with copious instructions on dealing with every kind of marine animal danger, octopuses did not rate a mention. Modern diving manuals are similarly dismissive. Yet the popular belief persists that big octopuses attack people, hold them under water or in rising tides, and drown them.

That belief is largely the creation of fiction writers, film makers — using ingenious mock-ups — and carelessly sensational journalists. But it is reinforced now and then by what seem to be solid facts. Early in 1986, for example, the

The octopus is a mollusc without a shell. It is a timid creature most often found on the rocky bottoms of coastal shallows, where it preys on crustaceans and smaller molluscs. It is rare to see one longer than about 75 cm, including the arms, in Australian waters. Some bigger types may be met at diving depths, but they are harmless provided they are left alone. They do not see human beings as potential prey.

natural resources minister of Kiribati, formerly the Gilbert Islands, commented on two deaths in the lagoon at Tarawa. The victims were said to have been held under water by octopuses 3-4 metres long. The minister called for a change in fishing methods.

Well he might. In common with many Pacific Islanders, the Tarawans traditionally catch octopuses by snatching them from their lairs or feeding grounds and inducing them to cling. Then they bring them to the surface and kill them by biting them between the eyes.

The men who came to grief must have picked on octopuses of unmanageable size. Conceivably the animals regained a purchase on coral outcrops and held the divers down long enough to drown them. In theory a man weighing 80 kg can be held under water by a pull of only about 4 kg — if he does nothing about it.

More likely, in attempting to bite their unwieldy catches, the men's faces were blanketed by the bodies and webs of the octopuses and they were suffocated.

Octopuses are molluscs without shells. To evade their many natural enemies, they rely on agility, the ejection of ink to cloud the water, and a ready access to hiding places. They are wary of any large body moving near them and, according to the underwater explorer Jacques Cousteau, pathetically frightened of humans.

Mistaken for a mate?

But once accustomed to the presence of a person close to its lair, an octopus may be emboldened to send out an exploratory arm. The touch is usually light, fleeting and — though perhaps repulsive — harmless.

Sometimes, however, if a startled person makes sudden movements, an octopus applies more of its eight arms and grips firmly. It is probably breeding time, when adults grapple as part of their preparation for mating. Waving human limbs may be mistaken briefly for the arms of another octopus.

There is no danger. If the 'victim' keeps still the octopus soon lets go. However big it may be, it certainly does not view a human as potential prey. Its body and parrot-like beak are too tiny — for nearly 95 per cent of the mass of an octopus consists of its arms. In the highly unlikely event of a bite, the puncture wound is slight. Salivary toxins are not known to be life-threatening except in our blue-ringed species. Where sucker pads have held the skin, reddened bumps may sting for a short time. Strong suction on the eyes could be damaging, but no such case has been recorded in Australia.

Frank Lane, a British author, has written the most comprehensive popular work on the life of octopuses. Human safety is not a major aspect of his book, *Kingdom of the Octopus*. But he went to considerable trouble to seek verifiable accounts of attacks. Lane received no evidence of fatalities. Nevertheless he offered a conclusion that octopuses were sometimes dangerous. He based this on stories from people who had been gripped in shallow water, needing the help of companions to break free.

These people believed that if they had been alone, and remained trapped when the tide rose, they would have drowned. Born of panic, it was an imaginary fear. Nothing known about octopuses suggests that they would hold on even for minutes, let alone for hours.

Some stories have been outrageously embellished to sell publications. Lane's enquiries uncovered such a case. An Englishwoman gave a simple account of having helped extricate her sister-in-law from the grip of an octopus in knee-deep water on the French Riviera. It had arms about 75 cm long. Later the women traced their tale through one newspaper after another. It became progressively more harrowing. Finally it told of the deaths of two young American bathing beauties, dragged into the sea by an octopus with 12-metre arms!□

Gustave Doré's engraving for Hugo's Toilers of the Sea.

THE POWER OF THE PEN
'Devil fish' horror was born in a novelist's brain.

Popular dread of big octopuses was fostered first and most powerfully by the best-selling French author Victor Hugo. In *Toilers of the Sea*, published in 1866 and set in the Channel Islands with an eye to the English market, the fisherman hero is shipwrecked and has to fight for his life against a huge 'devil fish'.

Hugo let his imagination run riot. The double rows of suckers on the arms of the octopus were like 'so many mouths devouring you at the same time'. Piling on the horror, the English translation goes on: 'You enter the beast... You become one. The tiger can only devour you; the devil fish inhales you.

'He draws you to him; and, bound and helpless, you feel yourself emptied into this frightful sac, which is a monster. To be eaten alive is more than terrible, but to be drunk alive is inexpressible.'

Splendidly dramatic, it was a flight of fancy that was eagerly taken up by other writers. But nothing like it could ever happen. When motion pictures came along, heroes and heroines were obliged to pretend to grapple with rubber models.

Squids are faster and fiercer than octopuses

Happily, we don't have to contend with the giant kind.

Compared with their octopus relatives, squids offer a more realistic threat of human injury. While an octopus is nearly all arms, a squid is mostly body, with a proportionately bigger beak. And instead of feeding sedately on bottom-dwelling molluscs, it is a ferocious hunter of free-swimming fish.

Squids belonging to the calamary family — the kind favoured for eating, common in bays and estuaries on most parts of the coast — are so avid in their pursuit of prey that they often leap from the water. Their streamlined speed and energy earn them the common names of 'sea arrow' or 'flying squid'.

Sometimes calamaries attempt to rob anglers of their catches, seizing and tearing at hooked fish. And if they are captured they take any opportunity to retaliate. Wounds from the slicing bites of bigger specimens have been likened to a snip with sharp scissors.

Calamaries can reach about 90 cm in the total length of their bodies and eight short arms. In addition, like all squids and cuttlefishes, they

Above: The kraken, an avenging sea monster of Norse mythology, has a real existence as the giant squid. Just nine years after this imaginative illustration appeared in a French novel in 1865, a 150-tonne schooner was reported sunk by a squid off India.

have two longer tentacles. Suckers are carried in two rows on the arms and four on the tentacles; their outer rings have teeth.

Deep-sea dwellers

Giant squids, the biggest invertebrate animals in the world, are probably responsible for ancient sea-serpent legends. They live where coastal shelves are deepest, surfacing rarely.

Dead or dying specimens have been washed ashore from time to time in New Zealand, but only one has been recorded in Australia. Found at Wingan Inlet, eastern Victoria, in 1948, its mutilated head and body measured 2.75 metres. The arms had been chewed off, but a total length of about 8.5 metres was estimated.

No one knows the maximum size of giant squids. Authorities do not rule out the possibility that they exceed 30 metres in total length and weigh many tonnes. Animals of such size could undoubtedly wreck boats and kill men.

One of the few authenticated attacks of modern times took place off West Africa during World War II. Twelve survivors from a sunken British troopship shared a small liferaft, some of them having to stay in the water while they clung to it. At night a tentacle wrapped around a sailor and he was plucked away, never to be seen again.

The report has credibility because another man, an army officer, was briefly grasped around one leg. When the party was rescued he needed treatment for ulcers where the squid's toothed suckers had gripped him. A biologist examining him two years later found circular scars the size of pennies.

As evidence of the immense proportions that hidden giants may attain, scientists point to sucker scars on the snouts of sperm whales. These deep-diving whales are specialised hunters of squids. Some bear scars 13 cm across — the size of bread plates.□

The southern calamary Sepioteuthis australis. *Squids belonging to the calamary family are also known as sea arrows or flying squids.*

Sting in the tail

Intent on a quiet life, stingrays glide over the sea floor browsing on shellfish. But woe betide people who step on them!

Rays are related to sharks, having skeletons of cartilage instead of bone. But in their evolution the rays took a course away from the sharks' life of predatory pursuit, with its demand for sustained tail-thrashing, energy-sapping swimming. With one exception — the horned 'devil' rays such as the manta — they settled for a humble existence as slow-moving bottom-feeders.

Rays' bodies became flattened, with the side fins enlarged to form lateral extensions. They developed twin holes on top, called spiracles, for drawing in water to pass over their gills and extract oxygen. Water taken in through their bottom-facing mouths, in the manner of sharks, would be full of sand and silt.

Instead of sharp, flesh-tearing teeth rays have grinding plates to crush molluscs and crustaceans. And having developed the ability to move by waving their sides, they have virtually lost their tails. At best a slender, puny-looking appendage remains.

But in many rays this vestigial tail has become a defensive weapon. It can be used like a whip, lashed upwards against threatening predators. In the types that we call stingrays or stingarees — the enormous manta ray is not one, fortunately — the tail is armed about halfway along with one or two barbed spines.

In about two-thirds of stingray species the spines are venomous, carrying glands under a sheath that is destroyed as a spine enters a target. Envenomation can be painful and sickening, and ulceration may follow. As is the case with many other fish stings there is a chance of fainting — hazardous in some situations.

Stingray venoms, however, do not directly threaten the lives of people in normal health. The initial physical damage, the high likelihood of infection, and the later foreign-body reaction

Above: The black stingray Dasyatis thetidis *is common in temperate waters from NSW to WA.*

Left: Barbs on a stingray's tail are usually about halfway along. In most species the sheaths covering them contain venom glands. But wounds are not venomous if the barbs have been recently used on predators, because the tissue and glands take a long time to regrow.

Bottom left: The eagle or bat ray Myliobatus australis *occurs in all Australian waters.*

Right: The common stingaree Urolophus testaceus *rarely exceeds 75 cm in length but can still be dangerous.*

if spine or sheath fragments stay in a wound, are usually of far greater concern.

Spines of the bigger species, as long as breadknives, can pass completely through limbs or tear deep, jagged gashes in the flesh. Often the greatest damage is done by barbs or serrated edges as the victim recoils from the attack and the spine is wrenched out. Spines penetrate wetsuits as easily as normal clothing, and even pierce leather footwear. Most injuries are to the legs, when people wading in shallow water tread on stingrays. Fishermen occasionally suffer wounds through inept handling of stingrays that are accidentally netted or hooked.

The people most vulnerable are divers cruising along over the sea bed. If someone takes a stingray unawares and frightens it by passing close above it, the lashing tail could ram a spine into their unprotected body. People swimming normally in shallow water are not usually in danger — they cause too much disturbance for stingrays to stay around.

Getting aquainted with a numbfish.

A SHOCK, PERHAPS — BUT NO HORROR
The electric ray can jolt you with a charge of 200 volts.

Four species of non-stinging rays in Australian waters defend themselves with electrical discharges. Paired organs behind the eyes consist of cells that present positive poles at the upper surface and negative poles on the underside of the body.

The most common species, *Hypnos monopterygium*, can pack a punch of up to about 200 volts when it is fully charged. If the ray is approached it does not flee but buries itself, sending out impulses at intervals. They travel through the water but are unlikely to be noticed by a passing swimmer. If the ray is touched, however, the shock is enough to cause a brief cramping of muscles. Hence its common names, numbfish or crampfish. Shocks can be received along wet sticks or boards that are used to move numbfish. They can be transmitted faintly across wet sand. Beached or netted, the distressed animal continues to discharge impulses — perhaps 50 or so in ten minutes. The voltage dwindles steadily.

Numbfish feed in sandy and muddy shallows or at depths down to more than 200 metres. Though they present a theoretical danger through cramping — if a perverse diver were to persist in embracing one, for instance — they and their less common relatives the torpedo rays have not been known to harm anyone.

AVOIDING TROUBLE WITH STINGRAYS

Wading on sandy bottoms, shuffle your feet instead of using a high-stepping action. If you dive and cruise along within 1 metre of the sea bed, look ahead with extreme care and remember that a stingray may be partly hidden in sand or rubble. If you see one in your path and cannot avoid it, disturb the water vigorously and it will quickly swim away. Unless you are expert in handling rays, cut free any that are accidentally hooked or netted. There is no safe place to hold some species — the tail can whip over to strike a wrist or forearm. Spines cut from stingrays remain dangerous and should not be placed in pockets or elsewhere in clothing.

WOUNDS Obtain urgent medical aid if the trunk has been penetrated, even if there is little external bleeding. Meanwhile give first aid appropriate for severe wounding and shock (page 358). Be ready to use resuscitation techniques. In the case of a limb injury, lay the victim down with the affected area raised. Wash the wound thoroughly with sea water. If the spine remains in the wound, remove it if this can be done gently, along with any noticeable fragments of sheath or glandular tissue. Lightly cover the wound with a clean dressing and seek prompt medical attention. Any stingray injury must be referred to a doctor because of the near-certainty of infection.

PAIN FROM ENVENOMATION Bathing wound and surrounding area in hot — but not scalding — water usually brings relief.

Stabbed in the heart
But Melbourne people were aghast in 1945 when an army sergeant was killed in the city's most popular sea baths, at St Kilda. Swimming strongly, he suddenly disappeared with a wave of his hands. He surfaced after nearly a minute, sank again, and when rescued by companions he was struggling to breathe.

Blood showed on a small wound in the soldier's chest. In spite of resuscitation efforts he was pronounced dead 20 minutes later. A postmortem examination showed that a stab wound had penetrated his ribs, left lung and heart. No manmade object capable of doing such damage was found in the baths. But the paling fence had gaps at a low level. The coroner attributed the man's death to penetration by a stingray spine. Presumably the animal, which in the open sea would have fled from a swimmer, was temporarily trapped.

Severe injuries are surprisingly rare, considering the abundance of stingrays around our coast. They are among the commonest types of big fish in almost any bay or estuary. James Cook in 1770 was so struck by their numbers that he originally gave Botany Bay the name Stingray Harbour. Coastal Aboriginal tribes relied on the rapier-like spines as heads for their hunting and fishing spears.

Beachgoers can be grateful on two scores. Many stingrays tend to be more active at night, often sheltering by day in crevices and caves. And they are not only timid but also extremely alert. Given any chance at all of getting out of our way, they will do so.□

The gentle giants

Forget the harrowing scenes of films and fiction. Enormous clams on the Great Barrier Reef close far too slowly to trap anyone.

Assertions have been made over and over — sometimes by reputable scientists though without a shred of authenticated evidence — that giant clams can and do trap people and hold them under water. It is a myth, an ancient tall tale of the sea that has been seized on happily by novelists and film makers.

Some *Tridacna* clams, all seven species of which are found in the Great Barrier Reef region, are indeed big enough to enclose a man's leg. And they possess overwhelming muscular power to lock shut their shells. A French naturalist tested a 250 kg specimen by anchoring it to a post and hooking on buckets of water. It took a pull of nearly 900 kg to draw it open. But although smaller species sometimes shut quickly if they are disturbed, the true giants, *Tridacna gigas* and *T. derasa*, are ponderous. They usually take minutes to close fully. It is inconceivable that anyone could have a limb in one of them and not know what was happening if it began to close.

Above: Varying colours and patterns in the mantles of giant clams are partly produced by millions of algae living in their tissues. The algae take carbon dioxide and nutrients from the clam, and the clam gains sugars and other carbon compounds that the algae make by photosynthesis. But this requires strong sunlight, so giant clams live as close as they can to the water surface.

Right: A species of Pinna *razor shell embedded in mud. If one of these is stepped on the fine edges cause deep cuts, and fragments breaking off delay the healing of wounds.*

It is also most unlikely that swimmers or waders, in daytime at least, could come upon giant clams without seeing them. Relying on strong sunlight for their growth processes, these animals fasten themselves to coral growths and rocks barely beneath the low-tide level. Their vividly coloured mantles make them perhaps the most conspicuous of marine creatures.

No predatory instincts

Closure of a clam's shell is purely defensive — nothing to do with any aggressive or predatory instinct. Clams are not even interested in fish that may swim through their open shells. They are filter-feeders, gaining nourishment by straining micro-organisms and vegetable particles from the water.

The biggest pair of *Tridacna* clam shells on record, taken from the Barrier Reef in 1917, are in a New York museum. They are 1.075 metres long and together weigh more than 260 kg. The Australian Museum in Sydney has a pair very nearly as big.

Giant clam flesh — especially the muscles — fetches high prices on Asian markets. Poaching by foreign fishing crews is prevalent on remote parts of the Barrier Reef. Specimens of leviathan size are encountered less and less often and total populations are declining. Though the clams are hermaphrodites, capable of fertilising their own eggs, they need the close proximity of others to ensure a good rate of reproduction through mutual spawning.

However big the clam that a reef visitor is lucky enough to find, there is nothing to fear from it. Waders concerned where they are putting their feet have more to worry about from slipping into coral crevices or gashing themselves on oysters.

In practical terms the most injurious molluscs, in sheltered shallows all around Australia, are the aptly named razor shells. Species of *Pinna* and *Atrina* feed by embedding themselves vertically in sand or mud with a cutting end jutting 2-3 cm from the surface. But in popular bays and estuaries, their threat does not rank with the menace of broken glass.□

Guard dogs of the reef

*Moray eels are ferocious defenders of their lairs.
They can be tamed, even patted — but mind how you offer
the hand of friendship.*

Divers paying regular visits to reefs or sunken wrecks in warm waters soon recognise the same moray eels, always lurking in the same places. Their confidence won with tidbits of food, the eels sometimes become so friendly that they allow handling and stroking with apparent enjoyment. Underwater feeding sessions featuring colourful morays have become the highlight of many a tourist trip on the Great Barrier Reef.

The behaviour of such eels is conditioned. Do not assume that others are as amiable. To attempt to touch a big specimen unaccustomed to humans would be as foolish as trying to pat a guard dog that doesn't know you. And the

Rocky reefs off NSW and the coral of Lord Howe Island conceal the mosaic moray Enchelycore ramosa *(right, above). The leopard moray* Gymnothorax flavimarginatus *(right) is a tropical coral species.*

Below: The green moray Gymnothorax prasinus *is common on rocky reefs off all temperate coasts.*

consequences could be as severe.

Reef eels are among the most sedentary of fishes, favouring the same confined habitats throughout their adult lives. By day they rest with only their heads protruding from crannies and crevices, occasionally snapping at passing fish or crustaceans. More active hunting takes place at night. Devotion to its home makes an eel a desperately ferocious defender against intruders. Savage retaliation is certain if someone tries to reach in after an eel that has retreated into its lair.

And although morays are not known for unprovoked aggression, a few people have been bitten when they merely lingered too long near an unseen eel, unwittingly threatening the animal's security. A NSW fisherman, cleaning his catch in a rock pool in the 1920s, had a finger bitten to the bone by a moray that had been concealed under a ledge.

Eels' teeth are designed for grasping prey that is too big to be engulfed whole. A moray's jaws, open to full gape, lock onto the body of a fish. Then the eel literally ties itself in a knot. A tight muscular contortion travels up from the tail, arriving against the fish's side with such force that the eel's jaws are wrenched away, carrying with them a huge chunk of flesh.

Blood loss and shock

Bites on human victims are deep and tearing, with most damage done by two heavy, oversized teeth at the front of the lower jaw. Wounds are usually ragged, and bleeding can be profuse. Shock from blood loss may result —

Long-finned eels
Anguilla reinhardti
(above and right) feed
at night in east coast
and Tasmanian rivers.

Biting reef worm Eunice aphroditois.

A WORM THAT BITES
It's as long as an eel, and has venomous bristles.

Smaller holes in coral reefs can conceal worms up to 1.5 metres long, capable of giving painful bites if they are interfered with. The biting reef worm *Eunice aphroditois* seldom exceeds 2 cm in diameter and its mouth is small, but it can push out its powerful jaws to inflict a fair-sized wound.

Eunice is a free-living, carnivorous marine worm — not a tube-dweller — related to the giant beach worms that bury themselves in tide-washed sandy shores and are prized as bait by anglers. While the beach worms hide, the reef worms defend themselves vigorously. They have an additional armament of venomous bristles that can cause rashes, and the sting of their bites suggests that salivary toxins may be injected.

AVOIDING TROUBLE WITH EELS

Don't put your hands into crevices or small caverns in coral or rocky reefs, and don't leave your feet near them unless they are heavily protected. If you see the head of a big eel keep clear — don't reach towards it.

SEVERE WOUNDS If bleeding is heavy try to stop it by applying local pressure (page 358). Tie a tourniquet if this does not succeed. Send for medical assistance and monitor victim for symptoms of shock. Be prepared to use resuscitation techniques if necessary.

Less severe wounds should be thoroughly washed in fresh water if possible, lightly covered with a dressing and referred promptly to a doctor. The chance of infection is extremely high.

Conger labiatus
(right) is a common
conger eel of east coast
waters.

highly dangerous if this should cause a victim to faint in the water.

Injuries inflicted by eels are notoriously prone to serious infection. There is a widely held belief that the bites of some morays are venomous, causing local paralysis, though scientists have found no venom apparatus. Those morays that live among coral may very well be dangerously poisonous to eat, being the likeliest accumulators of the toxin that causes ciguatera disease (page 248).

Morays of several species can exceed 3 metres in length, with bodies as thick as a man's thigh. Lengths of 1-2 metres are more usual. The fish are found mainly in tropical and subtropical waters, but they are not entirely restricted to reefs. The world record length for an eel, 3.7 metres, is claimed for a moray taken in a crab pot in the Coomera River, on Queensland's Gold Coast.

But muddy estuaries and the tidal reaches of rivers are more commonly inhabited by silvery-skinned pike eels — fearsome fighters if they are hooked by anglers and hauled into boats. At Currumbin, also on the Gold Coast, in 1976 a council worker was savaged and the heads of ducks and swans were bitten off by a giant said to measure 2.5 metres, inhabiting a stagnant creek.

Congers of the south
Around rocky reefs in cooler regions the eels of greatest size and presumed menace are the congers, often sought after as food fish. Much like morays but duller in colour and possessing side fins, the congers common along the southern mainland and Tasmanian coasts can exceed 2 metres in length.

Some danger from bigger specimens of the commercially fished and farmed freshwater eels, *Anguilla*, cannot be ruled out. They tend to be less confined in their habitats than reef eels, however, so they are more likely to avoid humans. Two species common in southeastern rivers can exceed a metre in length.☐

Big, bold and hungry

Barracudas and gropers don't deserve man-eater status. But they are impulsive — and far too powerful to make easy company down below.

Big barracudas are said in some parts of the world to be more feared than sharks. Their reputation seems to have originated in the Caribbean and is reinforced by some reported attacks on Americans at popular beaches in Florida. In the light of experience in tropical Australian waters, where the same species of barracuda abound, an excessive fear of them seems hard to justify. Proved attacks have been extremely rare, and they have usually involved fish that were speared or hooked.

The potential killing power of barracudas is undeniable, however. Their swift massacres of schools of prey are awesome. In flashing runs they slash from side to side with dagger-like teeth, then turn to feed on dead or dying victims at their leisure.

When fish of such an aggressive nature reach 2 metres in length and weigh perhaps 40 kg, they deserve to be given a wide berth. But big, solitary barracuda sometimes unnerve divers by stalking them, cruising to within a few metres and following even into knee-deep shallows.

Experts consider that there is no danger from this behaviour — unless a diver is holding or towing speared fish. Then a hungry barracuda's slashing attempt to hijack it could result in a grave though unintended injury. Since that possibility also arises in the case of sharks, sensible spearfishermen guard against it by disposing of their catches immediately.

Australian divers are generally counselled to avoid reef locations known to be frequented by barracuda, especially if the water is murky. In particular they are advised not to spear barracuda. And they should not wear or carry shiny instruments and equipment that could provoke a reflex-action attack.

The short-sighted groper

While the bad reputation of barracudas was imported to Australia, that of the giant groper seems to have started here and spread round the world. And such is the monstrous size of the animal — up to 3.6 metres long in the Persian Gulf — that people accept without question that it is a man-eater.

Growing to a maximum of about 2 metres in northern Australian waters, the Queensland groper or giant 'cod' *Promicrops lanceolatus* is among the friendliest of fish, coming to meet divers and taking offerings of food. Divers know not to flutter their hands or feet, because the groper is short-sighted and may snap at anything suggesting a fish. But injuries sustained in this way have been minor. The groper and the bigger species of *Epinephelus* rock-cods — similarly good-natured and also sometimes called gropers — are strongly territorial, however. A question mark is raised about how far they might go in resisting human intrusion on their home grounds. Divers have been rammed, but there are no authenticated cases of their being deliberately bitten.

Most stories of serious attacks are semi-legendary, dating from the rough-and-ready

Above: Queensland's Epinephelus tauvina is known as a brown-spotted, greasy or estuary rock-cod, and sometimes as a groper.

Right: The Queensland groper Promicrops lanceolatus is also called a giant cod or giant groper.

Below: Sphyraena barracuda, the biggest and fastest of their family, hunt on the Great Barrier Reef.

era of pearl and trochus shell diving and *bêche-de-mer* collecting. Unnamed workers were said to have been taken in remote places such as the Torres Strait Islands. Consequently the CSIRO in its early days described the Queensland groper in an official publication as 'vicious'. Maritime authorities also condemned it.

Some of the old stories mention divers having been seized by the head. It is probable, though not stated, that they were wearing shiny iron helmets. That provides a link with a more recent and better authenticated incident.

In 1943, naval salvage divers were repeatedly harassed by gropers attracted to their helmets. One fish took a helmet in its jaws and carried off the wearer, who was rescued by his mates. It was not really an attack — just another mistake by an impulsive but myopic species.

Savaged by a school of mackerel

Barracudas, gropers and rock-cods certainly do not present a danger comparable to that posed by sharks. Nor, if the experience of two Sydney teenagers is anything to go by, do they even rank with a frenzied school of small mackerel.

John Condron, 17, and Chris Leheane, 19, were exploring inshore waters at Caves Beach on the NSW south coast in January 1983 when they saw about ten pike darting towards them. In pursuit was a big school of slimy mackerel, all about 20 cm long. The youths found themselves between hunters and quarry and were indiscriminately bitten.

Hundreds of crazed mackerel beached themselves as the pair fled the water. Gashes on the youths' exposed arms, hands and legs required hospital treatment. Only their wetsuits saved them from more serious injury in a freakish event unparalleled in Australian fishing lore.

Among smaller fish, it seems that the ones giving us most to worry about are the various pufferfish. Not only are they lethally poisonous to eat, but they also bite with unusual power thanks to their heavy, fused teeth (page 244).

At least two paddling children, on opposite sides of the continent, have lost toes to these bold aggressors. At Kalbarri, WA, in 1977 a ten-year-old boy felt a sharp pain and saw a 35 cm northwest blowfish swimming away with one of his toes. Two years later at Shute Harbour, Qld, a five-year-old girl was deprived of the first joints of two toes by a 40 cm toadfish.☐

A blowfish's teeth can amputate a toe.

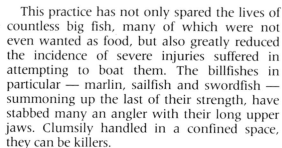

This practice has not only spared the lives of countless big fish, many of which were not even wanted as food, but also greatly reduced the incidence of severe injuries suffered in attempting to boat them. The billfishes in particular — marlin, sailfish and swordfish — summoning up the last of their strength, have stabbed many an angler with their long upper jaws. Clumsily handled in a confined space, they can be killers.

Mishaps still occur among casual fishermen. Often they are contestants in competitions that are thrown open to all comers and require that all catches be weighed. Participants in these light-hearted events sometimes let their enthusiasm — and the refreshments they take — get the better of them.

Boating even a small billfish is hazardous. A marlin that skewered a NSW man's thigh in 1985, crippling him for weeks, weighed a mere 45 kg. He would have been hoping for one ten times that size. Obviously anglers cannot foretell what they will hook. It can only be suggested that someone who is not expert, and whose reflexes may be slowed, should never attempt to bring in any billfish.

Greyhounds of the ocean

The elongated 'beaks' of these animals are used for stunning — rarely impaling — schools of prey fish, and are part of their streamlining for high-speed pursuit. Some of the various species of marlin, *Makaira*, have been timed at 80 km/h and can probably go faster in short bursts. Dashing runs by the swordfish *Xiphius gladius* have been estimated at 90 km/h. And some sailfishes, species of *Istiophorus*, are the fastest marine creatures in the world, capable of travelling at speeds of about 110 km/h.

Though they reach 3 metres, sailfishes are the lightweights among billfishes. The bulk of the other two types can be tremendous. Black marlin hooked by Japanese longline fishermen have exceeded 4 metres and 900 kg. Swordfish can reach 4.5 metres and over 500 kg. With

Fastest of all fishes, the sailfish (above) is relatively light. Catches in Australian waters seldom exceed 40 kg. Black marlin (right and bottom) may be 20 times heavier.

Opposite: Underwater researcher Valerie Taylor shows off the snout of a green 'sawfish' stingray. This giant species can exceed 7 metres.

Points in their favour

Badly handled by unwise anglers, billfishes can inflict terrible damage with the sword-like elongations of their upper jaws.

Game fishing has soared in general public esteem since clubs adopted an international code that recognises the need for conservation of the hunted species as important economic and social resources. It puts sportsmanship ahead of slaughter.

Under the rules of well-conducted clubs the lightest possible tackle is used. Catches exhausted by the most patient 'playing' are brought alongside boats — not to be gaffed and hauled aboard but in most cases to be tagged for scientific research purposes, then released. Exceptions are made, and fish landed for weighing, only if boat skippers believe that a record is about to be broken.

A single blade is carried at each side of the tail-base of the yellow surgeonfish Acanthurus olivaceus *(above). More prominent in a contrasting blue are the double blades of the long-snouted unicorn fish* Naso unicornis *(below).*

WELL-NAMED: THE SURGEONFISHES
Blades in their tails cut like scalpels.

Surgeonfishes are among the most gorgeously patterned reef-dwelling fish, commonest in the tropical zone but also occurring around rocky reefs to the south. They take their name from sharp blades in the bases of their tails. Normally these lie flat. But when the fish are threatened the blades are pushed out sideways, projecting like small keels.

Severe cuts are suffered if these fish, reaching 20-40 cm in length according to their species, are caught and allowed to lash their tails while being handled. The sharpest blades — one each side — are carried by species of surgeonfish *Acanthurus*.

Others in the group are species of *Zebrasoma* — generally called tangs — and *Naso*, which are known as unicorn fish because of their horned heads. The unicorn fish have two blades, fore and aft, on each side of their tail bases.

such masses moving at such speeds, it is no surprise that museums around the world show examples of old boat timbers impaled on swordfish and marlin bills. But there is very little evidence of deliberate running attacks on boats, even by hooked fish.

Most cases of boats being breached seem to have occurred simply as collisions, when the craft got in the way of billfishes in pursuit of prey. Among other game fish, it is hooked mako sharks that are most likely to make retaliatory attacks on boats.

A snout that saws
Certain big, stingless rays have an extraordinary snout adaptation that earns them the name of sawfish. The elongated top jaw forms a flattened blade with sharp teeth along each side. It is used to hack a path through packed schools of small prey fish, usually killing more than the ray can eat. Sometimes big fish are impaled, rubbed off on the sea bed then partly eaten.

The giant of the group, the green sawfish *Pristis zijsron*, is fairly common in warm, muddy-bottomed estuaries. Unusually for rays, it also moves far into fresh waters, having been found well upriver on the western side of Cape York Peninsula. This species can exceed 7 metres in length, nearly 2 metres of that being taking up by the saw.

Sawfish are a curse to net fishermen, destroying gear and posing a high risk of injury if attempts are made to boat them alive. There is one overseas account of a man cut in half. Incidents involving swimmers and waders are extremely rare, because sawfish are as timid as stingrays and faster-moving. The reported death of a 15-year-old boy, drowned after a massive loss of blood from an abdominal wound, is difficult to understand.

Also puzzling, as far as Australian authorities are concerned, are the many deaths and severe injuries suffered overseas from leaping needle-fish — species of *Tylosurus* that we usually call long toms or alligator gars. These slender, swift-moving creatures are closely related to flying fish. They often skim over the water surface in their pursuit of prey, and 'tail-walk' if something startles them.

In long toms, both jaws are extended and wickedly pointed. Swimmers and canoe fishermen have been struck and killed by them in Papua New Guinea — where a series of incidents was documented in a special medical study — as well as in Mexico and Tahiti. Long toms of several species are common in Australian tropical and subtropical waters, both in the ocean and in estuaries. Some enter the freshwater reaches of rivers. Giants among them may measure almost 2 metres. Yet they have never been known to harm anyone here.□

The sultan seals

Bulls competing for control over harems are savagely combative. People who go too near risk a bumping, biting punishment.

Seal breeding is marked by some of the greatest belligerence known in the animal world. Combat between bulls is no mere ritual: losers are battered, wounded and sometimes killed. The best fighters win control of the most cows, and guard their harems fiercely.

Approaching some of the bigger species of seals can be highly dangerous during their period of reliance on a shore base for pupping, quickly re-mating and rearing the young until they can fend for themselves in the sea. Seasons may extend from early spring to midsummer.

A lunging bull may simply bump an intruder, using its bulk with limb-breaking force. If thoroughly enraged it will inflict crushing, tearing bites. Cows can be just as aggressive if they are guarding pups, but in most common species they are much smaller than their mates.

Australians normally encounter only one type of seal of dangerous size. The endemic Australian sea lion *Neophoca cinerea* has permanent breeding grounds in South Australia on Kangaroo Island and at Point Labatt on the Bight side of Eyre Peninsula. Other sites are on little-frequented islets off WA and SA. Bulls of this species exceed two metres in length and weigh about 300 kg.

Far bigger leopard seals and elephant seals are occasional visitors. They do not come to mate — breeding takes place on islands far to the south — so they are fairly placid. And unlike

Under threat from a bull southern elephant seal Mirounga leonina *(above), a subantarctic research scientist backs off hastily. Penguins are treated more tolerantly at the breeding ground, on Macquarie Island.*

Below: An Australian sea lion Neophoca cinerea *of the Seal Bay colony, a major tourist attraction of Kangaroo Island, SA.*

the sea lion, which can use its hind flippers to move nimbly on land, they are restricted to sluggish crawling.

The enormous mass of the southern elephant seal *Mirounga leonina* demands respect in any circumstances, however. Bulls exceed 4 metres in length and weigh up to 4 tonnes. Although they cannot charge on land they can rear up. People interfering could be killed merely by being fallen on. Elephant seals used to breed on Tasmania's northwest coast and some Bass Strait islands, and it is in this region that most sightings are made these days. Two cows are known to have given birth to pups during their visits, but they did not mate afterwards.

The leopard seal *Hydrurga leptonyx* makes a solitary migration before winter from Antarctic pack-ice to subantarctic islands. Strandings of individuals that travel too far may occur in southwest WA, southeast SA and from Tasmania to northern NSW. Fully grown cows of this species are bigger than bulls, exceeding 3.5 metres in length and weighing up to 450 kg. Subantarctic research scientists who encounter the animals at their breeding grounds regard them as dangerous only if people stand in the way of their escape to the water.

Elephant seals on rare occasions have attacked small boats — whether through irritation or because they mistook the craft for predators, no one knows. When an aged bull elephant seal wrecked a fishing boat off Napier, NZ, in 1967, spilling five men into the water, it made no attempt to harm them.

Sea lions are said to have bitten swimmers off the South African coast. No such incidents have been reported here. Leopard seals often follow boats, but are not known to attack them.□

Collision course

Left alone, whales show no spite towards humans. They are just too big for us to be in the same place at the same time.

Early on a fine summer's day in 1963, staff at Sydney's South Head signal station rejoiced in the sight of a big hump-backed whale frolicking near the eastern suburbs beaches. For two hours the spectacle continued, the splendid animal sounding, surfacing, blowing, splashing about and basking.

Up since before dawn and oblivious to all the fun, two regular fishing companions were having no luck inshore off Coogee. They decided to try farther out. William Morris, 64, started up the engine of his self-built, 4-metre launch and with his 39-year-old friend Arthur Barrett headed for a spot about five kilometres offshore, between Coogee and Maroubra.

Anchored in about 40 metres of water, the pair had scarcely had time to throw out their lines when the whale surfaced beside the boat. Its tail slapped one side and the craft was tossed in the air, catapulting the men into the sea. Within seconds Morris called out: 'I'm going!' He suffered from a heart condition.

Barrett managed to reach his mate and hold his head out of the water until help came. The incident had been witnessed by another fisherman. But Morris was black in the face when he was hauled from the water, and did not respond to resuscitation attempts.

In May 1981 the yacht *Tahara'a*, a handicap favourite for the inaugural Perth-Bali race, was struck by a whale off Broome, WA. The boat

sank within three minutes but all six aboard were picked up by a following yacht. Exactly a year later, well out from the central Queensland coast, the Noumea-bound catamaran *Whiplash* collided with a whale, and one hull was destroyed. RAAF search aircraft directed a freighter to the aid of the couple aboard.

Incidents like that are not freakish, given the amount of small-boat traffic traversing Australian waters. The same waters are frequented by more than 40 species of whales. Some are small, many are rare. But enough of them reach colossal size for a greater frequency of collisions

Above: Marine growths adorn the enormous head of a right whale Eubalaena, *a slow (8 km/h) swimmer.*

Below: The hump-backed whale Megaptera novaeangliae, *an energetic swimmer, is often seen off the Australian coast during its annual migrations.*

to be expected. Further small disasters, perhaps involving loss of life, are certain to occur. No one will be to blame — least of all the whales. And there is no reasonable way of avoiding collisions, unless people are to desist altogether from putting to sea in small craft.

Where have the 'rogues' gone?

Old seafaring lore abounds with tales of vindictive whales, attacking boats on sight and sending their hapless crews to Davey Jones's locker. No doubt there was truth in some of them. In an era when whales were relentlessly hunted and harpooned by the crudest methods, vengeance by bereaved and perhaps wounded survivors would not have been surprising.

Nowadays, even where whale populations are showing upward trends, there is absolutely no evidence of 'rogue' behaviour. Whales are intelligent and apparently emotional, however. There can still be acts of vengeance.

In November 1979, after the New Zealand yacht *Dauntless* struck and killed a surfacing whale cow in the Tasman Sea near Norfolk Island, a bull rammed the 12-metre boat repeatedly. Terrified, a woman passenger began pelting it with everything she could lay her hands on, including a mirror. Perhaps frightened by a sudden flash of the sun's reflection, the whale retreated. By then the yacht was holed; after summoning help by radio the four people aboard took to a liferaft.

In bays where whales may come briefly to breed, maritime safety authorities can only urge that people in boats stay away from them. □

ORCA: A DOLPHIN DAMNED WITHOUT CAUSE
The making of the 'killer whale' myth.

As if the normal perils of the deep were not enough, seafarers have always shared a fondness for fanciful exaggeration. Legends of monsters were eagerly believed. And even this century, a new myth was created — that of the man-eating 'killer whale'.

The orca *Orcinus orca*, spectacularly coal-black on top and snowy white below, is the biggest of the world's dolphins. Exceptional bulls reach 9 metres in length. And it is the dolphin most tolerant of temperature variations. Though major populations are concentrated in the polar regions others occur in warmer waters, in every ocean. Orcas are the only marine mammals to seek warm-blooded prey. Their favourite food is seals. But hunting in fast-moving packs, orcas occasionally take bites out of whales and attempt to gobble up their pups. In bays too shallow for the whales to dive out of reach, they can be trapped and harassed to exhaustion. It is fair to call orcas whale-killers.

But the orca and its slightly smaller relative, the all-black, fish-eating 'false killer whale' *Pseudorca crassidens*, differ little from other dolphins in temperament. They are highly intelligent, and when accustomed to humans they are affectionate, playful and trainable.

There is simply no record of an orca having directly attacked a boat or a man, let alone having eaten one. But because hungry orcas sometimes batter polar pack-ice to get at seals and penguins, there is a theoretical risk that people could be mistaken for prey, tipped into icy water and drowned. In 1911 it almost happened. At McMurdo Sound, Antarctica, in preparation for Robert Falcon Scott's last attempt to reach the South Pole, husky dogs were tethered on floating ice. Herbert Ponting, the expedition photographer, was with them when eight orcas began bumping the ice from below. No harm was done — but the myth was born.

As the story was plucked from Scott's journals and told and retold, the involvement of the huskies was forgotten. This was represented as a deliberate, concerted attack on a man. A responsive chord was struck in North America, where suddenly it was discovered, or claimed, that Eskimoes were terrified of 'killer whales'. From South Africa came a report, accepted without question, that four men had been thrown from a boat and eaten.

The US Navy, among other authorities, swallowed the lot. Sailing directions for Antarctic voyagers included a warning that orcas would 'attack human beings at every opportunity'. The manual for naval divers rated the orca among the topmost animal dangers. In the 1950s, wide publicity was given to the survival of two men whose fishing dinghy was chewed by a 'killer whale' just off San Francisco. Little notice was taken of a later scientific finding that tooth marks in the boat's timbers were those of a shark.

A decade later, however, a methodical demolition of the whole myth began — much of it to the credit of Roger Caras, author of the book *Dangerous to Man*. Consulting authorities worldwide, he found no record of Eskimo deaths, no trace of the South African incident, and no shred of evidence of any other attacks by orcas.

MUCH TO DISCOVER

Dread of sharks stems from helplessness as much as horror. Of all dangerous animals, sharks are the ones we know the least about and can do the least to control. And they are masters of an element in which we are at our most vulnerable. They can produce effects of mind-numbing gruesomeness without even being seen.

Scientific understanding of shark behaviour is in its infancy. Little of practical use can be gained from captive specimens, or in conditions created artificially by offering food. The most menacing species are ever-roving, feeding where and when they can and defying consistent study.

It is clear enough, however, that sharks are not our natural enemies. Attacks are so rare as to be statistically negligible. And evidence is emerging that sharks are not the primitive, virtually mindless, automatic killing machines that we supposed. Along with extraordinarily sharp senses, the more active species seem to have considerable mental capacity. Their behaviour can be modified by processes of learning.

That suggests that continued efforts to find ways of repelling sharks rather than killing them will sooner or later pay off. Researchers also may be close to establishing exactly what it is about our own behaviour in the water that can trigger deadly attacks. If we too can learn, oceanic and estuarine waters will be safer for both kinds of creatures.

Above: A nurse shark at rest on the sea bed with a remora, or suckerfish, attached to its back. It was once thought that sharks had to swim all the time to stay alive.

Opposite: In the most advanced kinds of sharks – this one is a lemon shark Negaprion brevirostris *– the young are nurtured through umbilical cords and are born like mammals.*

HOMING IN ON THE SMELL OF THE SEA

A shark's nostrils and mouth constantly pass sea water to two big nasal sacs. These are lined with cells of astonishing sensitivity. At least one type of oceanic hunting shark has been shown to detect a trace of fish flesh so faint that it represented one part in ten billion parts of sea water. That is how anglers cleaning fish, spear-fishermen carrying or towing their catches, and perhaps people swimming with bleeding cuts attract trouble. When a hungry shark detects a stream of particles of animal origin, it faces into the current and heads for the source.

Sharks also have two systems for picking up sounds and other vibrations in the water, such as the erratic movements made by injured seals or fish — or accidentally, by people swimming or paddling on boards. One system works through pores on the head and body, containing sensory cells open to water pressure variations, and the other through the ears, on top of the head. Used together, they probably give sharks a very speedy and accurate 'fix' on the movements of likely prey.

Organs of yet another sensory system are scattered along a shark's body and around its lower jaw. Shielded by skin denticles, they resemble taste buds. No one has yet worked out what they are for.

Pores in the skin of the snout, just in front of the mouth, lead to a special network of cells and nerves that are sensitive to electrical fields. This is another way to detect schools of fish in the open sea, and it enables bottom-feeding sharks to find prey buried in sand. Some scientists conjecture that migratory sharks also use the sense for navigation, relating their positions to the earth's magnetic field.

Sharks also have good vision and an exceptional ability to distinguish moving objects from their backgrounds. As well as irises that contract or dilate to regulate light entry, sharks' eyes have remarkable mirrors behind the retinas. At night or in dark depths these bounce back the available light, increasing the visual stimulus. But if a shark feeds by day in bright sunshine, the mirrors temporarily cloud over with black granules.

Brainier than most other fishes

Sharks are sometimes said not to be 'true' fish. They and their close relatives the rays have skeletons of flexible cartilage instead of bone. But with a lineage going back at least 400 million years, they are highly evolved.

Nearly all sharks have greater proportions of brainweight to bodyweight than the bony fishes. The ratio also exceeds that of many birds and some mammals. Experiments on the few sharks that thrive in captivity show them capable of trained behaviour. They learn to press levers to obtain food, for example, as quickly as white rats.

In spite of their advantages in finding prey and their apparently superior intelligence, sharks have not achieved anything like the diversity and geographical spread of the bony fishes. While there are more than 20 000 known species of the latter, there are currently thought to be fewer than 350 shark species in the world, with a roughly equal number of rays. And very few sharks or rays are at home in fresh water, as so many bony fishes are.

The simple explanation for this disparity is that sharks, lacking an air bladder, rely much more on the buoyancy of salt water. Even with its help, and the added lift given by their oil-rich livers and aircraft-style tails, sharks have to spend more of their time swimming to stay at their chosen depth in the sea. They burn up more energy — and so they need more food.

It is not true, however, that sharks have to swim continuously to get enough oxygen through their gills, and can never rest. Researchers are finding more and more species that stop swimming, let themselves sink to the sea bed and lie there for long periods.

Temperature and movement

Different types of sharks live at different depths. The only groups with a bearing on human safety — and the only ones about which much is known, for that matter — are shallow-water

Grey reef sharks in a feeding frenzy.

HOW SHARKS DIFFER FROM BONY FISHES

Five to seven **GILL SLITS** can be seen on each side of a shark. A bony fish has only one on each side.

Instead of scales, sharks have small plates formed like teeth. Called **DENTICLES**, these can severely lacerate the skin of someone who is brushed by a shark.

A shark's **TAIL** is asymmetrical, with the upper lobe bigger. This gives the shark, which is heavier than water, lift while swimming.

While bony fishes spawn, sharks copulate. Males use a **CLASPER** on each of their pelvic or ventral fins.

Sharks' **TEETH** are replaced throughout their lives. If bony fishes lose any, they have to go without.

A shark's **LIVER** is big and especially rich in oil, lightening its mass in the water. Sharks lack the air bladders that give bony fishes buoyancy.

Most of a shark's **SKELETON** is made of tough but flexible cartilage. In nearly all other vertebrates this would become bone.

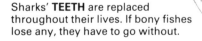

Twin claspers at the rear of a male white pointer.

A NEVER-ENDING SUPPLY OF TEETH

Each of a shark's visible teeth is part of a rotating battery containing five or more in various stages of growth. Whenever a tooth is lost a replacement swivels up from inside the jaw tissue. In fast-growing juveniles teeth are replaced every week or so; in adults the rate slows to every 4-8 weeks. The process continues throughout a shark's life.

Tooth shapes vary considerably between species. A mako's are long and narrow, for example, while a tiger shark's are broad with a saw-edged shoulder. In many of the dangerous species the teeth in the upper jaw are wider and more triangular than the lower set.

With the cutting edges of their teeth literally razor-sharp, sharks do not need particularly powerful jaws. Nevertheless the force concentrated at these fine edges can be enormous — three tonnes per square centimetre in the case of a moderate-sized whaler such as a dusky shark.

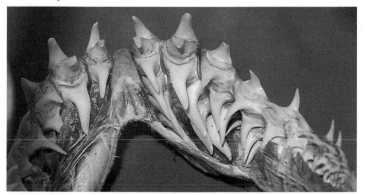

Replacements wait below each functional tooth.

sharks living above continental shelves, and pelagic sharks living at the upper level of the open sea, occasionally coming inshore.

Pelagic species migrate towards the poles in summer and back towards the equator in winter — not in uniform latitudes, but in directions dictated by warmer or colder currents. Shallow-water sharks also respond to changes in water temperature. Their appearances may be markedly seasonal around Australia's southern coastline. In the far north, where water temperatures fluctuate less, sharks cruise constantly but stay within home ranges.

Nearly all sharks are cold-blooded and slow-moving, rarely exceeding a leisurely human walking pace. The few warm-blooded exceptions, with more efficient muscles, significantly include the highly dangerous white pointer and the mako. With prey in sight they summon up short but powerful bursts of speed.

Every shark is carnivorous, preferring live, fresh prey, though some decaying flesh is scavenged. The diets of different species range from microscopic plankton to big seals and whales. Cannibalism of shark pups is common among the hunting species.

The feeding frenzy

When several sharks are attracted to the same prey, they often become excited to the point of insanity. They tear not only at the prey but also at one another — and even at their own bodies.

83

Objects such as cans and bottles, thrown into the water during such a feeding frenzy, are gobbled up indiscriminately. A person in the vicinity would have no chance among big sharks in this mood.

Because sharks' mouths are below their bodies, it is sometimes supposed that they have to turn upside-down to attack swimmers. But a shark's upper jaw is not fused to its skull. It can be disengaged and the whole snout pushed upwards, bringing the mouth to the front during an attacking run.

Sharks do not spawn like the bony fish. They copulate much like most mammals, the male depositing sperm in the female reproductive tract. Males are equipped with two organs, called claspers, for this purpose. They form extensions of the pelvic fins.

Very little is known of courtship or mating behaviour. It is apparently violent — females often bear the marks of extensive biting. In some species the females have much thicker skins in the area where they are held during mating, and the males have smaller teeth.

Gestation periods are surprisingly long — a minimum of nine months, and 22 months in the case of some dogfish. The young start life in three ways. Eggs are laid by the most primitive species and deposited in reef crevices, perhaps after being brooded for a while in the mother's mouth. In other species the eggs are nurtured internally, without attachment to the mother, until they hatch.

Born like mammals

The third and most advanced method of birth is just like that of most mammals. Pups are born live after developing in the mother's oviducts, where they are nurtured through umbilical cords. This allows the fewest pups to be born, but they are immediately self-sufficient.

Studies of the growth of sharks, through the recapture of tagged specimens, are in their early stages. So far they indicate a wide variation between different species, with some big pelagic sharks growing more than three metres in a year, and others adding less than 50 cm.

It is reasonable to assume that life spans vary just as widely. Several species are known to live for more than 30 years. Scientists think some may exceed 50, and there could even be 100-year-old sharks.☐

GIANTS, PAST AND PRESENT *Until only about 12000 years ago the dreaded white pointer shark had a cousin that could have swallowed it whole. Teeth of the now extinct* Carcharodon megalodon, *which are often dredged from sea beds, dwarf one taken from a 3-metre white pointer (inset, below), and indicate that this giant could have been up to 15 metres long and weighed about 20 tonnes. But an even bigger shark still roams the oceans. The whale shark* Rhiniodon typus *(the head of one is shown below, with Valerie Taylor) is thought to reach 18 metres and weigh up to 40 tonnes. Mercifully, it is a harmless filter-feeder.*

Who is at risk?

Odds against an average beachgoer being attacked by a shark are astronomical. But danger can be heightened in certain circumstances.

Of all instinctive human terrors, perhaps the deepest and most primitive is that of being eaten alive. A hangover of prehistory, it is still easily played on in advanced societies. Hence the success of morbid works of fiction about man-hunting animals like the film *Jaws*.

Fear of sharks is energetically fostered in Australia. Nowhere are people more dedicated to maritime leisure. For the news media, any means of highlighting the menace of sharks — even in the absence of attacks — is a guarantee of compelling reading or viewing. And just often enough, an actual attack occurs somewhere to reinforce the anxiety.

A spate of fatalities in the 1920s and 30s gave Australia's east coast a reputation as the most shark-infested in the world. Attacks in those days made up well over a third of the known global total. Seven or eight people were killed in bad years — just a day's road accident toll now, but somehow much more horrifying.

Reflecting a way of life

But global records were patchy, being virtually limited to data from English-speaking countries. And in making comparisons, people seldom realised how much more Australians used the sea. This country had then, as it does now, an extraordinary proportion of its population living within easy reach of warm coastal waters. Here more than anywhere else, harbour and ocean bathing was a pastime of ordinary people rather than a privilege of the elite.

Analysis of documented incidents, anywhere in the world, tells us almost nothing about the prevalence of sharks or their propensity to attack. In the main it merely confirms the obvious. Most attacks have occurred where and when most people spend most time in the water. On that score, Australia's recent record can be held up with pride. Many more people swim, ride boards and dive, through longer periods of the year, than ever did 50 or 60 years ago. Yet attacks involving serious injury average only about three a year, and the national death rate is well under one a year.

At popular beaches all around the country, millions of people enjoy near-total protection. For swimmers using those beaches during the recognised season, when shark-spotting patrols and measures such as harbour enclosures and ocean mesh trapping are in operation, the chances of an attack are infinitesimal.

Significant risks are run, however, by people of more independent spirit. In general the greatest danger lies in swimming or diving by

Seen from below, a board rider can look like a seal or a turtle.

oneself — not only because it may encourage bolder action from a shark, but also because it reduces the chances that the shark will be seen. And a lone victim is less likely to receive prompt assistance and first aid.

How figures can mislead

Raw statistics on the incidence of shark attacks can give a false picture. For example, a world-wide study funded by the US Navy revealed that 62 per cent of documented attacks occurred in depths of 1.5 metres or less. That suggests that shallow waters are the most dangerous. But it has to be remembered that 90 per cent of sea bathers do not go in out of their depth. So the more venturesome 10 per cent incurred 38 per cent of the attacks.

A similar misconception arises over the distance from the shore within which most attacks occur — 60 metres. But that is where the vast majority of bathers stay. Those who go farther out are subject to more than their share of attacks. Their peril may be heightened because they swim over channels or sharply increasing depths where sharks can approach unseen.

It is asking for trouble to swim where people are fishing. The bait alone may attract sharks. So will the struggles of hooked fish, and especially the guts and heads that are discarded when catches are cleaned.

Refuse from fishing and the remnants of people's meals contribute to a high incidence of attacks near wharves and jetties in harbours, estuaries and the tidal reaches of rivers. The most dangerous situation related to fishing, however, is that of divers who spear their catches and carry or tow them underwater, streaming blood.

Whether sharks are attracted by human blood, from injuries or menstruation, has been the subject of recent research and debate. Most overseas scientists accept experimental findings that sharks do not respond to traces of any mammalian blood.

But it is hard to ignore the fact that the most dangerous species, the white pointer and the tiger shark, deliberately hunt mammalian prey in the form of seals. People who are bleeding in any way would still be well advised to stay out of the water.

Danger in colder water

For many years it was thought that sharks rarely attacked in water temperatures below 20°C. The reality was that very few people used to swim, or at any rate stay long, in water colder than that. Since wetsuits have allowed divers and board riders to retain their body heat in chilly waters, it has been realised that some dangerous shark species are not only more abundant but also more active at lower temperatures.

THE CHANGING PATTERN OF TRAGEDY
Where fatal shark attacks have occurred.

Since the death of an Aboriginal woman was recorded in 1791, sharks have been blamed for killing about 150 people in Australian waters. No doubt many shipwreck victims, their ordeals unwitnessed, have met the same fate.

Old reports of attacks on swimmers and divers were often vague in location. Many failed to make clear whether or not the victims died. Fatal incidents summarised here are largely drawn from a compilation by the NSW photographer and writer Jack Green.

By far the most frightening toll was taken in the period between world wars. Fatalities averaged nearly three a year. This was the heyday of pearl and trochus shell diving in far northern waters. But more to the point, ocean bathing had completely taken hold in Sydney and Newcastle, where the greatest number of Australians lived close to surfing beaches.

Shark meshing off popular Sydney beaches changed the picture dramatically. For the 20 years after 1937 the national rate of fatalities fell to fewer than two a year. North Queensland emerged as the main focus of worry — due in part to the massing of troops training for the war in the Pacific, and in part to the rough-and-ready pioneering of tropical tourism.

Since 1957, fatal attacks have averaged fewer than one a year. The decline has occurred in the face of a heavy, migration-boosted increase in population and an even greater expansion of seaside holiday activity. It proves the value of further beach protection, along with improved first aid knowledge, communications and medical facilities.

NATIONAL TOTALS

Killed before 1918 ... **22**

Killed 1918-37 .... **57**

Killed 1938-57 .... **31**

Killed after 1957 ...... **26**

That discovery has a particular relevance here. The white pointer, a cold water shark, seems to be at its most prolific along Southern Ocean coasts and in the Great Australian Bight, where seals are also abundant. And wetsuited surfers and divers are entering those same waters, at times when once they would have been too cold for recreation.

Regrettably, the change is starting to show up in attack figures from Tasmania and South Australia. The remote surfing mecca of Cactus Beach, near Point Sinclair at the eastern head of the Bight, appears to be a particularly dangerous spot. As the recreational trend continues, the Southern Ocean coast is likely to inherit the east coast's ill-deserved notoriety as the world's worst for sharks.

Wetsuits themselves, along with the flippers that divers and snorkellers wear, may be to blame for some attacks. People wearing them, seen from below, can give the appearance of big seals. Board riders paddling with their arms can also look like seals or turtles.

The conjecture that sharks sometimes attack through mistaken identification will probably never be proved or disproved. A survey of wounds suffered by victims, however, supports a more general view that most attacks are made by mistake. Nearly all came as direct, hasty strikes, without any preliminary circling or passing to inspect the victim. And in most cases there was only one bite — as if the shark realised its error and lost interest.

Of more than 1000 international cases analysed in the US, only one in five — most of them fatal — involved more than one bite. Evidence of 'feeding frenzy' behaviour was noted in a mere 4 per cent of cases.□

Left: A diver's silhouette and movements may invite a tragic mistake.

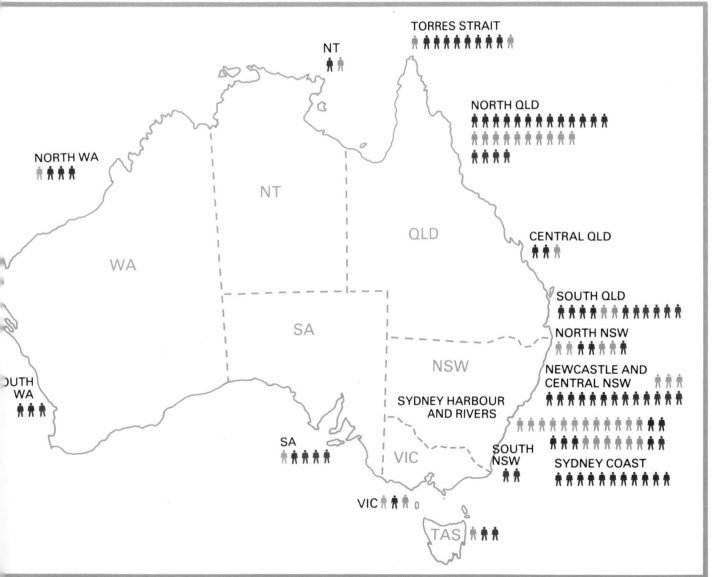

A streamlined form like a fighter plane allows deadly attacking speed.

The white pointer

*Huge, fast and aggressive, this is the world's most feared shark.
Yet scientists know almost nothing of how it lives.*

If any shark can fairly be called a man-eater, it is the white pointer *Carcharodon carcharias*. Although never found in great abundance, it accounts for more than a third of all human fatalities in which the attacking species is identified — despite it being most active in cooler waters where fewer people swim.

Americans know *Carcharodon* as the great white shark, or as the writer Zane Grey called it, white death. South Africans call it a blue pointer. Actually its predominant colour is grey — either light or dark — with an underside that is usually whitish but can have a rusty tinge. A clear distinguishing feature is a relatively small, conical head, coming to a sharp point in front.

White pointers probably have no greater taste for human flesh than any other shark. But they are omnivorous, devouring seals and even sea birds as well as fish, and unhesitating in their strikes. What makes them exceptionally dangerous to swimmers and divers is their sheer size and the force generated by their speed. One

exploratory bite can cut a person in two, even if the shark shows no more interest in the victim.

It was discovered in the 1970s that the white pointer, like the mako shark, is warm-blooded. In cold waters it can independently maintain a higher body temperature. Its muscles work more efficiently than those of most sharks, giving it greater attacking speed. But to gain its energy, it has to eat more voraciously. It is the most active of hunters, thrusting its head from the water to inspect the surface and to see what may be on reefs and beaches.

Nine metres long

The longest white pointer to be reliably measured was nine metres. Caught in the North Atlantic in 1978, it was estimated to weigh more than 4.5 tonnes. The jaws of a white pointer caught off Port Fairy, Vic, last century were displayed in the British Museum as having come from a shark 11 metres long. But recent comparisons with the jaws of sharks of known length indicate that this one probably did not exceed 5.5 metres — a good size for the animals caught in our southern waters.

White pointers are at home in all oceans, but may be most abundant in the Great Australian Bight. They are not commonly seen, however. Beyond some observations of their feeding habits, carried out by caged divers using baits, virtually nothing is known of their way of life.

Researchers can only guess how far white pointers travel, where they breed and how they reproduce. No pregnant females have been found. No one knows how big the pups are at birth, so their growth rates and likely life span are also a mystery.

Juvenile white pointers are sometimes trapped in shark meshing off the Queensland Gold Coast, but have never been seen in southern waters. There, from the observations of underwater photographers Ron and Valerie Taylor, a pattern of grouping has emerged.

Writing in the Reader's Digest book *Sharks — Silent Hunters of the Deep*, Valerie Taylor reported that white pointer populations near the South Australian coast seemed to be grouped according to sex and size. At Dangerous Reef, at the entrance of Spencer Gulf just east of Port Lincoln, most sharks seen were males 3-5 metres long. But on the other side of Eyre Peninsula, at Streaky Bay, most were females 5-6 metres long.□

Rodney Fox will bear the scars of his encounter with a white pointer for the rest of his life. In spite of his ordeal, he still expresses great admiration for sharks — and a certain amount of affection. More than two decades after the attack, he says: 'I probably have less fear of sharks than most people do, because I know more about them.' Much of his time in the intervening years has been spent studying sharks, and developing safeguards for researchers. Study of sharks is possible only in the open ocean. White pointers and most other dangerous species die quickly in captivity or if their movements in the sea are at all restricted.

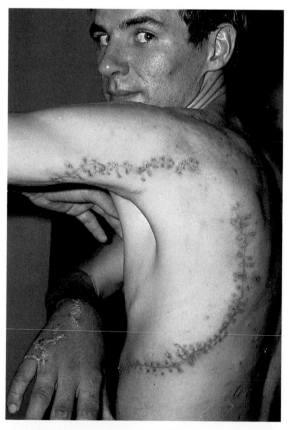

Diver's deliverance

After one bite, the shark decided to try something else.

Rodney Fox, adviser on underwater sequences for the *Jaws* films, gained his experience in the worst way. In 1963, aged 23, he suffered some of the gravest shark attack injuries that anyone has survived. Almost disembowelled, he was held together by his wetsuit.

A former South Australian spearfishing champion, Fox was competing at Aldinga Bay, 55 km south of Adelaide, to regain his title. He was about a kilometre from the shore, working the 'drop-off' where Aldinga Reef plunges from a depth of eight metres to 20 metres.

Without warning, while taking aim at a big morwong, Fox was hit and thrust through the water with enormous force. His face mask was wrenched off and the speargun was knocked from his hands. A white pointer had clamped its jaws over his trunk, from shoulder to stomach.

As Fox reached back with his free arm in the hope of gouging the shark's eyes, it released its grip — apparently realising already that it had chosen an unpalatable target. Fox's arm went into its mouth and its teeth shredded the flesh.

Fox kicked his way to the surface but the shark came too, lacerating his knees with the sharp denticles on its hide. Intent only on keeping away from the jaws, Fox did an astonishing thing — he wrapped his arms and legs around his attacker. Down they went together, the shark twitching to shake him off.

Running out of breath, Fox again broke away and gained the surface. The shark followed, rolling onto its side and surveying the blood-stained waters from a few metres away. Then it moved in to attack — but this time the target was not the diver but his marker bouy, attached by a rope to his weight belt.

Gulping down the buoy, the shark plunged at high speed. Fox was in more trouble, for he could not find the clip to release the rope. Mercifully, as he was dragged back into the depths, the line parted.

Shocked and faint from a huge loss of blood, Fox bobbed helplessly in the water. He was fortunate in that the spearfishing contest was under close observation. A boat was quickly at his side. As he was taken ashore his intestines spilled out of his slashed wetsuit, but were eased back by a bystander trained in first aid.

After a hectic dash by car and ambulance to Adelaide Hospital, Fox underwent four hours of surgery. Repairs to his abdomen, torso and arms required more than 400 stitches. But he was back in the water in five months, and a year later was victorious in a national teams spearfishing championship.□

The tentative tiger

*The tiger shark is hesitant in attack.
But a liking for warm inshore waters makes
it the main menace to summertime bathers.*

Usually smaller than the white pointer, slower-moving and markedly less impetuous in its hunting, the tiger shark *Galeocerdo cuvieri* has nevertheless been blamed for nearly as many attacks. It is a wide-travelling oceanic shark, breeding in the tropics but following warm currents south in the summer. Then it frequently enters bays and estuaries, at the very time that most people use them for swimming.

The world's biggest reliably measured tiger sharks were caught off Newcastle, NSW, in 1954 and off Mackay, Qld, in 1980. Both were 5.5 metres long and weighed more than 1500 kg. The average for adults in Australian waters is about three metres.

The species gets its common name not from any special ferocity but from dark back stripes that are prominent in juveniles but tend to fade with age. Adults are normally grey-green above and cream below. The upper lobe of the tail is long and finely tapered, with a small lateral keel.

Night stalkers

When they enter shallow waters tiger sharks seem to feed most often after sundown. Alert bathers or beach spectators have a reasonable chance of seeing them, however, because they are cautious and methodical in approaching a target as big as a human. They hang off potential victims, circling and trying to judge their suitability. The strike itself is not fast, and often entails only a small, sampling bite.

Tiger sharks are the least fussy feeders of their kind — the goats of the sea. It is quite usual to find tin cans, pieces of wood and other rubbish in the stomachs of captured specimens. In common with other sharks they can regurgitate their stomach contents at will. Tiger sharks seem also able to save up food. In 1950 a captured specimen was kept alive for a month at Taronga Park Zoo, Sydney. It was fed on horseflesh but regurgitated these meals. On its death the shark was found to be storing two dolphins, undigested.

It was a tiger shark at Coogee Aquarium in 1939 that initiated Sydney's celebrated 'Shark Arm' murder case. The ailing animal regurgitated a tattooed human arm, identified as having come from a minor figure of the local underworld, James Smith. The investigation led to a colourful exposure of criminal dealings, but no one was ever found guilty of the murder.□

Supreme sacrifice

To protect his shipmates, a trawlerman invited certain death.

Holders of the Star of Courage, the Australian government decoration for conspicuous bravery, include one young man who did not live to receive his medal. Dennis Murphy, 24, gave himself to a tiger shark to divert it from attacking two friends.

In 1983 Murphy and his 21-year-old girlfriend Linda Horton were working as deckhand and cook on the Townsville, Qld, trawler *New Venture*. On the night of 24 July they were off Broadhurst Reef, 90 km east of Townsville. Murphy was helping skipper Ray Boundy, 33, to secure a broken boom when the 14-metre boat was struck broadside by a big wave and capsized.

As the upturned hull slowly sank, the three collected a surfboard, a lifebuoy and some pieces of foam padding, lashing them together with fishing line, and set themselves adrift. They were not unduly worried: the steady southeasterly breeze would push them directly towards Keeper and Lodestone Reefs, where they knew other trawlers were working.

Without mishap, they were about 5 km from Lodestone Reef by the following evening. But soon after dark a tiger shark over four metres long began to follow, occasionally nudging the floats with its snout. It made one attempt to bite Boundy's foot, trailing beneath the surfboard, but he kicked it. The shark took fright and withdrew.

A few minutes later, however, a surging wave spilled the three from their floats. Almost immediately Murphy screamed that he had been taken by the leg. He was pulled under twice, trying in vain to punch and kick his attacker, before the leg was severed at the knee.

Desperately Boundy racked his brain for a way to staunch Murphy's bleeding and keep him out of the water. But 'Smurf', as his friends called him, summed up the situation quickly and calmly. 'Well,' he said according to Boundy, 'it looks like that's it. You and Lindy bolt, because he'll be back for the rest of me.'

With that 'Smurf' swam directly at the shark. He managed four strokes before there was a violent upheaval. Boundy and Linda, paddling as fast as they could in the opposite direction, saw his body thrust out of the water, upside-down in the shark's jaws.

The surviving pair paddled and drifted on, still on course to reach the safety of the reef by morning. But about 4 a.m. the shark — Boundy is fairly sure that it was the same one — cruised alongside. Linda was sitting in the lifebuoy, with her feet propped up on a piece of the foam padding.

The shark made a sudden turn and took Linda by the arms and trunk, shaking her like a doll. She let out just one squeal — Boundy believes she died instantly. The lifebuoy was hurled aside and the young woman and her attacker disappeared.

After daybreak, within sight of the breakers at Lodestone Reef, Boundy found that he was again being stalked. A shark was circling, some distance off but coming closer with each pass. With the last of his strength and one piece of foam he paddled frantically to the breakers and surfed in over the reef face. The crew of a search aircraft witnessed the last phase of his ordeal. He was rescued after about 36 hours in the water.□

Right: A plaque on a memorial wall beside Townsville's fishing port honours the victims.

IN EVERLASTING MEMORY
— OF —
DENIS PATRICK MURPHY   AND   LINDA BARBARA HORTON
(SMURF)   (LINDY)
WHO LOST THEIR LIVES IN TRAGIC CIRCUMSTANCES
ON 25TH JULY 1983 NEAR LOADSTONE REEF
FOLLOWING THE SINKING OF THE TRAWLER "NEW VENTURE"
"GREATER LOVE HATH NO MAN THAN THIS,
THAT A MAN LAY DOWN HIS LIFE FOR HIS FRIENDS" ST JOHN 15:13
ERECTED BY THEIR FAMILIES AND FRIENDS

A tiger shark's serrated teeth are highly distinctive – each is shaped like a rooster's comb.

Requiem sharks

The biggest family of active hunters is dominated by the whalers — including many potential killers in Australian waters.

Requiem was an ancient name given to any shark that followed a ship, because of a belief that it did so in the hope of receiving the bodies of dead sailors. In modern terminology the requiem sharks are carcharhinids, the most extensive family of hunting sharks. Worldwide there are nearly 50 species.

Most notorious of the carcharhinids, because it is by far the biggest, is the tiger shark *Galeocerdo*. But the family includes many others ranging in length from two to four metres. About a dozen species occurring in Australian waters are thought capable of killing people, though evidence that they have done so is available in very few cases.

The majority of the family are classed in the genus *Carcharhinus*, and some species are so alike that the only way scientists can tell them apart is by removing their backbones and counting the vertebrae. Cold-blooded, they range through tropical and temperate waters in a diversity of habitats. Some are oceanic wanderers. Others are reef-dwellers with a strong territorial sense and a resentment of intrusion.

Above: The blue shark Prionace glauca *frequents southern waters.*

Many like to feed and breed in nearshore shallows, bays and estuaries, and a few spend time upriver in totally fresh water.

Bigger requiem sharks frequenting bays and estuaries are commonly called whalers in Australia, because of the nuisance they caused the whaling industry by damaging catches being brought to land. They can be a similar headache in commercial and leisure fishing, destroying gear as well as mutilating catches.

Above and left: Bronze whaler Carcharhinus brachyurus.

The bronze whaler

Most feared of the whaler sharks seen around Australia is the bronze whaler *Carcharhinus brachyurus*, known in some other countries as the copper shark. Suited to cooler waters, it is common on the east and south coasts from southern Queensland to the Bight. Because this species is often active in surf it has been responsible for many attacks on swimmers and some fatalities. One authority, Neville Coleman, suggests that the bronze whaler has probably attacked more people in Australian waters than has any other single species.

The bronze whaler does not exceed three metres in length; its adult average is about 2.5 metres. More formidable is the black whaler or dusky shark *Carcharhinus obscurus*, which averages over three metres and may reach four metres. In spite of its wider range, into tropical as well as temperate waters, and its presumably greater killing power, it has been implicated in relatively few attacks.

Overseas, in heavily populated tropical regions, river whalers have a frightful reputation. One species, the bull shark, is blamed for the world's third highest toll of human fatalities, after the white pointer and tiger shark. So-called river and creek whalers in northern Australia are not of this species. They are said to have been responsible for some attacks but are seldom associated with swimmers. In the estuaries and nearshore shallows that they favour, greater menace is likely from crocodiles.

Near the surface of deeper waters all around the southern half of Australia, the blue shark or blue whaler *Prionace glauca* is abundant. A far-ranging migratory species, it moves out into the tropical Pacific so is less often seen near northern coasts. At more than three metres it is considered extremely dangerous.

Dangers to reef divers

Outer fringes of the Great Barrier Reef and other offshore coral structures are hunting grounds for big whaler sharks with lethal capabilities and proven records of aggression. These may include the oceanic whitetip shark *Carcharhinus longimanus*, reaching about 2.75 metres, or the somewhat smaller silvertip whaler *Carcharhinus albimarginatus*. The grey reef whaler *Carcharhinus galapagensis*, up to four metres long, is common around Lord Howe Island.

About nearshore reefs, including rocky formations in northern NSW, the grey reef shark *Carcharhinus amblyrhyncos*, exceeding two metres, provides inquisitive and often pugnacious company for divers. In coral lagoons and on reef surfaces the blacktip reef whaler *Carcharhinus melanopterus* has attacked in knee-deep shallows. But it seldom exceeds 1.5 metres and has not inflicted fatal injuries.□

Blacktip reef sharks Carcharhinus melanopterus.

THE PUZZLING HAMMERHEADS
No one knows the reason for their shape.

Apart from the curious lateral extension of their heads, hammerhead sharks of the genus *Sphyrna* are fundamentally little different from requiem sharks. They probably evolved from a carcharhinid ancestor. But scientists are at a loss to explain why they did so.

The eyes are mounted at each end of the mallet-like head and the various other sensory organs are distributed along it. Some biologists surmise that swinging the head from side to side gives the animal a superior ability to detect and locate prey. Others think the head shape simply aids the shark in making quick, precise changes of direction in pursuit of fast-moving prey such as squid and school fish.

Hammerheads abundant in inshore Australian waters include the great hammerhead *Sphyrna mokarran*, sometimes reaching six metres in length. It is mainly tropical but ranges into northern NSW. The scalloped hammerhead *Sphyrna lewini*, up to four metres, has a similar range. The species common in cooler southern waters is the smooth hammerhead *Sphyrna zygaena*, which can also reach four metres in length.

All big hammerheads are presumed to be dangerous, though evidence of actual attacks is hard to come by. Their mouths are small in relation to their overall size, and scientists attempting to study some species have found them unaggressive.

Great hammerhead Sphyrna mokarran.

Blue dynamite

The mako's fighting power and savagery are legendary. Beachgoers can be thankful that it usually stays well out to sea.

Exploding from the sea at close to 40 km/h, the mako summons up more energy and attacking power than any other shark. Built for speed, much like a tuna, it is from the same warm-blooded family as the white pointer. Australians often call it the blue pointer. Game fishermen, who rate its fighting qualities second only to those of the marlin, have nicknamed it 'blue dynamite'.

Makos live in all tropical and warm-temperate seas, from the surface to moderate depths. The species that roves farthest south, appearing in abundance off the NSW coast even in winter and reaching southern and south-western waters later in the year, is the shortfin mako *Isurus oxyrinchus*. Its longfin cousin is not known to come in from the Pacific.

Female shortfin makos — markedly bigger than the males, which is often the case with sharks — average more than three metres in length and weigh nearly half a tonne. Exceptional specimens have measured almost four metres. In spring, when most are caught, they are usually pregnant.

In common with a few other types of sharks, makos have a form of reproduction known as oviphagy. After pups are conceived the female goes on producing eggs. These are cannibalised

Fighting a hook, a leaping mako can land in a boat.

by the developing embryos while they are still in the uterus. From hundreds of eggs, litters of 4-14 well-nourished pups are born.

Makos prey mainly on oceanic school fishes, but sometimes take advantage of their speed to single out and pursue swordfish. They are unhesitatingly aggressive and undoubtedly dangerous to humans, but rarely enter nearshore waters where most people swim.

Vengeance on boats

Blue-water boat anglers are more at risk. Hooked makos are notorious for their attacks on small craft, ferociously chewing or stoving in hulls and even biting spinning propellers. More dangerous still, a big mako may leap as high as six metres out of the water in its efforts to dislodge a hook — and land among its persecutors on the way down. Even hooking other fish can bring trouble, if a mako is after the same quarry. South African fishermen in 1977 told a hair-raising tale of destruction and mayhem after a mako leapt into their boat in pursuit of a yellowtail. The havoc ceased only when the shark tangled itself in fishing lines. By then one man, severely bitten in the thigh, had thrown himself overboard — too terrified to be concerned that other sharks could be about.□

Many others will bite

Even placid wobbegongs may retaliate.

Dozens of species of sharks, while not equipped or inclined for the type of attacks that excite publicity, can inflict severe injuries. Most risks are associated with fishing. Many reef-feeding species are easily made aggressive by spearfishing activities, either because they are territorially jealous or because the struggles of target fish stimulate them.

Amateurs netting or hooking sharks should not underestimate the biting power of even small varieties if they are ineptly handled. And members of the squaliform order, most of them commonly called dogfishes, have a sharp spine in front of each of their two dorsal fins. Among the worst punctures are those received from the abundant piked dogfish *Squalus acanthius*. Some people are said to suffer a severe toxic reaction to material in the sheathing of the spines.

Wobbegongs, slow-moving bottom-feeders in tropical and temperate reef shallows, are generally regarded as harmless. But they are well camouflaged and easily trodden on. If frightened in this way they react savagely.

An aggressive species most surprisingly *not* implicated in any attacks, except in captivity, is the broadnose seven-gilled shark *Notorhyncus cepedianus*. Often simply called a ground shark, it favours colder waters in the southeast, including Tasmania, and can be active in shallows very close to the shore. With females averaging nearly 2.5 metres in length, this species should be presumed to be dangerous.

The grey nurse shark *Eugomphodus taurus* used to have an appalling reputation on the east coast, probably because it was confused with one or other of the many dangerous whalers. In fact it has been shown to be an extremely docile shark — but, like almost any other, it can be provoked to attack.☐

The vanquished mako dwarfs it jubilant captors.

ANGLERS UNDER ATTACK

A world record made up for the damage.

Off Long Reef, Sydney, in December 1975, gamefish anglers Jack Farrell and Pam Hudspeth were after the one that got away. The previous day Mrs Hudspeth had played a 4-metre white pointer for about 90 minutes before the line broke. Back in the same spot in Farrell's launch *Winamy II*, they baited their hooks for another try.

A mako struck instead — not as big, but a powerhouse of fighting energy. Almost as soon as it had taken Farrell's line it leapt from the water, landing just beside the boat. Then it began a series of retreats and attacking runs, repeatedly charging the boat and sinking its teeth in the timbers. The hull was chipped and battered, and metal stays supporting it were broken.

Farrell gaffed his 'blue dynamite' after almost two hours. When it was weighed by club officials at Watsons Bay he felt well compensated for the boat damage: his 349 kg catch was a world record under gamefishing rules, which forbid the use of tackle with a breaking strain of more than 60 kg. Farrell in fact used only a 24 kg line. The record was subsequently beaten in the United States, and in 1986 stood at 489 kg.

Spotted wobbegong Orectolobus maculatus.

Protecting our beaches

Offshore meshing depletes shark numbers in popular areas, but kills many harmless creatures. Selective methods are needed.

The only perfect protection from sharks is gained by staying out of the sea and any of its coastal inlets. Few Australians are prepared to settle for that as an end to their fears. State and local governments and community organisations have to find huge sums of money for attack-prevention measures — principally the periodical laying of mesh netting off the most popular beaches, the maintenance of harbour enclosures and the provision of aerial spotting patrols and warning systems.

Meshing inevitably traps harmless shark species, many fish of high commercial and ecological value, and — since it was introduced in Queensland in the 1960s — endangered animals such as dugongs and turtles. Yet there is very little encouragement for investigating more specific ways of keeping dangerous sharks away from people. Although Australia has had the world's worst reputation for shark attacks, nearly all research into deterrence has been funded and carried out elsewhere.

The efficacy of meshing was discovered by accident. A Sydney professional fisherman in the early 1930s made an experimental net with an unusually wide mesh, in the hope of trapping bigger fish by their gills. A day after he set it, the net had vanished. But a fortnight later, passing the spot, he found his marker buoys back in place.

The net had been dragged to the bottom by the sheer weight of scores of trapped sharks, and had not returned to its position until they had decomposed or been torn apart by their

Below: Cruising whaler sharks in shallow water are seen easily from the air. Spotting patrols by light aircraft and helicopter have added immensely to the safety of many popular beaches. If an impending threat is seen, a radio message to a lifesaving club base starts an alarm procedure, generally using sirens.

fellows. Presumably the struggles of the first to be caught had attracted more — no one knows to this day why shark meshing works so well.

Meshing as it is now employed off the beaches of cities and major tourist resorts along the east coast — and in parts of New Zealand and South Africa as well — is by no means a complete screening. Nets are set at any particular beach only four times a month, for periods sometimes as brief as 24 hours.

Sharks can swim over the nets, under them and around them. But local populations decline sharply — a fact that is simply demonstrated by the dwindling numbers that are found trapped. At the same time the fishes and squid that sharks prey on become more generally abundant, so there is less need for surviving sharks to hunt close inshore.

Game fishing and the use of sharks as food have also expanded in Australia, and some scientists suggest that these activities may now be sufficient to keep down the populations of 'man-eater' species to a reasonably safe level. It would take rare political courage, however, to propose the abandonment of meshing.

Gadgets and potions
Professional divers can protect themselves with armoured suits, cages and weapons. For people at risk from shipwrecks and plane crashes it is feasible — if someone foots the bill — to have a type of lifejacket that is a sharkproof envelope.

But efforts to find better ways of safeguarding mass populations of beachgoers have brought disappointing results. Great hopes were held out in the United States for a system of bubble-curtaining, with air pumped through perforated hoses. Sharks used in early aquarium tests were suitably frightened. But other species ignored the bubbles.

South African scientists have had more success by creating electrical fields beyond the surf zone of beaches. They rigged up cables and showed that if the current pulsing through them was strong enough, sharks would not pass. The cost of installing, operating and maintaining such a system, however, seems always likely to be prohibitive.

Since sharks are so extraordinarily sensitive to chemical traces in the water, the most fruitful quest ought to be for chemical repellants. If only they could be found, their use would be simple. They could be put on like a sunblock.

Military lifejackets during World War II were fitted with packets of a 'shark chaser' compound. It was later proved useless, though it may have helped to preserve the morale of some sailors and airmen awaiting rescue.

Much recent work has concentrated on natural chemical compounds — especially the secretions of some soles. These fish can dissuade sharks from attacking them by discharging a milky fluid with an active component resembling household detergent. Amounts needed to make a practical repellant for swimmers would be impossible to obtain naturally. But in 1986 chemists in Japan announced that they had analysed the chemical structure of the fluid and reproduced it synthetically.☐

Above: A white pointer trapped in nylon gill meshing. Nets 150 metres long but only 6 metres deep are set 300-500 metres offshore.

The worst thing divers can do is keep speared fish with them underwater. The faintest traces of blood or other juices will attract any shark that is down-current.

AVOIDING TROUBLE WITH SHARKS

● Whatever the time of year, do NOT wade, swim, dive or ride a surfboard where sharks are known to congregate. In a strange area, seek local advice. Choose patrolled beaches if possible, and stay near other people in the water. Do NOT swim at dusk or after dark.

● Do NOT swim where people are fishing, whether it is from the shore, a wharf or inshore boats. Avoid any area where untreated sewage or abattoir wastes are dumped in the water. Don't allow your own wastes, or blood from a wound, to spread in the water. And do NOT wear shiny jewellery.

● NEVER swim with pet dogs, or in the vicinity of other domestic animals. And in southern waters, NEVER swim or dive near a seal colony. The area will almost certainly be frequented by white pointers.

● Stay away from spots where the water abruptly becomes deeper — channels, for example, and the 'drop-offs' of reefs. Choose open beaches rather than river mouths or estuaries if you can, and avoid all places where the water is turbid and visibility is low.

● Don't become too intent on what is below, especially while snorkelling or diving. Take frequent looks around you and towards the open sea. Most big sharks can be seen in time for bathers to leave the water. In swimming away from a shark, do NOT splash about but use the smoothest, most rhythmical stroke you can.

● If you see fish, squid or turtles coming inshore in unusual numbers, or swimming in an erratic manner, get out at once. They are probably fleeing from predators, possibly sharks.

SPECIAL ADVICE FOR DIVERS

● If spearfishing, do NOT carry or trail your catch. Take it from the water immediately. Change your location frequently. Do your fishing on incoming tides, when blood and juices flow inshore instead of towards the open sea.

● Do NOT corner a shark — in a reef cavern, for example, or a lagoon. Keep a baton or some other tool handy to fend off a shark if you have to, but do NOT use it to molest or unnecessarily provoke a shark. Even small and docile species may bite savagely if they are harassed.

● If a shark's swimming action changes from a smooth motion to a stiffly jerky pattern it may be issuing a challenge. Back off immediately. But stay submerged, facing the shark and ready to defend yourself. Leave the water, moving gently and rhythmically, when it seems to have lost interest.

● If a shark of dangerous size approaches closely and seems intent on attacking, keep as calm and still as you can. Prepare to use whatever you have available to fend it off — but take no action until the last possible moment. Do NOT use a knife or spear against a shark unless all else fails. Many species are made more aggressive by injury. And sharks are cannibals: the wounds of one can attract others.

● If efforts to fend off a shark fail, try to strike it on the snout or eyes, or else jab it in the gills. Use your bare hands only as a last resort, because they will be lacerated by the denticles on the shark's skin, or worse still by its teeth.

HAZARDS IN BOAT FISHING

● Any big shark hooked from a small boat should be cut loose at once, unless the boat is equipped for subduing sharks before they are hauled aboard, or unless the shark can be confined in a pen separated from the rest of the deck.

● If fish are cleaned aboard a boat, the bilges should not be allowed to leak and the waste parts should not be thrown overboard while the boat is at anchor.

● In areas where white pointers and makos are known to occur, it may be advisable not to have brightly contrasting underwater paintwork, or a particularly shiny propeller.

FIRST AID: PAGE 358

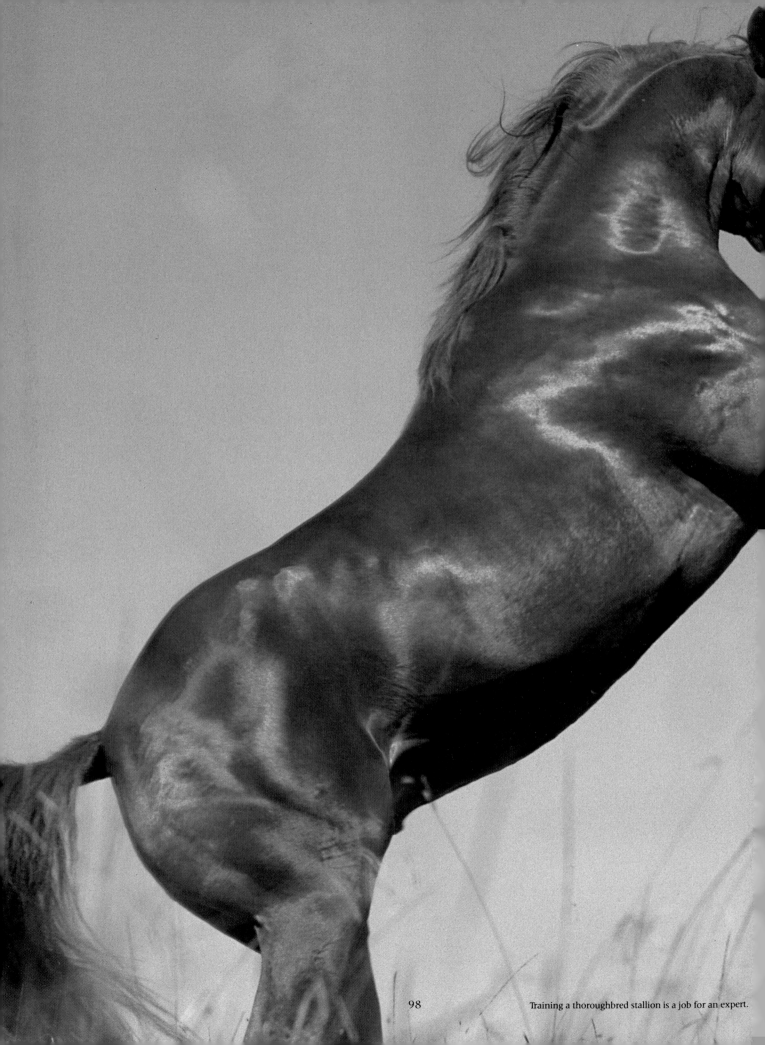

Training a thoroughbred stallion is a job for an expert.

Part 2

PERIL FROM PETS AND LIVESTOCK

*Attacks on owners as well as strangers
can stem from foolhardy choices and bad
treatment — or simply from a tragic failure to grasp the
complexity of some animals' personalities.*

Man's best friend?

Dogs account for a huge majority of the serious injuries inflicted by pets and livestock. Often it's not their fault.

Canberra Hospital staff in the late 1970s compiled a record of mammalian bite cases passing through their casualty department. More than 700 patients in 30 months suffered 800 bite wounds. The purpose of the study was to predict the scale of the medical emergency if rabies should break out. So only bites were analysed, not other types of injuries from animals. Even so, the figures were revealing.

Two-thirds of bites were inflicted by dogs. Cats accounted for 15 per cent, rats and mice for 7 per cent, and horses, donkeys and cattle together for 3 per cent — the same as bites by humans. No other mammal species was implicated to an important degree.

A quarter of the dogbite victims were children aged 12 or younger. In their cases 37 per cent of bites were to the face, head or neck. In the spread of ages notable peaks showed at two or three years — usually with the family pet involved — and among 10-year-old boys.

The national caseload

Estimating the size of the population that relied on their casualty service, the Canberra team concluded that dogbites severe enough to need hospital attention occurred at an annual rate of 184 per 100 000 people. That figure is low, compared with findings in some overseas communities where people are more rabies-conscious and report the most trivial injuries. But even the modest Canberra rate, if translated on a national scale, indicates that our hospitals could be handling 30 000 dogbite cases a year.

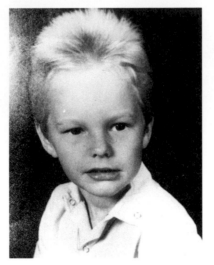
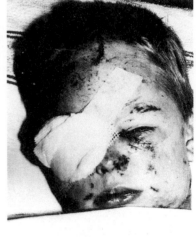

Above: Six-year-old Shane Spruce, of Newcastle, NSW, and (above right) how he looked after being attacked by a dog. Shane was savaged in July 1986 by his grandfather's guard dog as he was helping to unload a truck. The labrador-kelpie cross broke out of a cage and bit him on a leg and about the head. Microsurgery to repair eye, neck and facial wounds required more than 100 stitches.

Left: A NSW police dog being trained in the pursuit and apprehension of a suspected criminal. Police and army dogs are trained by experts. No member of the general public should attempt to train their dog in this manner. The owner of a dog that attacks another animal or person can be charged with an offence, and can be sued if an injury or damage to property arises.

Australia's estimated domestic dog population in 1986 was 2.6 million — roughly one dog for every six people. The density of distribution varies widely. In the Sydney municipality of Bankstown, for example, there was said to be a dog for every 2.8 people.

In the Canberra study the assumed dog population was 7000 in an area with 100 000 people — a ratio of only one dog to 14 people. On that basis it appeared that in a year, about one dog in 40 inflicted significant wounds. Perhaps a fairer assessment, based on the national ratio of dogs to people, is that one dog in 90 in a year seriously attacks someone.

Statistics aside, the problem is well enough illustrated by repetitious 'Pet Savages Toddler' headlines in newspapers, by prosecutions and civil damages suits, and by the frequency with which police and other authorities have to destroy dogs deemed to be dangerous.

By any measure, it is clear that a lot is found wanting in people's control of what are supposed to be obedient and amiable canine companions. The results are frequently injurious — the more so because big breeds are favoured over small by a majority of about six to four — and occasionally tragic. They are burdensome to the community as a whole. The important question is not how many incidents occur, but how many could be avoided.

What can go amiss

Aggression in dogs, big or small, can stem from senility or other pathological mental conditions. But far more often it is a reaction to cruel or thoughtless treatment. The greatest trouble is invited by denying dogs exercise and confining them for long periods — worst of all, on chains.

A sad mishap occurred in Newcastle, NSW, in 1984 — not through cruelty but simply because a dog's freedom was abruptly curtailed. Its owners' house was up for sale. To give estate agents easier access, the gates were taken down. So a docile German shepherd, which had had the run of the property for six years, was tied up for the first time in its life. A neighbour's little girl, a frequent visitor, was suddenly savaged. Lacerations to her head required 60 stitches. And the dog had to be put down.

Some owners, whether through lack of interest or out of a misguided indulgence, fail to instill adequate discipline. Dogs that are not taught the limits of acceptable behaviour, through the consistent use of short, firm commands, are easily confused. They may act unpredictably and perhaps violently if unusual situations arise.

Obedience classes are strongly recommended to the owners of unruly dogs. They make life easier for all concerned, and reduce the chance of danger to other people. Compulsory attendance at such classes has been suggested as a condition of dog registration. Unfortunately, compulsion would probably lead to an increase in the number of unregistered and poorly cared-for dogs.

Dogs and babies

Domestic dogs have an inherent pack sense. As puppies they learn their ranking in society, whether it is among their own kind or in a human household. Growing older, they take any opportunity to elevate themselves in the social order. If they are thwarted, or worse still downgraded, they are resentful. Preferment given a baby — especially in a previously childless household — is an affront. The newcomer is seen as a smaller creature that ought to be at the bottom of the heap.

Attacks on infants before they reach the crawling and fur-pulling stage are extremely uncommon, however. A disappointed and jealous dog may merely mope or display a general surliness. But it could bite or threaten someone else — perhaps a visitor or a passer-by.

Treated kindly, most dogs adjust fairly quickly to new family situations. Worried prospective parents can avert problems by thinking ahead. If the arrival of a baby will mean changes in a dog's way of life — particularly the loss of privileges and comforts — those changes should be made months in advance so that they are not directly associated with the family addition.

During preparations for any sort of household reorganisation, and especially when a baby is brought home, every effort should be made to show dogs the customary amount of

WHEN PREMIER JOH WAS TAKEN FOR A PROWLER
An over-zealous farm dog forgot who was boss.

A long, hard campaign in the cliffhanger state election of 1983 gave Queensland Premier Johannes (now Sir Joh) Bjelke-Petersen almost no time at home. When at last he returned to his Kingaroy property, a pet dog on the farm failed to recognise him. Sue, a huge seven-year-old German shepherd-labrador cross, meted out the treatment reserved for intruders. The elderly leader was seized by an ankle and flung to the ground. But he bounded back with typical vigour and quickly made the dog aware of her mistake.

Deep wounds to the ankle required stitching at Kingaroy Hospital, where the Premier was also given an anti-tetanus injection. Though forced to rely on a walking stick for a few days, he laughed off the incident — remarking that politics had taught him not to take even best friends for granted.

Premier Joh recuperating while a chastened Sue (top) ponders her error.

101

HEADLINE HUNTER

A dingo took a piece of newshound Anna.

When word crept out in July 1984 that a little girl had been severely bitten by dingoes at a private fauna park near Auckland, New Zealanders were agog. The Azaria Chamberlain controversy was at its height, and attracting just as much attention across the Tasman as it was here.

The victim, two-year-old Debbie Cowlishaw, had strayed from her family and reached into a wire-netted enclosure to pat two dingoes. Both attacked, seizing her by the forearm and trying to snap at her throat. Luckily the toddler's mother was not far away and managed to pull her daughter to safety, but even so gaping wounds in the child's arm required nearly 60 stitches. She was a week in hospital, and had to return later for plastic surgery.

Police and medical authorities voiced grave concern, fuelling a public outcry. The park proprietor protested that the publicity was unfair. The dingoes had never attacked before — they were completely tame. And it was up to parents to make sure that children did not interfere with the exhibits.

To make his point the proprietor urged 18-year-old Anna Storey, a cadet reporter on the *Auckland Star* newspaper, to get into the cage with the 'tame' dingoes and pose for photographs. Anna complied. The male of the pair immediately rushed at her, fangs bared. She was bitten on the hand as she tried to brush the dingo aside, then on a buttock and thigh as she turned to flee. The red-faced owner undertook to build an extra fence.

Cadet reporter Anna Storey on the retreat from a 'tame' dingo.

attention and affection. Childless dog-owners are advised to take similar care if they have other people's children to stay as guests.

By the time children reach an age of crawling or toddling, contented house dogs are normally highly tolerant of them. If a child's attentions become excessive or annoying, a dog may chastise it as it would a puppy, with a low growl and a slight, harmless nip.

Exceptions can be found, however, with elderly or ailing dogs and certain breeds that are usually trained as hunters or guard dogs. If a dog utters teeth-bared snarls at a child of its own household, keep them apart. And make sure that visiting youngsters do not pester the dog. Typically it is they, rather than children that a dog knows best, who will receive the full brunt of its anger.

Comparing the breeds

Veterinarians consider that emotional volatility in dogs, leading sometimes to unexpected outbursts of savagery, goes with very high intelligence. For this reason they regard the dobermann pinscher as a breed requiring exceptionally careful treatment. The often-feared German shepherd, by comparison, has a stolid temperament. Animal welfare organisations such as the RSPCA are glad to advise on the suitability of breeds for particular households.

Bigger breeds require more space and more exercise — along with more food, obviously. But unless they are trained to savagery, big dogs are no more inclined towards it than are small dogs. It is simply that the triggers of aggression are different. Bigger breeds are more likely to rise to what they see as a challenge. Little dogs react from fear.

Aggression can be incited accidentally, especially by strangers. A sudden movement can spark it. A broad smile can be seen as a snarl. And an attempt to pat a dog's head, by anyone but its owner, may be taken as an attack. If you want to befriend a strange dog, hold your hand near the ground, palm up.

Fright was probably the trigger for an appalling attack in Melbourne in 1926, when three young bulldogs killed an 80-year-old woman. Mrs Christiania Olsen was not a stranger to the dogs — she boarded with their breeder and they had known her since they were puppies.

Mrs Olsen went to hang washing in the backyard, where the dogs ran freely. Neighbours heard her screaming, and through the fence saw her lying face down with the dogs astride her, tearing at her head. She was able to crawl into the house when they tired of the attack, but died in hospital six hours later. Her face was wounded beyond recognition, according to a press report. Her left arm and hand as well as the back of her head were also mangled.

The dogs were known to be happy and friendly. The only conjecture that could be offered to explain the attack was that they misinterpreted some movement of Mrs Olsen's — most likely with the clothes prop. Perhaps they thought they were to be struck with it.

Even so, had Mrs Olsen provoked only the one dog she would not have come to much harm. The frenzied way in which the three pursued their onslaught and mutilated her was a frightening example of pack behaviour. Pet dogs, just as much as feral dogs (page 44), are capable of exciting one another into excesses of bloodthirstiness.

Legal responsibilities

Owners who let their dogs roam free, rather than taking the trouble to exercise them under supervision, should contemplate the possible consequences. A pet of the mildest temperament can show another side to its character, especially if it runs with other dogs. And the owner could face ruinous legal proceedings.

If a dog makes an attack in public or on someone else's property — against other animals, as well as people — or creates other nuisances, the owner can be charged with failing to keep it under proper control. If someone is injured or suffers any material loss, the owner can be sued. Damages awards are especially high if it can be shown that the owner knew the dog was dangerous.

Anyone who is attacked by a dog can do whatever is necessary to defend himself, including striking and kicking the dog. The owner has no comeback if the animal is injured or killed — though to shoot it in a public place would be a firearms offence. Savage dogs on the loose in public can be seized and impounded, but such actions are best left for the police to arrange.

Even on your own property you do not enjoy absolute legal protection if your dog hurts other people. And you cannot use a savage guard dog to patrol your front garden or yard. The law says access must be available to your front door. 'Beware of the Dog' signs have no validity.

Guard dogs can be used in backyards or inside houses — but recent legal decisions are establishing that even unlawful intruders have a right to redress if excessive force is used against them. In some states it is a specific offence to allow a dog to attack anyone, and a more serious offence to urge it to do so.

Dingoes as pets

Dingoes may not be kept without special permission, generally granted only for purposes of exhibition or scientific research. The prohibition stemmed from old declarations of dingoes as noxious animals because of their interference with pastoral agriculture. Its continuance in

David Fleay at the Gold Coast reserve he founded, with Puggles the echidna.

BITING THE HAND THAT FEEDS
How a dingo lost its 'coward' tag.

David Fleay, veteran of six decades of looking after wildlife, can fairly claim to have experienced the world's oddest range of animal dangers. In at least one respect his record is never likely to be surpassed — not only did he photograph the last thylacine or Tasmanian tiger in captivity, in 1933, but he was also bitten by it.

After early years of field research Fleay founded the native fauna section of Melbourne Zoo in 1934, then set up the Sir Colin Mackenzie Sanctuary at Healesville, Vic. There he became the first and only person to breed platypuses in captivity. In 1952 he put all his savings into developing a private sanctuary at West Burleigh in southern Queensland. At the age of 75, in 1982, he gave the 23-hectare reserve to the state's national parks and wildlife service.

Among the pioneers of snake 'milking' for research and the production of antivenoms, Fleay has been bitten by most of the deadly species. He has been stung by platypuses, kicked and scratched by marsupials, pecked and clawed and slashed by birds of prey. Flesh is missing from the back of one hand — the memento of a struggle with a goanna.

The saddest of Fleay's encounters, and the most significant in its effect on public opinion, was the attack that ended a years-long friendship with Brindle, the top dog among the seven dingoes he kept at West Burleigh at the time. He describes the clash graphically in his book *Living with Animals*.

During the autumn breeding season, Fleay noticed that Brindle was in a highly irritated state. Children had teased him the day before, and now Ginger, one of his sons from last year's litter, was doing the same. Fleay entered the enclosure to remove the younger dingo to a separate pen.

Without a sound, Brindle leapt and bit both of Fleay's forearms to the bone. The naturalist clutched the dingo and beat its head against a gatepost but the onslaught continued. He was bitten in the chest and on one thigh and knee before an attendant came to his rescue and pulled him away.

Though Fleay felt sure that the animal would have returned to good temper, he could not allow any risk to his staff. Brindle was shot. But by his action he did some justice to his breed, giving the lie to the popular belief that no dingo would have the courage to defend its rights against a grown man, whatever the circumstances.

urban areas has support now on grounds of public safety.

State laws have been widely flouted since the 1970s, when many people openly demonstrated their success in raising and keeping dingoes as obedient, affectionate and loyal pets. But the incidence of mishaps, related to the numbers kept, is far higher than that experienced with domestic dog breeds.

After a spate of attacks on children in Melbourne in 1981 the president of the Australian Dingo Foundation, which campaigns for the preservation of the breed, was obliged to point out publicly that dingoes were not and never could be truly tame household pets.

Unlike domestic dogs, dingoes do not have a strong social instinct. Most of their life in the wild is solitary. So they do not easily accept a ranking within a household. The greatest success in taming them is usually achieved by solitary bush workers. A dingo can become a good 'one-man dog', ranking itself second only to its master and remaining wary of anyone else — much like police dogs and many guard dogs.

That was exemplified in Barcaldine, central Queensland, in 1981. A mining prospector came to town to visit friends, bringing his dingo companion. Tethered in a yard, it was admired by neighbours as a placid pet. But when four-year-old Toni Plumb walked within reach, it leapt on the little girl from behind and bit her on the neck. She managed to roll clear as it tried to seize her throat.

Cats avoid trouble

Australia is said to have the same number of cats as domestic dogs — about 2.6 million. Though capable of inflicting severe wounds they rarely do so. More independently inclined than dogs, they withdraw from situations of serious conflict with humans.

Older family cats often punish small children who tease them persistently or handle them roughly. But their bites and scratches seldom cause serious wounds. Infection is a greater worry — all scratches must be disinfected, and wounds should receive medical attention.

Occasional examples are reported of pathological behaviour by cats. Usually big toms, they terrorise neighbourhoods by leaping at people whom they evidently see as intruders on their breeding territories. Authorities are ill-equipped to deal with such cases, but owners should bear in mind that they could be sued for damages. Short of having such a cat put down, desexing ought to reduce its aggression.

Family cats may well resent any withdrawal of privileges and attention caused by the arrival of a baby. They will take no revenge, however. In a disturbing case reported in 1973 from Newcastle, a month-old boy in his bassinet was severely slashed and bitten about the face and head. The cat responsible, a Siamese, was an even more recent arrival, having followed the parents home a week before.

Infants can be smothered

A greater danger can arise peacefully, from a cat's love of warmth and comfort and its readiness to go to sleep. From time to time a baby is suffocated because the family pet jumps into its pram or cot and lies on it. Trickles of burped milk may increase the attraction. Such deaths are extremely rare, but should be guarded against by keeping cats out of babies' sleeping areas or netting babies' cots.

Other conventional household pets cause little trouble — though parents should warn children that rabbits, rats and mice can inflict severe bites if they are interfered with and have no escape. Big cockatoos and their relatives may bite maliciously and unpredictably. They are best bought from reputable shops, to be sure that they are raised to be content in a cage. Fly-by-night market traders and hawkers sometimes sell drugged birds that have been caught in the wild and will become unmanageable.

Unscrupulous traders allegedly promote grave risks by offering tranquillised, unbroken brumbies as children's ponies. Parents should be suspicious of any extraordinary bargains. Riding accidents aside, enraged horses can be exceedingly dangerous. As well as kicking and trampling they can tear out big chunks of flesh with their bites — and sometimes they go for the throat.

Farm livestock do remarkably little harm in Australia, compared with the thousands of killings and hundreds of thousands of injuries reported each year in Europe and North America. Because of the spacious nature of agriculture here, and a climate in which there is generally no need to bring animals into shelter in winter or pen them in feedlots, our farmers and their families are far less often at close quarters with their stock.

The price of trespassing

But people visiting rural areas should not assume that there is no danger — especially if they enter farm paddocks without permission. Fierceness is not confined to bulls and stallions. Mares protecting foals may also resist interference, and so may cows. Rams, goats and pigs are all capable of charging at people with limb-breaking force.

New risks have arisen with the recent development of deer farming in Australia. The stock become fairly docile at most times of the year — but not because they are tame. They simply become accustomed to humans and lose their natural fear. During the breeding season, stags

Above: A western Queensland stockman prepares to vault for safety from yarded longhorns.

Right: Red deer stags compete for breeding dominance. Farmed deer can be highly dangerous because they are not truly domesticated — simply relieved of their natural fear of man.

Below: A rearing stallion can inflict terrible injuries.

'in rut' may see people as rivals and attempt deadly attacks with their antlers and hooves.

Does and juvenile deer can make charming pets. So can their marsupial equivalents. Male kangaroos and wallabies reaching maturity, however, may become highly dangerous. Many attacks by kangaroos in confinement have been reported — including some in fauna parks with public access. Young children should be warned not to approach kangaroos and wallabies that are bigger than they are.

It is beyond the scope of this book to consider the risks run by zoo staff, circus handlers and so forth in the artificially forced contacts they have with potentially dangerous animals. From a public point of view it should simply be stressed that warning signs are there to be heeded — even if no menace is apparent.

At Sydney's Taronga Park Zoo in the 1950s, a visitor was intrigued by some small jungle fowl in a high-fenced enclosure. He climbed in for a closer look. Unseen were a pair of nesting cassowaries. They knocked the intruder down, slashing and stamping on him. He was critically injured with punctured lungs and about 30 other wounds to his face, torso and legs.☐

A wild-eyed Jethro, caged at Marong.

Jethro goes to town

Police manhunt methods drove a pet emu mad.

Until late in April 1983 Jethro the emu was a placid family favourite on the farm of Mr Bob Brown, near the central Victorian township of Lockwood. Then a murder was committed in the district. Police sent in tracker dogs and a surveillance helicopter in search of the offender. With the chopper clattering overhead, Jethro went out of his mind.

The Browns watched in despair as the six-year-old bird dashed blindly about the property, running into trees and fences and even into the dam. Eventually the locality returned to its normal peace — but not Jethro. He began to make frequent escapes, defeating a 2-metre fence. Always he made for the main road. 'He just liked people so much,' Mrs Brown recalls. 'And he simply loved traffic.'

In mid-May Jethro turned up in Marong, ten kilometres north of Lockwood. Pensioner Gus Steel, 77, was walking to the post office when

Right: Mr Gus Steel, the emu's victim. Stunned and bleeding after the attack, the 77-year-old pensioner told a reporter that he simply hadn't seen the emu coming for him when he returned from the post office. His injuries included a skinned nose and deep cuts in both arms.

CAMELS ARE LOSING THEIR LAST SANCTUARY

Only in Australia do 'ships of the desert' roam free.

Australia's 'Red Centre' and the arid regions of South Australia have the world's only significant populations of free-ranging camels. An estimated 20 000 dromedaries of Arabian descent are scattered in groups over an area of nearly a million square kilometres. Their ancestors were imported as pack and riding animals between 1840 and 1907. Survivors were turned loose when road and rail transport made them unnecessary.

The camels have thrived in their desert fastnesses, eating mulga, saltbush and porcupine grasses, and have remained enviably disease-free. Now the eyes of oil-rich North African and Middle Eastern governments are on them as sources of meat and milk for their people and as riding mounts.

With negotiations for huge contracts well advanced in the mid-1980s, and talk of young, saddle-trained bulls and heifers fetching up to $2000 a head, the mustering and farmed breeding of camels began in many inland localities.

Managing camels can, however, carry a high degree of risk. Even when well treated the males — especially in their breeding season — are the most cantankerous of traditionally domesticated animals. And if overburdened or brutally handled, they are capable of a fearful vengeance.

A mature male stands as tall as 2.5 metres at the shoulder and if well fed can weigh half a tonne. It can kick forwards as well as backwards — or sideways if it is lying down. It can trample a persecutor, or kneel on him and crush him with its breastbone. Its bite is said to be powerful enough to sever a limb. Deaths from injuries inflicted by camels have been recorded everywhere that they have been employed, including Australia during the pioneering era.

Freedom is over for a western Queensland camel, descended from the one-humped Camelus dromedarius *of Arabia.*

the bird pecked at him from behind a fence. On his return, it launched a full-blooded attack. A leaping, two-footed kick in the chest flattened the old man. Suffering deep cuts in one arm and the opposite armpit, he lay doggo in the hope of avoiding further injury.

Would-be rescuers were chased off by Jethro. But he was sufficiently distracted to leave Steel alone and run off up the road back towards Lockwood. Shire council workers diverted him into their equipment compound, where he attempted to attack one of them but was stunned with an iron bar.

With Jethro back in their care but still highly agitated, the Browns tried to have him accepted by Melbourne Zoo. Officials were not prepared to recommend the tranquillising needed for his safe transportation. Bob Brown took him deep into a nearby forest, but the cars on the Calder Highway lured the bird back.

Jethro gradually settled down on the farm, but about a year later took to escaping again. To prevent the road death that seemed bound to come sooner or later, the Browns found him a new home behind the high, reinforced fences of a deer farm.□

'A leaping, two-footed kick in the chest flattened the old man.'

Eastern tiger snake *Notechis scutatus*

Part 3
ANIMALS USING VENOM

Their variety is unmatched anywhere else on earth. Australian scientists have worked wonders to find antidotes. But our best defence is to be alert to the dangers, and to leave venomous creatures alone.

Paralysis or pain

*Constituted in a confusing variety, venoms
share simple purposes: either to rob victims of the power to move, or to inflict agony.*

Venoms are poisonous fluids produced by animals to subdue or repel other animals. They vary in composition, not only between a snake and a wasp or a spider and a jellyfish, but also between different species of the same type of animal. And they are complex. Eastern tiger snake venom, for example, contains three separate toxins to attack nervous systems, and about a dozen more protein compounds that act in other ways.

Most venoms are intended to help predatory animals to obtain food. For those creatures that are poorly equipped to grasp or chase their prey, the best way to stop the prey resisting or escaping is to induce rapid paralysis. So the primary aim of a predator's venom is to sabotage the machinery of movement.

Some animals use venoms solely in defence. Toxins injected by their stings or spines are intended to dissuade predators, mainly by inflicting instant pain. Apart from stonefish venom, most of the toxins that are purely defensive present little direct threat to human life. At worst they may cause collapse in hazardous situations.

But if people are stung by the same kind of animal on many occasions, a devastating allergic reaction can develop. Sting allergies are discussed later in relation to the animals that most often cause them — bees, wasps and ants.

Snake venoms too can promote severe allergies, rare in the general community but common among collectors and exhibitors. Immature snakes and weak-venomed species that are normally harmless can become dangerous to people who are oversensitised in this way.

A few animals secrete venoms that are not used in biting or stinging, and concern us only if we eat them. For that reason pufferfishes, the giant toad and venomous frogs are not found in this section but in Part 4, as animals that are poisonous to eat. Stingrays are venomous but they pose a greater risk of maiming, so they are in Part 1 among animals that wound. Paralysis from tick saliva is an accidental effect, not the result of a venomous attack; tick paralysis is covered in Part 5 as a disease.

Neurotoxins block the brain's commands

Most dangerous of all venom components are the neurotoxins. They besiege the nerves that convey the brain's commands for movement, concentrating on or about the interfaces between these nerves and the muscles. Chemical blockades stop the passage of signals.

Normally, an electric impulse from the brain is carried down whichever nerve reaches the appropriate muscle group. The message travels by a chain reaction of ions — positively charged molecules — of sodium coating the nerve. These enter the nerve in turn, like heads of corn bobbing progressively as a puff of wind crosses a field.

Cells towards the end of the nerve constantly manufacture a chemical, acetylcholine. It is stored in little sacs on the surface of the nerve ending, which flattens out to spread along the muscle wall. But the nerve and the muscle do not touch — there is a tiny gap between them.

When the electric impulse from the brain arrives at the nerve ending — called the motor-end plate — it triggers the discharge of the sacs. The acetylcholine crosses the gap and is collected in special receptors on the muscle wall opposite. There it sets off a reaction that makes the muscle contract.

This is a once-only event. The acetylcholine is immediately destroyed by an enzyme and the muscle relaxes to await the next shower. Meanwhile, on the other side of the gap, the empty sacs retreat into the nerve to reload with acetylcholine.

If a toxin disrupts any part of this process — the activity of the sodium ions, the release or reception of acetylcholine, or the recycling of the sacs — the muscle cannot function. The result is paralysis. If the volume of toxin is great enough it reaches most parts of the body, including the breathing muscles. And if they fail, so will the heart.

Each toxin finds its own target

Where precisely a neurotoxin sets to work and how long it takes, once it has found its way in circulating body fluids, depend mostly on the size of its molecules. Among different toxic compounds, sizes can vary to the same degree as those of marbles and footballs.

Tetrodotoxin (TTX), the important component of blue-ringed octopus venom and also of pufferfish poison, has tiny molecules that invade the nerve fibres before they reach the motor-end plate. The sodium ions are denied entry. No brain commands are transmitted and paralysis is absolute.

TTX acts speedily and its effects may last for many hours. Yet it does no actual damage to the nerve fibre. If a victim is kept alive by artificial breathing aids, the nerve function is completely restored once the toxin's effects wear off.

Tetrodotoxin from the blue-ringed octopus (above) and atraxotoxin from the funnelweb spider (above right) prevent electric impulses from passing down the nerve.

Postsynaptic membrane ———

Acetylcholine sacs

Purified TTX extracts from pufferfish are used worldwide in research into how nerves work.

Nerve surfaces are also the target of a rapidly occurring attack by atraxotoxin, from funnelweb spider venom. With molecules about eight times bigger than those of TTX, it triggers false impulses that cause uncontrolled twitching of the muscles. Not only the motor nerves are affected but also many other types. Their abnormal stimulation produces gross sweating, salivation and dangerous rises in blood pressure and pulse rate.

Alpha-latrotoxin, the main component of redback spider venom, works its way into the nerve endings and causes the slow release of most of the stored acetylcholine. This just trickles out, in amounts too small to cause the muscle to contract. Paralysis can develop as the supply is depleted.

A similar action at the ends of other types of nerves leads to sweating and pain in the patchy areas that can change from hour to hour. Redback spider venom also raises the blood pressure and accelerates the pulse, but more slowly and mildly than funnelweb venom. Alpha-latrotoxin has relatively huge molecules;

WHERE ANIMAL NEUROTOXINS ATTACK

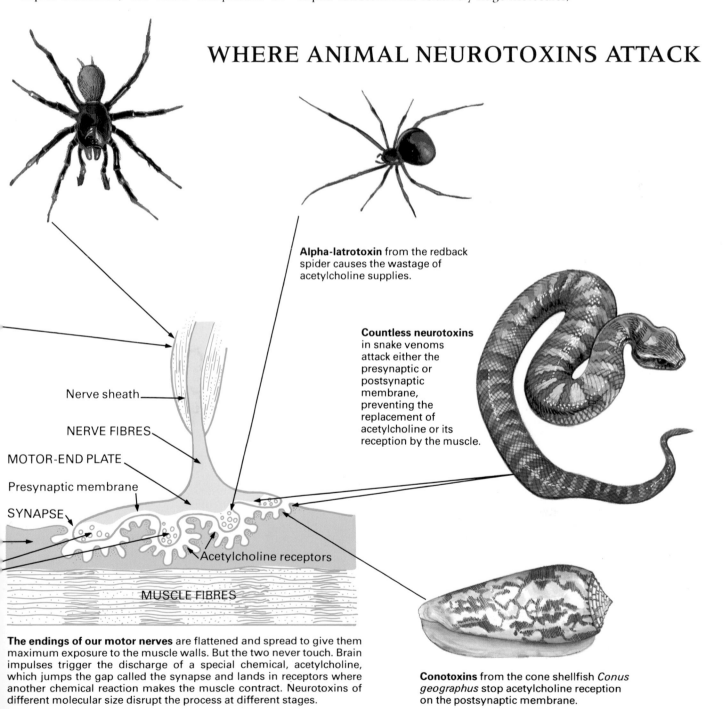

Alpha-latrotoxin from the redback spider causes the wastage of acetylcholine supplies.

Countless neurotoxins in snake venoms attack either the presynaptic or postsynaptic membrane, preventing the replacement of acetylcholine or its reception by the muscle.

Nerve sheath

NERVE FIBRES

MOTOR-END PLATE

Presynaptic membrane

SYNAPSE

Acetylcholine receptors

MUSCLE FIBRES

The endings of our motor nerves are flattened and spread to give them maximum exposure to the muscle walls. But the two never touch. Brain impulses trigger the discharge of a special chemical, acetylcholine, which jumps the gap called the synapse and lands in receptors where another chemical reaction makes the muscle contract. Neurotoxins of different molecular size disrupt the process at different stages.

Conotoxins from the cone shellfish *Conus geographus* stop acetylcholine reception on the postsynaptic membrane.

their size, and the fact that they actually have to get inside the nerves, explain why illness does not set in for hours after a bite.

The versatile venoms of snakes

Most of our medically important snake venoms contain at least two different neurotoxins that go to work on each side of the nerve-muscle gap. Scientists call the gap a synapse, and refer to the two sides as the presynaptic and postsynaptic regions.

Snake neurotoxins acting on the presynaptic or nerve side of the gap apparently enter the surface membrane. As sacs of acetylcholine move out and discharge their contents, big toxin molecules stop them from going back into the nerve to reload. Paralysis takes hours to develop, but it is most serious — and difficult to

reverse with antivenom because the toxins are almost buried in the membrane.

Postsynaptic muscle regions are put out of action by a wide range of poisons. Any substance attracted to the acetylcholine receptors will block the access of the chemical. Best known of the postsynaptic paralysers is curare, the plant extract that South American Indians put on arrows and blowpipe darts. Modified curare is often used to relax the muscles of surgical patients.

Paralysis from the postsynaptic neurotoxins of our dangerous snakes sets in much more quickly than the presynaptic effects. The molecules are generally far smaller, and they need only attach themselves to the acetylcholine receptors — there is no requirement for time-consuming penetration. But antivenoms reach

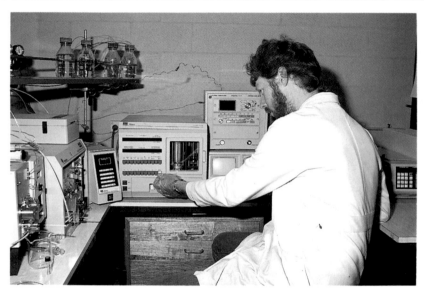

A chromatograph separates venom components within each minute sample.

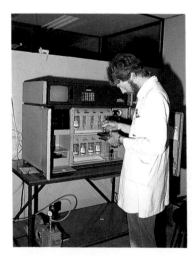

The sequencer analyses individual venom components.

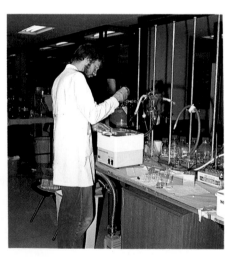

Funnelweb venom being placed in a centrifuge for drying before analysis.

TRACING THE PROFILE OF A POISON

*Laboratory detective work
unravels the mysteries of toxin structure.*

Advanced equipment at Sydney's Macquarie University can operate day and night, programmed for the automatic analysis of venoms. Liquid chromatographs separate all the components of tiny samples of venom, then detect them and register their amounts by scanning them with ultra-violet light. Each run of venom takes only about 25 minutes. The information can be graphed simultaneously, allowing scientists to interrupt the process at any stage to draw off a particular component for further investigation. Otherwise the data can be stored in a computer and printed out later.

Individual protein toxins, isolated in this way, are placed in the sequencer. It identifies their molecular structures. The university's school of chemistry research group, under the leadership of Associate Professor Merlin Howden and Dr David Sheumack, is the first in Australia to have discovered the complete structure of an animal toxin.

Their special interest is the venoms of various funnelweb spiders, but they have also made full analyses of toxins in brown snake and death adder venoms.

A funnelweb spider being 'milked' for its venom.

these toxins easily. Reversal of the paralysis can be dramatically swift. Neurotoxins in the venoms of cone shellfishes also have a fast-acting postsynaptic effect. Three of these conotoxins were identified early in the 1980s in the potentially lethal *Conus geographus.*

Why some victims took days to die
Some of the snake neurotoxins have recently been found to be capable of damaging muscle fibres. In severe cases a lot of the muscle protein — myoglobin — escapes into the bloodstream, eventually clogging the kidneys and causing their failure. Tiger snake venom is the most common cause.

This discovery accounts for the blackened urine of many snakebite victims. And it explains why, before antivenoms were available,

some victims died days after they were bitten rather than within the usual 24 hours. But scientists still cannot understand why one victim suffers severe muscle damage but only mild paralysis while another, bitten by the same kind of snake, experiences the opposite effects.

Few venoms have muscles as a primary target, however. With the exception of protein loss by snakebite victims, and perhaps a reduction of muscle contractability in stonefish victims, the machinery remains in perfect working order. But the starter button is disconnected.

Snake venoms — but strangely none of the others — often disturb the process of blood clotting. Bite punctures or scratches, and any other wounds a victim may have, bleed steadily. A sample of blood may take over an hour to clot, instead of the usual 4-10 minutes.

The purpose of this property of snake venoms is a mystery. It is easily overcome with antivenom and human blood products if necessary, and unlikely to cause problems unless a victim already has an internal complaint such as a stomach ulcer.

Destruction of red blood cells was once thought to be an important effect of snake venoms. It was blamed for the darkened urine of bite victims — now known to be caused mainly by muscle damage. While the venoms of the copperhead and the *Pseudechis* group in particular do have cell-damaging factors, they are of minor medical significance. Chances of severe anaemia or kidney failure from this cause are remote.

Venoms that destroy the skin
Contending with so many venomous creatures, Australians are fortunate in one respect. Few of the animals do much damage around the site of their attack. The most important exceptions are turning out to be spiders, though what property of their venoms could be responsible for massive destruction of skin remains unknown.

The fastest and most agonising tissue damage by an Australian animal is produced by the chironex box jellyfish. The skin-killing component of its venom acts almost immediately, but long-term damage is slight if antivenom is given within about 15 minutes. Other marine injuries such as stonefish stings often feature some tissue destruction.

But no snake venoms in this country cause widespread skin damage. A bitten area appears normal within a few days. It is this harmlessness of snake venoms at the bite site that makes pressure/immobilisation first aid such a success. In contrast, some foreign snakes wreak terrible tissue destruction. American rattlesnake venom, for example, can cause the loss not merely of all the layers of skin in the bite area, but also of big blocks of flesh below.□

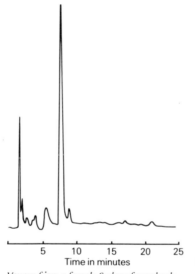

In the venom profile of a male Sydney funnelweb spider, scanned by chromatography, the lethal component shows up at 20 minutes, near the end of the run. Many other components may be toxic but not deadly. The areas under the peaks show the relative amounts of components. The height of the peaks does not necessarily indicate a particular toxicity, but is rather a reflection of the rate at which these components come through.

Venom from a female Sydney funnelweb shows a very different make-up. Most significantly, there is no peak of lethal toxicity at the 20-minute mark.

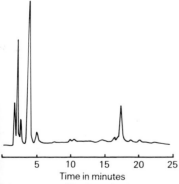

The venom profile of a male Blue Mountains funnelweb has a worrying peak near the end of the sequence.

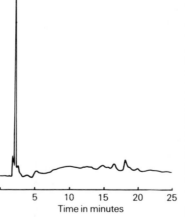

As individual as a signature, the venom profile of a northern tree-dwelling funnelweb differs yet again.

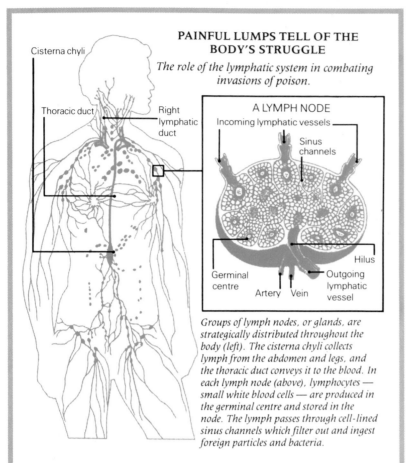

PAINFUL LUMPS TELL OF THE BODY'S STRUGGLE

The role of the lymphatic system in combating invasions of poison.

Cisterna chyli

Thoracic duct

Right lymphatic duct

A LYMPH NODE

Incoming lymphatic vessels

Sinus channels

Germinal centre

Artery

Vein

Hilus

Outgoing lymphatic vessel

Groups of lymph nodes, or glands, are strategically distributed throughout the body (left). The cisterna chyli collects lymph from the abdomen and legs, and the thoracic duct conveys it to the blood. In each lymph node (above), lymphocytes — small white blood cells — are produced in the germinal centre and stored in the node. The lymph passes through cell-lined sinus channels which filter out and ingest foreign particles and bacteria.

Venoms circulate in all of our body fluids — not just the blood. A lot of venom may not find any target. It is simply discarded as liver or kidney waste and excreted in urine. Some is carried by tissue fluids into the lymphatic system, where the body's natural defences start an attempt to neutralise it. But this is a slow process — to no avail if neurotoxins are causing paralysis.

The lymphatic system is a network leading to strategically placed filtering stations called lymph nodes. These are rubbery glands about the size and shape of beans. They are concentrated mainly around the jaw, neck, collarbones, armpits and groins, and at greater depth around major organs and the roots of the lungs.

Lymph nodes manufacture and store lymphocytes, white cells whose job it is to filter out any foreign matter in the circulating fluid. They in turn start manufacturing proteins to counter the foreign proteins — they make antitoxins in the case of envenomation and other poisonings, just as they make antibodies in response to bacterial infection. Once neutralised, the invading particles are destroyed by other white cells in the blood.

Whenever lymphocytes are called on to resist heavy invasions of toxins or bacteria, the lymph nodes swell up. They become tender to touch and sometimes very painful. Their enlargement may be visible at a glance, especially in children. Involvement of the deep nodes around internal organs probably accounts for some of the abdominal pain that often occurs after snakebites.

If a person is bitten or stung by some unidentified creature, swelling and pain in the lymph nodes nearest the wound may offer confirmation that a substantial quantity of venom has been injected. It is not necessarily a danger sign, though — venoms that provoke this reaction may not have any paralysing neurotoxic components. And the same reaction can be caused by many other kinds of poisoning and by infectious diseases.

Identifying a venom

Australian ingenuity has cut the time from days to minutes.

Rapid identification of snake venoms is a triumph of 10 years of Australian research and development. A simple test with a portable kit ensures that a bite victim gets the right care. Here it saves just a few lives, because relatively few people are seriously bitten. But overseas, where some 30 000 people a year are killed by snakes, the kit's potential is limitless.

In 1972 it became possible for the first time to detect and actually measure snake venom at the site of a bite or in the victim's blood or urine. The technique, involving radioactive iodine, was highly complicated and could be performed only at the Commonwealth Serum Laboratories in Melbourne. Results took days to obtain.

This type of test, called a radioimmunoassay, gave CSL researchers their first chance to see how venoms travelled in animal systems. It allowed the comparison of various methods of delaying the spread of venom, leading to proof that the pressure/immobilisation first aid method works best.

A speedier radioimmunoassay technique was put to use in dramatic circumstances in 1975, when a 10-year-old Melbourne girl was found unconscious near a suburban railway line. She was thought to have suffered a head injury or a drug overdose. But strange aspects of her condition led hospital doctors also to consider snakebite. Samples from a scratch on her leg and from a sock, as well as a urine specimen, turned out to contain tiger snake venom. Quickly given the right antivenom, the girl made a full recovery.

Radioimmunoassay is still the most sensitive technique for research purposes. But the Melbourne girl's case, along with some others in the mid-1970s, spotlighted the need for a more simple, bedside method of detecting and identifying snake venom.

Uncertainty was risky — and costly

In 90 per cent of Australian snakebite cases, the snake responsible is not positively identified. Without knowledge of the venom type, a seriously poisoned victim may have to be given a combination antivenom including all types. That requires the injection of a large volume — increasing the risk of an adverse reaction. It is also very expensive. Immediate identification of the snake type offers other advantages in treatment. It gives doctors a better idea of special problems to watch for. For example, tiger snake venom can cause muscle damage leading to kidney failure. That worry does not arise from the bite of a brown snake.

The first type of venom detection kit was issued free by the CSL in 1979 to hospitals that regularly treated snakebite cases. It clearly indicated if any venom was involved and if so, into which of the five antivenom groups it fell. But the test still took hours and was best performed by a skilled laboratory technician.

Once the value of the test had been established researchers started to devise an even simpler, faster system. The result in 1982 was an ingenious miniaturised kit that can identify venom in a sample within 30 minutes. It can even be used by outback Flying Doctors when they and their patients are airborne.

The kit works on the principle that toxin molecules will attach themselves to the appropriate antitoxins. Its core is a string of six narrow tubes. Five of them are coated inside with the different antivenoms — tiger snake in the first, brown snake in the second and so on. The last tube is coated with a venom mixture as a control measure.

A sample of blood or urine, or fluid swabbed from the site of a bite, is drawn through the tubes by a syringe and squirted out. Then four different solutions are drawn in for short periods. The last one contains an indicator chemical that changes colour from orange to dark purple in a tube that has venom. The control tube also shows purple, confirming that the test has worked properly.

The CSL kit is distributed to hospitals and bush nursing centres, but not to the general public because its materials require storage within a closely controlled temperature range. It has attracted worldwide attention. Modified for different venoms, it is likely to be adopted in many countries.☐

Animals to the rescue

Production of life-saving antivenoms relies on a peaceful four-legged workforce, tenderly cared for and kept super-immune.

A horse being bled to produce serum.

Antivenoms lend us a swift resistance to the worst effects of venomous attacks. They are natural proteins, imitating those that the human immune system is capable of producing for itself. But when neurotoxins paralyse the breathing muscles, our bodies have no time.

Instead the neutralising agents are created in animals that have been made highly immune to venoms. Serum is separated from their blood and purified. The animals, carefully husbanded because they have to be in perfect health, remain unharmed.

Horses are the principal donors. Dozens of them, chosen for their size, graze on a special farm in Victoria. CSL staff give them courses of venom injections, starting with tiny amounts and building up to doses that would kill any other horses.

Each fully immune horse is bled three times in a week, yielding a total of 24 litres. Then it is rested for several months. Later, after some booster doses, more blood can be taken. The advantage of using horses lies not only in the great blood volume that goes with their size, but also in the long lives that they enjoy under good care. One horse in its career provides serum for many thousands of antivenom injections.

Horse serum is the basis of all snake antivenoms, along with redback spider and stonefish antivenoms. Sheep are immunised to make chironex box jellyfish antivenom. Tick antitoxin is made from the blood of infested dogs. Funnelweb spider antivenom comes from rabbits. These different animals are used either because horses did not respond to the immunisation or because the quantities of antivenom required are small.

Pigeons were the pioneers

Venom immunity was first induced in animals in 1887. An American physiology professor, Henry Sewall, worked out the smallest amount of rattlesnake venom that would kill a pigeon. Injecting pigeons with only a fraction of that amount at first, then gradually increasing the doses, he eventually found that the birds survived volumes of venom many times greater than the original lethal dose.

By 1894 a French scientist, Albert Leon Calmette, had immunised horses in the same way, using cobra and viper venoms. His *serum antivenimeux* was marketed worldwide. Calmette believed it could be used for any type of snakebite. But at Sydney University in 1897 it was shown to be ineffective against tiger and red-bellied black snake venoms.

A year later Dr Frank Tidswell of the NSW Health Department began immunising an ambulance horse that had been retired because of a shoulder injury. Its first dose was a mere 0.5 milligrams of tiger snake venom. By 1901 it was tolerating 600 mg.

Antivenom from this horse was used in experimental studies but apparently not given to bite victims. No further work was done for over 20 years. Luckless people continued to die from venomous attacks in all parts of Australia.

In 1928 the director of Melbourne's Walter and Eliza Hall Institute of Medical Research, Dr Charles Kellaway, initiated studies that led to a rapid improvement in the understanding of venoms and the care of victims. His interest was sparked by Sir Neil Hamilton Fairley, who had extensive experience of Indian snakes and made significant contributions on treatment, first aid, biting mechanisms and venom yields.

The breakthrough in Australia

With the basic research done, Kellaway teamed up with Dr Frank Morgan, head of the Commonwealth Serum Laboratories. Their collaboration produced a CSL tiger snake antivenom for general use in 1930. At last Australian doctors had a means of saving the lives of at least some dying bite victims.

But it was a long time before other snake antivenoms were available, mainly because of difficulties in obtaining reliable supplies of fresh venom. Taipan antivenom was distributed in 1955, brown snake in 1956, death adder in 1958. Black snake antivenom, made from a New Guinea species but used against bites of the mulga snake, was issued in 1959. A sea snake antivenom followed in 1961.

Polyvalent antivenom, combining all the land snake venom groups, was released in 1962

Above: Robyn Worrell coaxes venom from a black tiger snake at the Australian Reptile Park near Gosford, NSW. An eastern tiger snake (right) and a taipan (below) also contribute.

THE RISKY ART OF 'MILKING' A KILLER

Handling dangerous snakes is all in a day's work for venom suppliers.

Unerring judgment and great dexterity are needed to obtain snake venoms for antivenom production. The snake has to be made to bite through a latex membrane — usually a piece of rubber glove — stretched over a glass container. But the head must be grasped in a way that allows the snake to breathe comfortably. And the venom glands are massaged to gain the maximum yield. All this must be done with one hand, for the other is busy restraining the snake's muscular, writhing body as well as holding the container.

The Commonwealth Serum Laboratories take venoms from only the most careful and knowledgeable collectors. Yet every regular supplier has needed antivenom for at least one bite. Most have been badly bitten three times or more. Handling the same species so frequently, venom suppliers run an additional risk of developing allergies that make even slight bites immediately dangerous.

Dr Charles Kellaway

Sir Neil Hamilton Fairley

Dr Saul Wiener

Dr Frank Morgan

to treat poisonings by unidentified snakes.

Meanwhile the CSL was occupied with other dangerous venoms. An antitoxin to combat tick paralysis was in use by 1938. Dr Saul Wiener's redback spider antivenom, issued in 1956, put an end to deaths from that cause. The world's first marine antivenom, against stonefish stings, was also prepared by Wiener and released in 1959. Meticulous work by Dr Harold Baxter and his team culminated in the debut in 1970 of chironex box jellyfish antivenom. The problems of developing a funnelweb spider antivenom were solved in 1980 under the leadership of Dr Struan Sutherland.

Nothing but the best

To make effective antivenoms in the large quantities that are distributed in case of snakebite, the best of everything is required. Venoms must be of the highest quality — accepted only from a few collectors who know a great deal about the identification and care of the snakes that they 'milk'. Donor snakes, as well as the animals to be immunised, must be kept healthy. And the CSL needs the most advanced facilities for purification and testing.

Venom collected from identical snakes can be pooled immediately, or mixed after it has been dried. Bacteria may grow in liquid venom and destroy it, so any that cannot be put in a drying machine quickly must be frozen. Every batch arriving at the CSL in Melbourne is tested to confirm that it is from the right species and has the expected toxicity.

Converted into a dry powder and sealed from the air, venoms can be kept indefinitely. When

needed for animal immunisations they are put into injection solutions in carefully measured amounts. Laboratory workers wear masks when they handle the powdered venoms — inhaling them would be highly dangerous.

After immunisation, the whole blood that is taken from the horses contains many components other than antivenom. Purifying the product entails a step-by-step process of fractionation to isolate the single component needed to treat bite victims. At last a clear, straw-coloured liquid, tested to verify its antivenom activity, is dispensed into glass ampoules.

How the doses are measured

Ampoules vary in size, according to the amount of venom that the snakes involved are expected to inject with a bite. Each ampoule contains the least amount needed to neutralise the average 'milking' yield of a snake — generally more than would be delivered by biting. In most cases it is enough, though occasionally a patient may require as many as six doses to regain health.

Antivenoms are measured in units of activity, each unit being able to neutralise one-hundredth of a milligram of venom. Bites by the low-yielding brown snakes are counter-acted by just 1000 units. But initial dosages in the other antivenom groups are 3000 units of tiger snake, 6000 units of death adder, 12 000 units of taipan and a massive 18 000 units of black snake. An ampoule of polyvalent antivenom contains the equivalent of all those units of activity but not their total volume, because there is some overlapping effect.

Supplies for human use are issued through state health departments, and dispensed with-

---

WANTED: SNAKEBITE VACCINES

Prevention is better than a cure, but venom immunity is a long way off.

An antivenom cannot be used preventively. It works only when someone is already in danger from the effects of a venom. What is needed, to avoid those effects altogether, is a vaccine that confers longterm immunity.

Just as the CSL's horses can be immunised against snake venom by gradually increased dosing, so can people. The trouble is that the degree of protection falls off rapidly as soon as the dosing stops. And in the first place, it is only good against one kind of snake or group of snakes.

Snake handlers who receive minor bites from the same species over a period of years can accidentally develop some natural immunity to that species. But they are just as likely to develop a severe allergy that threatens their life when they are bitten by even a baby snake.

In countries where one species of snake presents a particular threat, some newly developed vaccines are being tested. For Australia, higher hopes lie in the advances of biotechnology — genetic engineering. One day antivenoms could well be produced in laboratories rather than in animals, and venoms could be subtly altered to become perfect immunising agents.

out charge to victims. Hospitals are advised to have on hand four ampoules for each of the snake groups in their region and — except in Victoria and Tasmania — some polyvalent antivenom. A mixture of tiger snake and brown snake antivenom will neutralise the venom of any Victorian snake. In Tasmania the only snake antivenom required is tiger snake.

Bush nursing centres, church missions and some individual doctors in outback areas also receive emergency supplies free. Of the total distribution of thousands of ampoules, very few are actually used to treat bite victims. Even though snake antivenoms are kept under re- frigeration, they start to lose their effectiveness after three years and have to be replaced.

For veterinary use to save pets and livestock, antivenoms are made available at the full cost of production. But their purchase for other private purposes — by doctors accompanying remote expeditions, for example — is not encouraged. If antivenom injections are called for, resuscitation equipment, supporting drugs and laboratory facilities should also be on hand.

Avoiding the risk of a bad reaction

Antivenoms are not administered as a matter of course, but only if symptoms of general envenomation that could be a threat to life are observed. Because the proteins are of non- human origin they can produce adverse reac- tions in the systems of some patients. Their use often has to be accompanied by other medi- cation to avoid the worst effects, and needs careful monitoring.

Until 1976 it was customary to give antivenom in 'neat' doses. Serious reactions were not uncommon, especially among people born before 1940. They were likely to have been sensitised to horse protein as children, through injections of relatively crude prep- arations to guard them against tetanus and diphtheria.

Today's highly refined antivenoms are in- fused slowly in a diluted form if time permits, after premedication with drugs such as antihistamines and adrenalin. Steroids are given to older people and those with histories of allergy. Bad reactions during antivenom treat- ment have become extremely rare.

But there is still a chance of a later reaction — delayed serum sickness. It is a disorder of the immune system that may show itself within a week of an antivenom injection, or take as long as a fortnight. The effects are generally mild. A fever, a rash and some soreness of the lymph nodes all resolve themselves within 48 hours.

In severe cases of serum sickness the symp- toms are painful and debilitating. Sufferers may take months to recover their strength. Such cases are usually associated with the heaviest

doses of antivenom. With venom detection kits reducing the use of high-volume polyvalent injections, they are increasingly rare. But they highlight the need for caution in using any antivenom.☐

Note: Older publications may use the terms 'antivenin' or 'antivenene'. These terms have been obsolete since 1979, when the World Health Organisation chose antivenom as the sole form for international use.

A spitting cobra from Africa.

ON GUARD AGAINST EXOTIC SNAKES

Customs officers, zoo personnel and snake collectors are all at risk.

Small amounts of foreign snake antivenoms are held in Australia, chiefly in case Customs officers need treatment. They occasionally encounter dangerous types such as cobras and rattlesnakes, either deliberately smuggled or as accidental stowaways in cargo. Keepers of exotic collections, too, are sometimes bitten.

Supplies are kept at Sydney's Taronga Park Zoo, the Royal Melbourne Hospital, and the Australian Reptile Park near Gosford, NSW — where the proprietor was gravely poisoned by his favourite king cobra in 1985.

Australia in turn is a substantial exporter of antivenoms. The Commonwealth Serum Laboratories make snake antivenoms for Papua New Guinea, and CSL marine antivenoms — especially those against sea snakes and stonefishes — are in wide demand in Southeast Asia and the Pacific Islands.

POTENCY TO SPARE

Most Australian snakes are venomous, and some of their venoms are astonishingly potent. At least 10 species are more lethal than an Indian cobra. America's most feared rattlesnake would barely rank in our top 20.

Killing the single rat that may satisfy its hunger for a week or two, the fierce snake injects a dose of venom sufficient to finish off thousands of rats. Such a capability for overkill, common among our snakes, has no apparent purpose. They have no designs on bigger prey.

Nor is the country crawling with snakes. Bites from unseen snakes happen now and again, but few people suffer seriously from them. And with advances in first aid and antivenom science, death is the least likely outcome.

Rather than fear or hatred, snakes call for caution — and a certain amount of concern. Left to themselves they are incapable of malevolence towards us. Instead, in the dwindling territory that we allow them, they perform a valuable service in controlling vermin.

SUPERBLY ADAPTED TO LIFE WITHOUT LIMBS

Nature may seem to have given snakes a raw deal. Deprived of their ancestral legs, they are cast into the lowliest of roles — squirming on their bellies while their prey and their enemies can run or hop or fly. They are limited in vision, virtually deaf, and their activities are ruled by the whims of the weather.

But snakes represent one of the great successes of specialised development. Descended from the same stock as lizards, they have adapted to claim their own ecological niche in habitats as varied as forests, deserts and oceans. One of their great strengths, compensating for many disadvantages, is the ability to slip through small openings. It allows them to hunt by stealth, to invade the burrows and nests of prey animals, and to enjoy shelter and rest where no danger can enter.

Along with the loss of limbs the main evolutionary change was an extreme narrowing and lengthening of the body. Internal organs were modified and rearranged to fit. The spine, extending until it comprised hundreds of vertebrae, became unusually supple. A unique system of movement developed around it.

Every vertebra in a snake's trunk is connected with a movable rib, through a ball-and-socket joint. Muscles run in all directions — from vertebrae to ribs, from one rib to the next, between ribs that are several vertebrae apart, and from the ribs to the skin. Other muscle tendons are linked to one another in chains spanning up to 30 vertebrae.

Two methods of travelling
All of this intricate mechanism is aimed at exerting backward pressures on the surface over which a snake wants to move. When resistance is met the snake is thrust forward. If it can apply pressure on many points at the same time or in quick succession, it goes faster.

Squirming is the basic method. The body is flexed in a series of waves that travel backward as the muscle chains are contracted in sequence. Each wave stops where the snake's skin is in contact with an obstacle — a rock or a tree root, or merely a bump in the ground — and the body is thrust forward a few millimetres.

But many land snakes, including our venomous species, can also creep along in a straight line. For this purpose they use their wide belly scales, each of which is connected to a vertebra and rib. Controlling opposed pairs of muscles that link each rib with the skin of the belly and lower flanks, they hitch the scales forward, tilt them and pull them back with the trailing edges forced against the ground.

A snake can lift the first third of its body clear of the ground to seek a higher position — or of course to attack. All land snakes can climb trees, as well as swim. But none can jump, though an angry one may hurl itself forward for up to a third of its length.

Acceleration is rapid when a startled snake is fleeing for cover. But its speed over level ground does not exceed 7 km/h. Even that can be kept up for only a short distance. A normal rate of travel is about 4 km/h — an easy walking pace for humans, but still an impressive effort by creatures that have no limbs.

A tongue that tracks down prey
Unable to chase prey, a snake relies on catching its victims unawares. It finds them by following scent trails — using its long, forked tongue. Constantly flickering in and out through a notch in the lips, the tongue samples particles from the ground and the air and passes them to sensory organs in the roof of the mouth.

The eyes are useful only in detecting movement at close quarters, and closed inner ears can pick up little more than ground vibrations. But they provide a remarkable sense of balance — snakes traverse swaying vegetation with ease, and smaller ones can rest on things as thin as fencing wire.

Venomous snakes seldom eat anything except live or freshly killed animals. In Australia rodents, other small mammals, frogs, lizards and sometimes birds are the usual prey — along

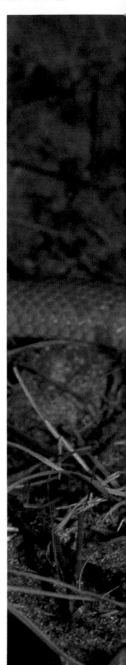

with other snakes. A hungry snake may gorge itself on a succession of small kills but then it rests: taipans in captivity are content with about 20 good meals a year.

Whatever is to be eaten has to be swallowed whole — nearly always head-first so that fur, limbs or feathers go in easily. No snake can chew its food. Small, backward-pointing teeth are used only to draw prey into the mouth.

Often a snake's kill is wider than its head. But its jaws come apart at the back, and other bone connections allow an enormous stretching of the mouth. To avoid choking, the windpipe is reinforced and its opening can be thrust out at the side of the mouth. Once the prey has been worked past the jaws, the body skin is elastic enough for the bulkiest meal.

Activity is ruled by the climate

Like all reptiles, snakes are cold-blooded — they cannot compensate internally for temperature changes as mammals and birds do. Their body temperatures rise or fall with the air around them. They enjoy full activity only within a range of about 20-35°C. At higher or lower temperatures they become sluggish, and below about 3° or above 40° they die.

In the freezing winters of northern Europe and North America snakes hibernate in a state of unconscious, helpless torpor. But true hiber-

Mobile ribs flatten the neck of a threatening eastern brown snake as it arches its body in a characteristic S shape.

THEY SKIN THEMSELVES ALIVE

Fresh skins are needed frequently to accommodate the growth of juvenile snakes and to remedy wear and tear in maturity. The process of shedding an old skin — called ecdysis or sloughing — takes place four to six times a year in the land-based species.

A snake's colours become dull, then turn milky, and its normally transparent eye scales cloud over. This is because chemical changes are taking place in the outer layers of the skin. Partly blind, the snake goes into hiding for a week or so.

Restoration of sight signals that the new skin is ready. Within a few days the snake starts to break the old one by pushing its snout against rough objects. The skin peels away around the lips. Pushing harder, the snake forces the facial skin back over its head and wriggles forward with its body pressed to the ground. The old skin is pulled inside out, right to the tail, and the snake emerges in bright new colours.

A sloughed (pronounced sluffed) skin is stretched and weakened, and of no commercial value. But it can be useful in identifying the type of snake that shed it and sometimes, by counting the scales, its exact species.

Colouring is brightest immediately after sloughing.

A Northern Territory death adder captures (above) and paralyses (below) a dtella gecko. The prey will then be pushed into position to be eaten head first.

nation is rare in Australia. In adverse weather snakes wait in shelter by choice — not because they have lost all vigour.

More Australian snakes have problems with excessive heat, which can damage some of their tissue proteins. In the tropics and the hot outback they take shelter from summer sunshine, and usually convert from daytime to dusk or night hunting. In prolonged spells of fierce heat some inland species find deep cracks in the ground and become completely inert. Again it is a voluntary retreat, and such snakes can retaliate if they are disturbed.

Three ways of giving birth

Mating usually takes place in spring. Males track the females by scent and pairs start a courtship routine of gentle twining and tail-twitching. At its culmination the female's anal plate is pushed down and the male inserts one of two penises. Pairs of snakes may sometimes be seen intertwining vigorously — these are more likely to be both males, in ritual combat.

About half of the dangerous species are viviparous. They give birth to live young, though some may be encased in light membranes from which they emerge in a few hours. All are classed as live births in the species descriptions given later. The remainder are oviparous, laying leathery eggs that require weeks of incubation.

Neither the litters of live young nor the clutches of eggs are tended by the adults. The young have to fend for themselves and are able to kill tiny prey within days of birth or hatching. Their mortality from cannibalism and the predation of birds and big lizards is often high. But snakes that reach maturity are thought to have a life span of 10-30 years.□

Above: A pair of male copperheads writhe in ritual combat.

Above: A gwardar lays its eggs.

Left: A new generation of taipans is born. Eggs from captive snakes of this species take about 15 weeks to hatch.

Below: Young of the bardick Echiopsis curta are born live. Found in dry parts of the south, it may be slightly dangerous.

BITES — AND HOW TO AVOID THEM

Snakebite statistics are sketchy. It has been suggested that about 3000 people are bitten each year in Australia but the vast majority of them feel few effects. The only firm figures for serious poisoning relate to antivenom usage: in a typical year 200-odd cases are reported back to the Commonwealth Serum Laboratories. About as many again are thought to go unreported.

Wittingly or not, we provoke all snakebites. No snake is normally inclined to attack anything that is too big to swallow. Our size is a threat, as well as a waste of venom, and most snakes will do their utmost to stay out of our way. Even when they enter buildings, it is without knowledge of our presence. They retreat hurriedly when they discover their error — unless their escape is cut off.

Unseen snakes, trodden on in grass or inadvertently touched in their hideouts, are said to account for a minority of bites. More attacks occur, according to professional handlers, as a result of deliberate acts of bravado or fear. People try to kill, catch or harass snakes when all they should do is let them slip away.

While that may be an apt summary of adult behaviour, it is probably unfair to children. A recent study of child victims in Brisbane showed that 75 per cent were bitten at play — walking, running or climbing.

A disturbing aspect of the Brisbane study was the inclusion of three infants, found in their cots clutching snakes. It is a reminder that the usual human repugnance to snakes is taught, not instinctive. And a baby is especially vulnerable because a snake inside a house, encountering

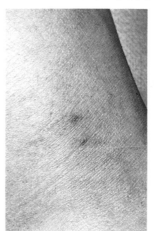

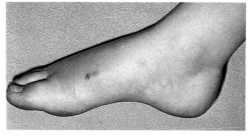

Above: Typical twin puncture marks on a child's foot were accompanied by brief bleeding, some bruising from the impact of the snake's head, and general reddening and swelling. Left: Fainter marks with no evidence of bleeding — but the poisoning, from an unidentified snake, was severe in this case.

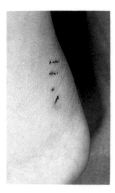

Evidence of a double strike above the heel of a 9-year-old Brisbane girl. The snake was an eastern brown.

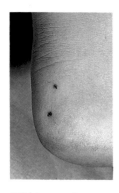

Widely spaced punctures on the heel of a younger child indicate a snake of great size — again an eastern brown.

such a tiny form, may 'freeze' beside it while pondering the likelihood of danger.

Keeping trouble at bay

Snakes can be kept away from homes, schools and farm buildings by denying them cover and eliminating rats and mice. Grass around buildings should be cut and garden refuse turned over frequently. Boxes, cartons and tree branches should not be left lying about. Building materials and firewood should never be stacked on the ground but raised on bearers.

If a snake enters a building, don't poke at it or try to corner it. Move about, giving it a wide berth, and it will probably go out the same way it came in. Even if it crawls towards you it is not attacking — just step quietly out of the way. If a frightened snake hides in an inaccessible spot, call the police or state emergency services.

Children should be instructed in the ways of snakes and warned against playing in long grass or among fallen trees — especially if they go barefoot. Making pets of baby snakes and harmless species is a hobby that ought to be discouraged unless children have a full understanding of the possible danger of other types.

In bushland, where snakes are more at home than we are, the safety rules are simple:
- Wear boots and thick socks if you walk through grass, or over ground you can't see.
- Use a torch around campsites on warm summer nights, and shake out sleeping bags.
- If a snake blocks a trail, back away. Make plenty of noise — stamp your feet — and let it go where it wishes.
- NEVER blindly plunge your hand into a hollow log or a chink among rocks.☐

CURES THAT KILLED

Distilleries did best out of snakebite last century. Many doctors believed that the actual cause of death was fright, and recommended alcohol to overcome it. Worse still, rest was forbidden. Drunken victims were made to walk up and down, and sometimes whipped, until they collapsed.

Dangerous chemical antidotes were regularly proclaimed, and not easily discredited. Among the substances that Australians were urged to take were liquid ammonia, mercury, chloride of lime, potassium permanganate and even strychnine — which killed a child as late as 1912. A more natural cure, seriously proposed in 1903, was eating the venom gland from the same kind of snake.

FANGS THAT QUEUE UP TO KILL

Wear and tear are heavy on the slender fangs of venomous snakes. They are soon blunted — if not wrenched out in the struggles of prey animals or broken when they strike bone. But fresh fangs are always held in reserve. Each bides its time, poised to move into a position where it can kill.

Fangs are teeth that are modified to carry venom. In the front-fanged elapid snakes and sea snakes — the only kinds that are dangerous in Australia — each fang is curved around a deep groove running from near its base. The lips of the groove meet, enclosing a tube along which venom is pumped. The tube opens almost at the tip of the fang.

A snake normally has one active fang on each side of its upper jaw, attended by at least one and sometimes a cluster of three or four spares in different stages of growth. The biggest ones may receive venom but they jut out at awkward angles, not as far as the main fangs.

A mobile bone holds each active fang and its successor.

So unless a snake fastens onto a victim's flesh and gnaws at it, the reserve fangs are unlikely to do more than scratch the skin.

Every few weeks a worn-out fang works loose. It is usually swallowed, embedded in prey. The senior reserve fang soon takes its place and the other spares move closer. If a fang is lost prematurely its replacement takes longer, but the snake is not seriously handicapped. It eats infrequently and can bite with a single fang if it has to.

Poorly armed to attack humans

Except for the brown snakes and the rough-scaled snake, all of the dangerous Australian snake species keep their mouths shut until the final instant of the strike. A swivelling bone in the jaw allows the fangs to be pushed forward at the moment of biting. But the degree of rotation is extremely slight in all species other than death adders.

Against bigger targets the attack is often inefficient. The fangs — if they do not glance off — may enter the flesh at an acute angle. The venom injection is then shallow and less quickly effective.

This restriction of mobility also places a limit on the length of fangs. Foreign vipers can swivel their fangs not only well forward but also backward, to fold them away. So they can accommodate fangs that are far bigger than their closed mouths — sometimes longer than 30 mm. An Australian brown snake's fangs, by comparison, do not even reach 3 mm. A taipan's fangs, the longest of any snake in this country, rarely exceed 12 mm.

How the venom is squirted

The bases of the functioning fangs, and often the first reserve fangs as well, are penetrated by ducts that lead from big glands behind each eye. These glands — modified salivary glands — are surrounded by muscles reaching from the jaw.

Biting is a voluntary action. A snake can choose to launch a mock strike with its mouth closed, or open without any ejection of venom. But if the attack is wholehearted, a snake will tighten its jaw muscles in a way that squeezes the venom glands.

A clear or yellowish fluid is forced along the venom ducts and down through the fangs, squirting out under pressure as if from a pair of hypodermic syringes. The squeezing is repeated with each bite, so if a snake makes a series of strikes or holds on, shifting its grip, every wound may contain venom.

How much venom — if any — is discharged depends on the snake's previous activities. If it has just been 'milked' in captivity there may be none. If it has made a recent kill the output may be slight. But if a snake in the wild has gone hungry for long, the ducts as well as the glands are loaded. Venom spurts at the instant of striking and a copious amount is injected into even a glancing wound. If the target is missed, venom can squirt three or four metres through the air.

Immunity is not complete

Snakes are not totally immune to the venom of their own kind, but can withstand amounts that would be received from a bite or while cannibalising smaller snakes. Their resistance is thought to stem mainly from the nature of their nervous systems, rather than from the development of antibodies.

Australian snakes also enjoy a substantial immunity to the venoms of related local species. Big doses of death adder venom, for example, have no effect on a tiger snake. But there is no protection against foreign venoms — the same tiger snake would be killed by a tiny amount of Indian cobra venom.□

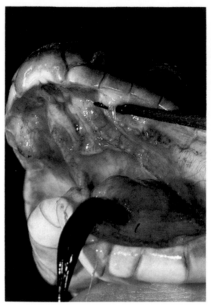

The tip of a pointer is slipped behind the tiny fang of a tiger snake.

A fang is not hollow, but curled around a deep groove.

A cutaway diagram shows one of a death adder's paired set of venom glands, ducts and fangs. Other teeth, small and backward-pointing, are used only for drawing in prey. The forked tongue is a harmless sense organ.

ATTACKED BY A SEVERED HEAD

Vital impulses — including the killing mechanism — can continue long after a snake has been mortally injured. It is probably this phenomenon that gave rise to the popular fallacy that snakes do not die until sundown.

A housewife in Alice Springs, NT, visiting a neighbour in 1977, found that he had just killed a gwardar. He had chopped it in two with a spade, then cut off the head along with about five centimetres of the neck and body.

Seeing a dog about to sniff at the head, the visitor reached down to move it. As she touched the neck end, the head shot round and she was bitten on a finger. Punctured by one fang and scratched by the other, the unfortunate visitor required antivenom treatment and 18 hours in hospital.

In a gruesome American study, 13 rattlesnakes were decapitated to test the later reactions of the heads and bodies. One snake's heart went on beating for 59 hours. The heads remained dangerous for 20-50 minutes, with the fangs pushing out and the eye pupils contracting whenever a hand was moved near them. One head bit a stick and discharged venom after 43 minutes.

Is it dangerous?

Colours are the least reliable guide in identifying snakes. Experts observe subtle differences in scale formation.

Of more than 100 species of land snakes in Australia, fewer than 20 are capable of killing even a baby. The others cannot bite humans effectively, or their venoms lack potency, or they have no venom at all. But harmless snakes are often feared — and regrettably some are killed — by people who have no way of knowing which ones are dangerous.

Only herpetologists who spend their lives studying snakes can recognise all of the different types. To be sure of the particular species, they usually have to make a careful examination of the scales, and often the teeth. In some cases they may have to dissect the animal to study its internal organs. Untrained people can at least be confident that most snakes shorter than their hand or thinner than a finger present little threat. Confronted by a bigger snake, the only way they can be sure that it is harmless is by contrasting its looks with those of the possible killers.

All of the highly dangerous species found in Australia are illustrated in the following pages, along with some that do not kill people but are considered capable of causing significant illness. You need not keep a mental picture of every one — just those that the maps indicate occur in your locality.

BODY AND UNDERSIDE SCALE PATTERNS

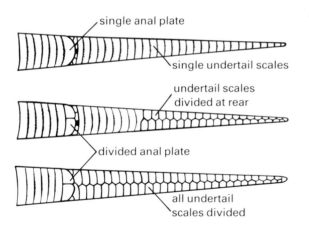

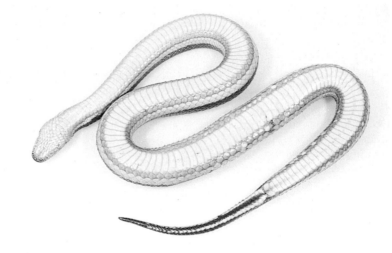

Anal plate and undertail scale formations
Variations in anal plate and undertail scale formation help distinguish snake types.

Ventral view of a land snake, showing belly and undertail scale formation.

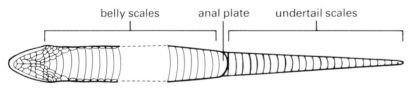

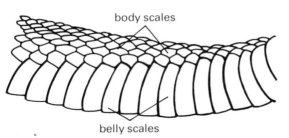

Body and belly scales
All of the venomous snakes have belly scales many times wider than their body scales. But so too do the harmless pythons and colubrids.

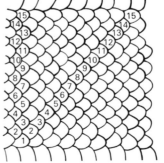

Underside scale formation
The underside scale formation typical of most land snakes. In this diagram the belly section has been shortened by about 90 per cent. The true ratio can be seen in the photograph above.

How to count body scales
Body scales can be counted in rows straight across the back, but a diagonal count is easier. It should be done halfway along the snake.

Rating the risks

Eleven of our most dangerous snakes have been 'ranked' by Queensland Museum staff.

The most fearsome of the dangerous snakes, in practical terms, is whichever one you step on. Comparing the menace of different species is difficult because so many factors are variable. The fierce snake, for example, has the most potent venom — but it isn't especially fierce and hardly anybody goes where it lives.

Queensland Museum staff, however, have attempted a ranking. They scored snakes 1-5 points on each of five criteria: venom toxicity, venom yield, fang length, temper, and frequency of attacks on people. By their reckoning the taipan comes out well ahead, with 21 points out of a possible 25.

Scale 0 1 2 3 4 5

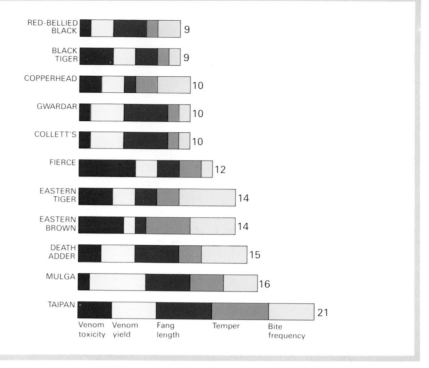

RED-BELLIED BLACK				9
BLACK TIGER				9
COPPERHEAD				10
GWARDAR				10
COLLETT'S				10
FIERCE				12
EASTERN TIGER				14
EASTERN BROWN				14
DEATH ADDER				15
MULGA				16
TAIPAN				21

Venom toxicity · Venom yield · Fang length · Temper · Bite frequency

In Tasmania, for example, people need only be familiar with the appearance of the copperhead and black tiger snake. Anything else on the island will be harmless. In the far north a big, vividly banded snake may look like an eastern tiger snake. But it cannot be one — there are none in the tropics. It is most likely a python, and non-venomous.

Remember the general look of a dangerous snake, rather than any particular colouring or marking. In many types the colours and patterns vary from place to place and between snakes of different ages. And the one snake can change its appearance in weeks: bright hues and prominent markings become dull or faded as the time comes for its skin to be sloughed.

What the scales can tell

Experts confirm their identification of most snakes by counting the scales on various parts of the body and observing certain peculiarities in their structure. Only an accomplished handler can do this while a snake is alive. It is not suggested that any untrained person should kill a snake simply out of curiosity to know what it is. But if you should come across a snake that has been accidentally killed, or find a recently sloughed skin, a surprising amount can be learnt from the scale patterns.

All dangerous land snakes, and many that are harmless, have wide scales running crosswise on their undersides and staggered rows of smaller scales reaching over their backs. Towards the rear underneath, they have a noticeably bigger plate shielding the anus and the genital openings.

Scientists call the underside scales forward of the anal plate ventrals or gastrosteges, and those behind it subcaudals. For simplicity in the descriptive notes in the following pages, the terms used are belly and undertail. The smaller scales on top are called body scales.

Different types of snakes can be distinguished by counting various scales. The body scales are counted in rows across the back, halfway along the snake. Belly and undertail scales are counted along the snake, from the throat to the anal plate and from the plate almost to the tip of the tail. Counts vary within a known range among snakes of the same species.

Further distinctions are made by observing whether scales are smooth or ridged, and whether the anal plate and undertail scales are single or divided. Other important differences can be detected in the shape or arrangement of individual scales on the head and face.

Ruling out the harmless types

If a snake has 23-30 rows of body scales ranged across its back and a divided anal plate, it belongs to the colubrid family. If it is venomous, its fangs are at the back of its mouth, so it cannot strike at people. Most Australian colubrids are water snakes of the far

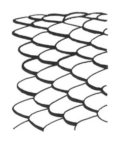

Body scale variations Some or all of a snake's body scales may be ridged (top) rather than smooth and shiny.

north. Two other aquatic species of the tropics have sharply ridged and pointed scales reaching all or most of the way round their bodies, with no obvious belly scales. Appropriately named file snakes, they are non-venomous. A heavy-bodied snake with wide, distinctive belly scales, but 30 or more rows of body scales above, must be a python. The bigger pythons can deliver a damaging bite if they are interfered with, but they have no venom. Also non-venomous are the blind snakes — rarely encountered because they live underground. They have small, smooth scales all around their bodies. Their eyes, no longer functional, appear only as black spots under the head scales.

All of the other land snakes are elapids. Their belly scales are many times wider than their body scales, and the body scales are ranged across the back in not more than 23 rows — fewer in most cases. All elapid snakes are venomous and front-fanged. But by no means all of them are dangerous to humans.☐

Taipan *Oxyuranus scutellatus*

Australia's most notorious snake is not common — but a fondness for farms and dumps can bring its menace close to home.

Before a taipan antivenom was developed in the 1950s, any full bite meant certain death. The only known survivor was attacked through boot leather and a thick sock, which prevented the fangs from penetrating far into his flesh. Friends quickly hacked open the wound with a chisel and bled it — not a recommended treatment these days, but one that probably saved his life.

The taipan is Australia's longest venomous snake and its fangs are by far the longest. They can grow to more than 12 mm — about the length of an index fingernail. Though the taipan's venom is not the most potent, the amount delivered in a bite and the depth it reaches make this our most formidable snake.

Stories of taipans chasing people are mythical. They are as shy as any other snake, and quicker than most to slither away from something as big and inedible as a human. But if cornered and harassed they are easily angered and will attack ferociously. They hurl themselves at a tormentor and deliver a quick succession of snapping bites, not attempting to hold on.

Well-grassed tropical woodlands are the natural habitat of the taipan. But its favoured diet is rats and mice, along with a few birds and bandicoots. So although it is a relatively uncommon species, it is often drawn to farm outbuildings, canefields and even urban rubbish dumps because of the vermin that infest them. It is mainly active by day or at dusk, but in the hottest weather it may hunt at night.

The southern limit of the taipan's range, around Grafton, NSW, is also the limit of sugar growing. To the chagrin of cane farmers, however, the taipan seems never to eat the other bane of their lives, the giant toad.☐

Body colours vary widely but at least part of the head always shows a creamy contrast.

APPEARANCE Long, narrow head, thinner neck, slim body. Back uniformly brown, from fawn to russet or nearly black in different localities. Snout or whole head cream; eye orange. Belly cream to yellow with reddish marks towards throat.

LENGTH To 3.3 metres, average 2.5 metres.

SCALES Mid-body 21-23 rows, slightly ridged at least near neck; 220-250 belly scales; 45-80 undertail, divided; single anal plate.

BIRTH Eggs, clutches of 10-20.

ANTIVENOM Taipan.

*Above: A wide
mouth, reaching
back to the neck,
accommodates
the longest fangs
of any Australian
snake.*

*Left: Back scales
on the slender
body are tiny but
numerous.*

For science's sake

A man-killer was allowed to live and was used for the first taipan venom studies.

'Its orange-red
eyes glittered
as it lunged
and hissed in
furious anger.'

Taipan antivenom was among the last of the snakebite antidotes to be developed, though it was the most desperately needed. Virtually nothing was known about the nature of the venom until 1950. In July of that year a young Sydney collector, Kevin Budden, spotted a taipan among rubbish at the city tip in Cairns, Qld. He seized the 1.9 metre snake and carried it to his car, but as he was stuffing it into a sack it gave him multiple bites on a thumb.

Witnesses tried to kill the taipan but Budden stopped them. He was taken to hospital and given tiger snake antivenom. That evening,

before he became too paralysed to speak, he stressed that the snake must be kept alive and sent to the Commonwealth Serum Laboratories in Melbourne. Budden died the following day.

When the snake reached Melbourne in its air-freight box, the naturalist David Fleay was called in to milk it. He had not seen a taipan before, let alone handled one, but he was all too aware of its fearsome reputation.

'Its orange-red eyes glittered,' he recalled later in his book *Talking of Animals.* 'The lightish nose and upper lip gave the mouth a grinning appearance, especially now as it lunged and hissed in furious anger.'

One milking yielded a dried weight of 128 milligrams of venom — four times what could be expected from a tiger snake. The CSL had plenty on which to start its research. And within five years, the first life was saved by the new antivenom.☐

A taipan's big eyes are bright orange around the pupil.

Death in 10 minutes

Six bites gave a little boy no chance.

Taipans snap repeatedly when they strike, pumping venom time after time. And because they are such long snakes, they often strike high. A four-year-old boy, attacked near his seaside home north of Townsville, Qld, in 1979, was bitten again and again on his upper thigh and buttock.

The child collapsed almost immediately. Mouth-to-mouth resuscitation, given within 10 minutes of the attack, failed to revive him. A post-mortem examination revealed no fewer than 12 puncture marks. The boy's death was among the fastest on record from snakebite in Australia.☐

The barefoot farmer

His feet were so tough he didn't feel the bite.

'The farmer
firmly denied
any contact with
snakes.'

Even a deadly bite may not be noticed. In 1980 a tobacco farmer went to hospital at Mareeba, on the Atherton Tableland in North Queensland, after an hour-long bout of vomiting. Paralysis set in and he was transferred to Cairns with suspected food poisoning, from canned mince he had eaten two days before. A venomous bite had been suspected, but the farmer firmly denied any contact with snakes.

Soon he was unconscious, and so paralysed that he needed mechanical aid to breathe. Convinced he was a snakebite victim, doctors

examined his body for puncture marks. But his feet were such a mass of cuts and scratches — he worked barefoot — that they could not be sure.

The farmer was given all-purpose antivenom. It saved his life, according to a report in the *Medical Journal of Australia.* Analysis of blood and urine samples eventually proved that he had been bitten by a taipan. After nearly three weeks on life-support systems he made a full recovery.

And the good name of the canned mince company was restored.☐

Fierce snake *Oxyuranus microlepidotus*

The taipan's inland cousin kept out of sight for 80 years. It's back — but the extent of its range is anyone's guess.

Venom yielded in an average milking of a fierce snake could kill 100 000 mice. The supply from a big specimen that had not fed recently could dispose of 250 000. It is by far the most potent land snake venom in the world. In spite of that the snake's common name seems undeserved. It is a retiring animal, not readily roused to anger, that seems to have retreated from the spread of inland pastoral agriculture.

Fierce snakes were first found in 1879 near the junction of the Darling and Murray Rivers, and in the 1880s at Bourke, NSW. No more were seen until 1967, when one turned up in the far southwest of Queensland. The species has since proved to have a stronghold in this 'Corner Country', reaching into South Australia. Whether it survives over a wider range is uncertain. As this book was reaching completion, venom in a snakebite victim in central NSW, indicated taipan or fierce snake.

Habits of hunting and drought-avoidance keep the fierce snake out of sight, so it could be more numerous and widespread than people suppose. In its presently known range it feeds on the plague rat *Rattus villosissimus* and often occupies the burrows of its prey. It waits out

OTHER NAMES Small-scaled snake, western taipan, inland taipan.

APPEARANCE Narrow head, slim body, no distinguishable neck. Back brown with dark-edged scales giving speckled look, sometimes faintly banded towards tail. Head dark; eye black. Belly cream or yellow with darker edges.

SIZE To 2.5 metres, average 1.7 metres.

SCALES Mid-body 23 rows; 212-237 belly scales; 54-66 undertail, divided; single anal plate.

BIRTH Eggs in clutches of 9-12.

ANTIVENOM Taipan.

The fierce snake is so different from the taipan that some scientists prefer to assign it a genus of its own, Parademansia.

drought by descending to the cool, moist depths of cracks in the ground.

The rediscovery of the fierce snake in 1967 was also the occasion of its only recorded bite on a human. The naturalist who found it, Mr Athol Compton, was snapped at twice on a thumb. He received immediate first aid and was flown to hospital, but because he mistook his attacker for a brown snake was given the wrong antivenom. He was critically ill, his heart stopping twice, and took a month to recover.□

Eastern tiger snake *Notechis scutatus*

Provoked to strike, it clamps onto its victim and gnaws with a vigour that makes up for the shortness of its fangs.

Stripes tend to be more prominent at the sides than on the back scales.

People receive severe bites from eastern tiger snakes more often than from all other Australian species. This heavily built snake is not especially aggressive but it is a prolific breeder with a range that covers the most densely populated corner of the country. And it is more likely than most species to be trodden on. Instead of fleeing when it first detects human footfalls, it often allows people to approach within a step or two.

If cornered and aroused, a tiger snake flattens its neck in cobra fashion and hisses threateningly before striking. The attack is aimed flat and low. Taking a grip if it can, the snake works its jaws in a chewing motion that pumps extra venom through its fangs. These are relatively short for a snake of such size — seldom longer than 3.5 mm — and the venom yield is not great. But tiger snake venom contains one of the world's most potent combinations of toxins. Before an antivenom was developed the death rate from bites was high. And still today, a fairly heavy toll is taken of pets and smaller livestock.

Eastern tiger snakes inhabit upland rainforests, river valleys and the floodplains of the major river systems of the southeast. They favour swampy ground, for their favourite food

is frogs. But if they are forced out by flooding they may climb low trees to prey on nestling birds, or hunt rodents around houses. They are active mainly by day or at dusk, but in hot weather they turn to hunting at night.☐

OTHER NAMES Mainland tiger snake, common tiger snake.

APPEARANCE Broad head, heavy build. Wide, shield-like plate between eyes. Back usually has 40-50 cross-bands of yellow or cream on background of grey, green or brown — sometimes reddish or nearly black. Belly cream, yellow, olive or grey.

SIZE To 1.8 metres, average 1.2 metres.

SCALES Mid-body 17-19 rows; 140-190 belly scales; 35-65 undertail, single; single anal plate.

BIRTH Live, average litter 30.

ANTIVENOM Tiger snake.

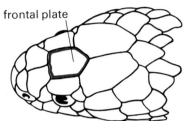

frontal plate

Above: A widening head merges into the eastern tiger snake's body without an obvious neck. Right: The eastern tiger snake's shield-shaped frontal plate — the centre scale between the eyes — is unusually elongated, twice as long as it is wide.

Little boy lost

He preferred catching snakes to seeing a princess.

Snake collecting was the favourite pastime of a nine-year-old farm boy in Gippsland, Vic. During Princess Margaret's tour in 1975, when schoolchildren were given time off to see her, he took the chance to go hunting. He was not missed until nightfall. Searchers found him in bushland late next morning, lying comatose. He was dead on arrival at hospital, from what proved on later analysis to be tiger snake venom.

The boy's movements were backtracked in a zigzag pattern of running and crawling. It seemed that the toxin's effects had overtaken him, and destroyed his sense of direction, before he made any attempt to get home. Doctors reporting the case in the *Medical Journal of Australia* surmised that he did not seek help sooner because he was afraid of getting into trouble for 'wagging it'.☐

The bad dream

Deadly intruder came in through a cupboard.

On a summer night in 1977 near Delegate, NSW, a seven-year-old girl woke with vomiting and diarrhoea. As her mother remade the bed, the girl told her that a big snake had just gone into a cupboard. Assured it was only a dream, she went back to sleep.

Gravely ill by the following evening, the child was treated for snakebite with an all-purpose antivenom. Tests later confirmed tiger snake envenomation. She recovered in a few days. Meanwhile her father checked the bedroom cupboard — and found a knothole accessible from the ground.☐

'The girl told her mother she had seen a big snake.'

Black tiger snakes *Notechis ater*

Genetic variations, evolved in long isolation, include an imperturbable giant that fasts for 10 months every year.

Colour banding on the black tiger snakes is absent or barely discernible in most adults, but often more prominent in juveniles.

OTHER NAMES Island tiger snakes. Subspecies *N. a. ater*, Krefft's tiger snake (Flinders Ranges, SA); *N. a. humphreysi*, King Island tiger snake; *N. a. niger*, peninsula tiger snake (Yorke and Eyre Peninsulas, Kangaroo Island and islets, SA); *N. a. occidentalis*, western tiger snake or norne (WA); *N. a. serventyi*, Chappell Island tiger snake.

APPEARANCE Broad head, heavy build. Back black or dark brown; faint cream or yellow bands on some western snakes. Belly usually grey, sometimes cream, yellow or blue.

SIZE Chappell Island to 2.4 metres, average 1.8 metres; Krefft's less than 1 metre; others 1.5 metres.

SCALES Mid-body 17-19 rows; 155-190 belly scales; 40-60 undertail, single; single anal plate.

BIRTH Live in litters of 20-30.

ANTIVENOM Tiger snake.

Dark tiger snake breeds, seldom noticeably striped, occur in Tasmania, on some smaller islands and in isolated pockets across the southern mainland. They include the biggest and smallest of their kind, and a type that is adapted to semi-desert survival.

Scientists have found enough distinctions to classify the black tiger snakes in five subspecies, not including the main population in Tasmania. Odd differences arise even in a distance as short as 300 kilometres, between islands at the western and eastern ends of Bass Strait.

Both Bass Strait subspecies prey on migratory muttonbirds. They raid nesting burrows and eat the chicks. But that provides them with food for only two months of the year. On King Island and neighbouring islets, the snakes turn cannibal for the rest of the time. To the east, on

The orange flanks of this Western Australian norne N.a. occidentalis (above) are unusually vivid. Krefft's tiger snake N.a. ater (above right), from the Flinders Ranges in SA, shows a subtle cross-banding. But the peninsula subspecies N.a. niger (right), from farther south in SA, is funereal by comparison.

Chappell and Badger Islands in the Furneaux group, juvenile snakes survive on tiny lizards. The adults eat nothing — they live off their own fat for 10 months.

In spite of that deprivation the Chappell Island breed is the biggest of all tiger snakes. Muttonbirders run a considerable risk when they plunge their hands into burrows. But otherwise this snake is remarkably placid. The Queensland collector Michael Cermak, writing in *Australian Natural History* magazine, describes how he walked round the island among snakes as they sunned themselves outside the burrows 'taking absolutely no notice'. He lifted one high in the air by the tail, then put it down. It moved just 2 metres, then curled up again to rest. □

Mulga snake *Pseudechis australis*

Not surprisingly, it used to be called the king brown. But brown snake antivenom is no help to victims of its paralysing bite.

APPEARANCE Broad, flat head, slightly distinct from neck; heavy body. Back most often copper but varies from light brown to russet, dark-olive brown or chocolate. Belly pinkish or yellowish cream, sometimes with pink or orange blotches.

SIZE To 3 metres, average 1.5 metres.

SCALES Mid-body 17 rows; 185-225 belly scales; 50-75 undertail, single at front but usually paired towards end; divided anal plate.

BIRTH Eggs in clutches of 10-20.

ANTIVENOM 'Black snake'.

Habitats of the mulga snake vary from northern vineforests and woodlands to the barest deserts.

Mistaken identity in the case of the mulga snake could easily cost the life of a bite victim. Though its skin is always some shade of brown, and though it occurs in the same regions as several members of the brown snake group, its copious venom is differently constituted.

Antivenom prescribed against bites by brown snakes of the genus *Pseudonaja* is altogether useless. The mulga snake has to be countered with black snake antivenom from its own kind, *Pseudechis*. Confusingly, bites by the smaller black species of *Pseudechis* are treated with tiger snake antivenom.

Head shape and build are the important guides in distinguishing a mulga snake, without taking the risk of trying to kill it. The heads of *Pseudonaja* brown snakes are narrow, merging into the body without a noticeable neck, and their bodies are slim in relation to their length. The mulga snake is more heavily built and its broader head stands out slightly from the neck.

The mulga snake is the most massive of the venomous Australian species and its output of venom is the greatest of all. It does not rank high in toxicity but the volume delivered in a bite — often increased by a chewing action — makes the snake extremely dangerous. The action of the venom is curiously different from that of other species, having its major effect on the muscles, causing paralysis, rather than directly on the nervous system. It often causes pain and swelling at the site of the bite — rarely experienced from other snakes.

Mulga snakes live on rodents, lizards and birds, and often on other snakes. They shelter under fallen trees, or in rabbit and goanna burrows in the drier regions of the outback. Active mainly by day in the south, they are nocturnal hunters in the tropics.□

Trying everything

The tragic case that showed the need for another antivenom.

Sleepless on a hot night in the late 1960s, a young Western Australian farm labourer reached under his verandah bed for cigarettes. He was bitten on the hand by a mulga snake measuring more than 1.8 metres.

A local hospital did all it could for him. He was injected with brown snake antivenom, death adder antivenom and tiger snake antivenom — each of them twice in the course of the following two days. Nothing worked, and his condition declined steadily until he died, 37 hours after the bite.

This case led to the use of Papuan black snake antivenom against bites by the mulga snake, and the discouragement of its alternative name, the king brown snake. Some mulga snake constituents are now included in polyvalent (all-purpose) antivenom.□

Right: The tongue is used to track prey, taking particles from the air or ground to sensory organs in the mouth.

Death adder *Acanthophis antarcticus*

Its calmness is a worry. Camouflaged as it lies in wait for prey, it refuses to budge even when someone is about to step on it.

APPEARANCE Wide, flat head tapering to blunt snout, thin neck. Fat body tapers abruptly to thin, curved spine at end of tail. Back colours vary widely from light brown to reddish and black, sometimes banded. Belly creamish grey with pink or brown blotches.

SIZE To 1.1 metres, average 65 cm.

SCALES Mid-body 21-23 rows, smooth or slightly ridged; 110-130 belly scales; 35-55 undertail, mostly single but some divided at rear; single anal plate.

BIRTH Live, average litter 15-20.

ANTIVENOM Death adder.

The death adder is the least nervous of our venomous snakes, but its placid nature makes it all the more dangerous. Instead of slipping away at the approach of a human, it lies motionless. For that reason it used to be called the 'deaf' adder. People have jumped up and down near this snake without provoking any response. But if trodden on it strikes low, with lightning speed and unusual efficiency.

It is not one of the true adders. They are members of the viper family, which has no representatives in Australia. But the death adder has some viperish characteristics, noticeable in its big, angular head and more importantly in the structure of its upper jaw. It has the ability to push its fangs well forward when it strikes, so that they penetrate deeply into a large target instead of hitting at a shallow, glancing angle. And the fangs are of generous length — usually more than 6 mm.

The death adder is a subtle predator that wastes little energy. It wriggles into sand, gravel

This boldly banded young death adder was found on the western side of Cape York Peninsula. Some scientists regard it as a separate species, Acanthophis praelongus.

Below: The tail of a more southerly specimen is part-buried in sand but the worm-like tip emerges, wiggling as a lure.

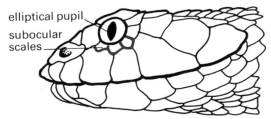

elliptical pupil
subocular scales

Only death adders have tiny extra scales under their eyes. The oval pupil signifies a nocturnal hunter.

or leaf litter so that most of its body is concealed. Looping around, it lies with its head close to the worm-like tip of its tail. Flicked back and forth, this lures small rodents, lizards and birds to within striking distance. The snake is active mostly at night in sandy scrubland.

Scientists make a distinction between the common death adder and a rarely-seen variety that inhabits arid regions of Central Australia and the northwest. This 'desert' death adder *Acanthophis pyrrhus* is more often reddish in colour and has many more belly scales. Death adders of the far north, from Cape York to the Kimberleys, are sometimes classed as a third species, *Acanthophis praelongus*. Such distinctions are not important in recognising the snake or understanding its habits.☐

Eastern brown snake *Pseudonaja textilis*

The colouring may be nothing like brown. But when this snake is angry, rearing high with jaws agape, its threat is unmistakable.

OTHER NAME Common brown snake.

APPEARANCE Narrow head, slender body; no distinguishable neck. Back a uniform colour in shades of grey, brown, orange, russet or nearly black. Belly cream, yellow or light brown with darker blotches.

SIZE To 2.4 metres, average 1.4 metres.

SCALES Mid-body 17 rows; 185-235 belly scales; 45-75 undertail, usually all divided; divided anal plate.

BIRTH Eggs, average clutch 20-30.

ANTIVENOM Brown snake.

Provoked to fury, a brown snake thrusts a third of its body off the ground, tightly folded and ready to spring almost like a jack-in-the-box. Jaws wide open, it rears high and strikes high. This attacking posture earned it the generic name *Pseudonaja* — false cobra.

The unusual open-mouthed strike helps to compensate for undersized fangs. So does a chewing action if the snake manages to get a grip on its target. Even then the output of venom is low — but in toxicity it is second only to that of the fierce snake. One component, acting on the nervous system, is the most potent neurotoxin known in a land snake. It accounts for only 3 per cent of the venom quantity, but 70 per cent of its killing power.

Sheltering among rocks, in hollow logs or in the burrows of other animals, the eastern brown snake is common in a variety of habitats. They range from dry scrublands to damp coastal and upland forests. But this species avoids the boggy ground favoured by tiger snakes. It hunts by day, even in hot weather, feeding mainly on mice and lizards. After an attack it throws a coil or two around its prey, securing it until all struggling has ceased.

The brown snake's liking for mice and small rats frequently brings it close to farm outbuildings. It is the leading cause of serious snakebite in Queensland, and second only to the tiger snake in the southeast. The venom is fast-acting — stomach pains, vomiting and dizziness often set in within minutes — but victims respond particularly well to antivenom injections. □

Left: The neck is not discernible when a brown snake is relaxed. If angered, it spreads it like a cobra.

*Above: In the more
sombre colour
variations of the
eastern brown snake,
the belly is often the
brightest part. This
one was found in the
MacDonnell Ranges,
Central Australia.*

Up and running

The treatment was costly — but the dog repaid the bill.

Prized pets and valuable livestock can be saved from the effects of snakebite in the same way as humans, with proportionate doses of antivenom and supporting drugs. Dogs are the animals most often treated, partly because they are more likely than other animals to seek out their owners when they start to be affected by a bite. Treated quickly, their recovery rate is high — but so is the cost.

When a well-bred young greyhound was bitten near Horsham, Vic., in 1979, no expense was spared. Along with a shot of brown snake antivenom — the snake had not been identified but the veterinarian correctly summed up the symptoms — the dog was given oxygen through a mask and three different drugs. It made a good recovery and became a substantial stakewinner. Its name: Full of Venom.□

Gwardar *Pseudonaja nuchalis*

Even juveniles, less than 30 cm long, can inflict symptoms severe enough to require antivenom and intensive care.

Colour contrast is exceptionally bold in this strongly banded gwardar, photographed near Alice Springs, NT.

As ferocious as the eastern brown snake, though perhaps less easily roused, the gwardar fortunately has relatively little contact with people. But it is common throughout its vast range, in habitats that embrace deserts as well as seasonally watered tropical woodlands and vineforests. It hunts by day in the south but at night in the north, preying mainly on mice and lizards.

The venom is less toxic than that of the eastern brown, but it is effective at a surprisingly early age. In cases recorded in Western Australia in 1960 and 1980, juvenile snakes no more than 30 cm long caused severe envenomations. One victim was a collector who ignored the bite and eventually collapsed, losing consciousness for 15 minutes.□

OTHER NAMES Western brown snake, collared brown snake.

APPEARANCE Narrow head, slender body; no distinguishable neck. Back colouring extremely variable from olive-grey to many shades of brown or nearly black. Often black on head or in V or W pattern behind head. Occasionally dark-banded on body. Belly yellowish cream with grey or orange with darker blotches.

SIZE To 1.8 metres, average 1.4 metres.

SCALES Mid-body in 17-19 rows; 180-230 belly scales; 50-70 undertail, divided; divided anal plate.

BIRTH Eggs, average 20 in clutch.

ANTIVENOM Brown snake.

Gwardars of varied appearance photographed in Western Australia (centre) and South Australia (left and right).

Dugite *Pseudonaja affinis*

Rapid development around Perth and in the extreme southwest heightens the risk of wayward snakes entering people's homes.

Excitable and aggressive, the dugite has a bad name for invading sheds and houses in search of mice. It favours sandy country, where it is active by day, and has long been the commonest cause of snakebite in Perth. The rapid growth of population and increasing tourist interest in the Swan coastal region and the southern ports may bring this snake into greater prominence as one of Australia's most dangerous.

Dugite venom, along with that of the gwardar, interferes most markedly with the clotting properties of human blood. It can lead to unusual problems of haemorrhaging, even after brown snake antivenom has reversed other symptoms of a bite.☐

Black specks distinguish a dugite, seen near Esperance, on the south coast of WA.

OTHER NAME Spotted brown snake.

APPEARANCE Narrow head, slender body; no distinguishable neck. Back olive or dark brown, speckled with black. Belly yellowish cream or olive with grey or pinkish spots.

SIZE To 1.8 metres, average 1.5 metres.

SCALES Mid-body in 19 rows; 190-230 belly scales; 50-70 undertail, divided; divided anal plate.

BIRTH Eggs, up to 20 in clutch.

ANTIVENOM Brown snake.

A dugite of the arid hinterland, basking on sand near Norseman, WA.

Copperheads

Austrelaps superbus

Kings of the mountains, they bask even after snow falls. Southerners are the biggest and most deadly of the breed.

Copperheads are the hardiest of the highly venomous snakes, actually preferring a cold climate. They are the last to become dormant in winter and the first to resume activity in spring. Altitude is no bar — copperheads occupy the highest parts of the Great Dividing Range including Mount Kosciusko, where they have been seen above the snowline in winter. Only in freezing conditions do they hibernate, using animal burrows.

In NSW the copperhead is confined to high country, but in Victoria and Tasmania it spreads to many lowland areas. From western Victoria, where it is most common, its population reaches to Mt Gambier in SA. Isolated pockets occur farther west in the Mount Lofty Ranges and Adelaide Hills and on Kangaroo Island.

The biggest copperheads are found in Victoria, southeastern SA and Tasmania. Scientists recently have moved towards a division of species, reserving the original name *Austrelaps superbus* for this larger breed. Highlands copperheads in NSW, significantly shorter and usually darker in colour, are expected to be renamed *Austrelaps ramsayi*. The separate communities closer to Adelaide — even smaller,

A big copperhead on Flinders Island, Bass Strait. Tasmanians and the strait islanders have only this and the black tiger snake to concern them.

and sometimes called pygmy copperheads — may prove to be a third species.

Austrelaps superbus is undoubtedly capable of inflicting a fatal bite on humans, although it is not easily aroused and its strike often misses. In the smaller types, not only are the fangs shorter and the venom yields lower, but the venom also seems to be less toxic. Any copperhead bite should be taken seriously, however.

Copperheads favour swampy country, where if food is plentiful scores of them may congregate among dense clumps of grass. Hunting by day except in the hottest weather, they prey on frogs, small mammals and reptiles. They also frequently eat other snakes, including their own young.□

OTHER NAMES Southern copperhead, lowlands copperhead, proper copperhead, diamond snake (Tas.), superb snake.

APPEARANCE Narrow head, heavy build. Back grey, brassy, copper, russet, chocolate or black; cream, yellow or red along sides. Light scales above lips often give striped appearance. Belly yellowish cream or grey.

SIZE To 1.8 metres, average 1.3 metres (highlands copperhead average 1 metre).

SCALES Mid-body in 15-17 rows, enlarged at sides; 140-165 belly scales; 35-55 undertail, single; single anal plate.

BIRTH Live, litter of 10-20.

ANTIVENOM Tiger snake.

Smaller, darker copperheads in the NSW high country — this one is from Lake Eucumbene — may be a separate species, Austrelaps ramsayi.

Striking back

Vengeance on a snake made the venom work faster.

A young farmer's first mistake was to hop off a tractor with only thong sandals on his feet. He trod on a copperhead and it bit him on the ankle. Incensed, he jumped up and down on the snake to kill it. His vigorous movements almost cost him his life, because they hastened the passage of venom to his nervous system.

And the farm, in Gippsland, Vic., was three hours' drive from a hospital. Paralysed, barely conscious and his circulation failing, he was treated with antivenom only just in time.

The first rule for snakebite victims is to avoid any unnecessary movement. Least of all should they go chasing the offending snake.□

'His vigorous movements almost cost him his life.'

Black snakes

Reluctant attackers, they rank low in killing power. But the bigger ones pack enough punch to put a child's life at risk.

Even when cornered, the familiar black snakes of the east and south seldom bite. Low in venom output and potency, they are unwilling to waste their armament. Displays of aggression are usually bluff. But it is unwise to count on that. A good-sized black snake can inflict a severe poisoning — potentially fatal in the case of a child.

Black snakes have suffered more than most other species from the advance of European settlement and agriculture, because their range exactly matches Australia's belt of greatest fertility. The red-bellied species has a particular liking for damp coastal forests and inland river margins. The blue-belly is also at home on natural grasslands.

Both are daytime hunters except in very hot weather. The blue-bellied black snake eats small mammals, lizards, frogs and sometimes other snakes. The red-belly's choice is wider because it takes readily to water, often lying still and easily passing for a stick. It supplements its diet with fish — especially eels — and crustaceans. The red-belly is also a persistent cannibal, eating snakes almost up to its own size.□

Blue-bellied black snake
Pseudechis guttatus

OTHER NAME Spotted black snake.

APPEARANCE Narrow head, heavy body. Back usually glossy black (sometimes dark brown) with single cream spots on some scales. Belly blueish grey, sometimes with yellow spots.

SIZE To 2 metres, average 1.25 metres.

SCALES Mid-body in 19 rows; 175-205 belly scales; 45-65 undertail, single towards front but most divided; divided anal plate.

BIRTH Eggs, average clutch about 15.

ANTIVENOM Tiger snake.

BLUE-BELLIED

Left: A blue-bellied black snake of the Macquarie Marshes, NSW, towards the western limit of its range. This species also inhabits grasslands.

Red-bellied black snake
Pseudechis porphyriacus

OTHER NAME Common black snake.

APPEARANCE Narrow head, heavy body. Back shiny blue-black, lower sides bright red, orange or pink. Belly dull red or pink; undertail black.

SIZE To 2.5 metres, average 1.25 metres.

SCALES Mid-body in 17 rows; 180-215 belly scales; 40-65 undertail, single towards front but most divided; divided anal plate.

BIRTH Live, average litter 12-20.

ANTIVENOM Tiger snake.

Redness showing on a NSW Pseudechis porphyriacus *is on the flanks, not the actual belly. In northern types the flanks are pink or creamy.*

RED-BELLIED

A baby red-belly emerges from its birth sac.

Rough-scaled snake *Tropidechus carinatus*

What it lacks in size, this hot-tempered breed makes up for in belligerence. Many a handler has been taught a sharp lesson.

OTHER NAME Clarence River snake.

APPEARANCE Big head, distinct from neck; heavy body. Back greenish brown with dark, narrow cross-bands. Belly creamy yellow or olive with blotches of darker green.

SIZE To 1 metre, average 75 cm.

SCALES Ridged; mid-body in 23 rows; 160-185 belly scales; 50-60 undertail, single; single anal plate.

BIRTH Live, litter size unknown.

ANTIVENOM Tiger snake.

Most known victims of the rough-scaled snake have been collectors or exhibitors, accustomed to dealing easily with much bigger species. This one's pugnacity takes them by surprise. It seems more eager to attack, with less provocation, than any other snake. And while many species eventually adjust to captivity and handling, the rough-scaled snake remains excitable, hissing explosively and lunging whenever it is approached.

The fangs are relatively long — 5 mm in a specimen of less than 1 metre — and the venom has an extremely rapid neurotoxic effect. Healthy adults have lost consciousness within five minutes of a bite. Though only one death is definitely blamed on this snake, it is a leading suspect in killings by unidentified snakes in its region. Blood and urine tests developed so far are no help in identifying it — they indicate tiger snake venom — unless a bite occurs well outside the range of tiger snakes.

Rough-scaled snakes are usually found close to water in moist, forested country — often at high altitude. Hunting frogs, small mammals and lizards, they are active more often at night than by day, especially in warm weather. Their habitat is shared by the keelback, a harmless semi-aquatic snake that looks almost identical. Mistaking the species, collectors have sometimes ignored dangerous bites.☐

Even expert collectors can mistake the risky rough-scaled snake (below) for the innocent keelback (right). They favour the same habitats. The keelback or freshwater snake, known until the mid-1980s as Amphiesma, *is now classified as* Styporhynchus mairii.

Life of the party

*Prankster's 'python' took five minutes
to kill him.*

Venom can have an exaggerated effect on someone whose muscle control is already weakened by heavy drinking. In the 1950s a Sydney man sought to enliven a social club gathering by producing what he thought was a baby python. As he took it from its bag it bit him three times on the hand. Five minutes later he had a fit of coughing and fell dead. The snake, just over 60 cm long, was identified as *Tropidechis*. The coroner's finding: snake poisoning while under the influence of alcohol.□

A rough-scaled snake's big head juts from a slender neck.

The longest battle

A bite victim spent 10 weeks on life-support systems.

A rough-scaled snake is believed to have caused the most prolonged case of severe snakebite illness ever recorded. In 1978 a nine-year-old Queensland boy suffered cardiac arrest, kidney failure, muscle wastage and paralysis that kept him in hospital for 18 weeks — dependent for 10 of them on mechanical ventilation and life-supporting drugs. He was not cleared to return to school for a further six months — still with a throat constriction that hampered his eating.

The boy was unaware that he had been bitten on a finger while playing in long grass near Noosa. It was more than two days before doctors suspected snakebite as the cause of his critical condition. Diluted all-purpose antivenom was administered 60 hours after the bite, according to the *Medical Journal of Australia*. It had no effect. But full-strength antivenom, even injected 90 hours late, was some help in restoring muscle power.□

Broad-headed snakes

To an untrained eye, the biggest of these pugnacious climbers may masquerade as a harmless diamond python.

Vivid mouth markings relieve the sombre look of a pale-headed snake.

Ridged scales on the bellies of the broad-headed snakes distinguish them as the only habitual tree-climbers among Australia's dangerous species. Active mostly at night, they prey on lizards, small birds, mice and frogs. The pale-headed and Stephens' species also rest in trees, sheltering in hollows or under loose bark. The Sydney broad-head by day is more likely to be found among rocks in dry, broken sandstone country.

People familiar with the diamond python *Morelia spilotes* — a harmless creature that is sometimes made a pet — could mistake the Sydney broad-head for one of these. Authorities now doubt an old story that a man was killed through such an error. But there is evidence to

suggest that the bite of any well-grown *Hoplocephalus* species could cause acute illness and collapse.

The broad-headed snakes are bad-tempered if disturbed, and eager to attack. They rear in the manner of brown snakes, their bodies arched in tight S-shapes, and strike high, repeatedly and accurately. The venom of the Sydney species has been likened in potency to that of the southern copperhead.☐

Stephens' banded snake
Hoplocephalus stephensi

APPEARANCE Wide head, stocky body. Back light grey or brown with wide black cross-bands. Belly creamish yellow with black blotches towards tail.

SIZE To 1 metre, average 60 cm.

SCALES Mid-body in 21 rows; 220-250 belly scales, ridged; 50-70 undertail, single; single anal plate.

BIRTH Live, litter size uncertain.

ANTIVENOM Tiger snake.

Pale-headed snake
Hoplocephalus bitorquatus

APPEARANCE Wide head, stocky body. Back uniformly light grey or brown, sometimes dark-spotted; cream or white band behind head. Belly creamy grey, sometimes darker spots towards rear.

SIZE To 1 metre, average 50 cm.

SCALES Mid-body in 19-21 rows; 190-225 belly scales, ridged; 40-65 undertail, single; single anal plate.

BIRTH Live, litter size unknown.

ANTIVENOM Tiger snake.

Sydney broad-headed snake
Hoplocephalus bungaroides

OTHER NAME Yellow-spotted snake.

APPEARANCE Head much wider than neck, fairly heavy body. Back mainly black with narrow, irregular yellow bands. Face yellow. Belly grey, sometimes with yellow blotches.

SIZE To 1.5 metres, average 75 cm.

SCALES Mid-body in 21 rows; 200-230 belly scales, ridged; 40-65 undertail, single; single anal plate.

BIRTH Live, litter of about 12.

ANTIVENOM Tiger snake.

Increasingly rare, the Sydney broad-headed snake is a favourite of collectors.

Stephens' banded snake is almost always found in trees.

Theoretical threats

Beautiful but shy, these outback neighbours are classed as likely killers — though there's no proof they have ever bitten anyone.

COLLETT'S

SPECKLED BROWN

Colour bands on the beautiful Collett's snake are sometimes creamy rather than pink.

On the Mitchell grass plains of central-western Queensland live three of Australia's most attractively marked snakes, rarely seen but presumed to be highly dangerous. No records exist of people having been bitten by them — the verdict is based on venom studies and a knowledge of related species.

Collett's snake venom has been likened to that of the mulga snake. The species is so uncommon that it was not shown to be an egg layer, rather than a bearer of live young, until the 1970s. It hunts by day, feeding on small mammals, lizards, frogs, and birds.

Ranging across the Barkly Tableland into the Northern Territory, the speckled brown snake *Pseudonaja guttata* is adapted to seasonal extremes of drought and drenching rain. It preys by day on frogs, small mammals and lizards. When the soil is parched it shelters in deep cracks in the ground.

The northwestern part of the speckled brown snake's range is shared with the rare Ingram's brown snake *Pseudonaja ingrami*. This species has a confusing variety of colours — sometimes chocolate or reddish brown, but often very like its neighbour or like the eastern brown snake *Pseudonaja textilis*. It has 17 rows of back scales at mid-body, rather than the speckled brown's 19 or 21, and achieves a much greater size.☐

Collett's snake
Pseudechis colletti

APPEARANCE Narrow head, heavy body. Back chocolate or black with irregular bands of cream or pink; sides often predominantly cream or pink. Belly creamy yellow or orange.

SIZE To 2 metres, average 1.25 metres.

SCALES Mid-body in 19 rows; 215-235 belly scales; 50-70 undertail, single towards front but most divided; divided anal plate.

BIRTH Eggs, clutch of about 12.

ANTIVENOM Tiger snake.

A speckled brown snake of the Barkly Tableland gives an angry warning.

Ingram's brown snake, seen on a Northern Territory cattle station.

Speckled brown snake
Pseudonaja guttata

OTHER NAMES Spotted brown snake, downs tiger snake.

APPEARANCE Small, flattened head, slender body. Back creamy brown to salmon, flecked by many dark-edged scales; sometimes with dark cross-bands. Belly creamy yellow, blotched with orange.

SIZE To 1.4 metres, average 1.25 metres.

SCALES Mid-body in 19-21 rows; 190-220 belly scales; 45-70 undertail, divided; divided anal plate.

BIRTH Eggs, clutch size unknown.

ANTIVENOM Brown snake.

Some other

Their bites are unlikely to threaten human life or require antivenom treatment — but the effects could still be severe.

Collectors, bitten previously and made oversensitive to venom, can suffer drastic reactions to the bites of snakes that are not generally regarded as dangerous. Other people may experience symptoms that are extremely unpleasant, though not life-threatening. Dogs and cats are certainly at risk from the species shown here.□

Above **SMALL-EYED SNAKE** *Cryptophis nigrescens.* Great Dividing Range to coast, from Cape York to eastern Vic.

Left **DUNMALL'S SNAKE** *Glyphodon dunmalli.* Ranges and western slopes, southeastern Qld.

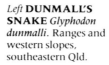

Left and below left **YELLOW-FACED WHIP SNAKE** *Demansia psammophis.* All mainland except coastal SA, southwest WA, and far north from Gulf of Carpentaria to Kimberleys. Has wide colour variation from olive-green (top) to grey-brown (bottom).

Right **DE VIS' SNAKE** *Denisonia devisii.* Inland from central Qld to central NSW.

snakes to avoid

Above **MARSH SNAKE** or **BLACK-BELLIED SWAMP SNAKE** *Hemiaspis signata.* Coastal, northern Qld to southern NSW.

Above **BLACK WHIP SNAKE** *Demansia atra.* Ranges to coast from Gympie, Qld, to Cape York; Gulf of Carpentaria to E. Kimberley, WA.

Left **CURL SNAKE** or **MYALL** *Suta suta.* Inland plains of Qld, NSW, SA, NT, northern Vic., eastern WA.

Right **GREY SNAKE** *Hemiaspis damelii.* Central NSW to Qld coast at Rockhampton.

ORNAMENTAL SNAKE *Denisonia maculata.* Rockhampton district, Qld.

GOLDEN-CROWNED SNAKE *Cacophis squamulosus.* Ranges to coast, mid-north Qld to southern NSW.

Sea snakes

Netted along with prawns or fish, sea snakes can be a boatman's nightmare.
But swimmers and divers have little to fear.

Most sea snakes in Australian waters are limited to the tropics but a few ride warm currents south in summer, some as far as Tasmania.

Close to shore in tropical waters, sea snakes are a constant hazard to trawler crews. In the Gulf of Carpentaria particularly, any haul is likely to include potential killers. Seasoned deckhands in seaboots and gauntlets nonchalantly flick them overboard. Serious bites are unheard-of. But amateurs handling nets anywhere in warmer waters should be wary, especially at night.

More than 30 of the world's 50-odd sea snake species have been identified off the Australian coast. Most are limited to the tropics but a few ride warm currents south in summer, even to Tasmania. The yellow-bellied *Pelamis platurus*, fairly common year-round as far south as Sydney and Perth, is the only ocean traveller. All the others stay inshore or near reefs.

Frequently inquisitive, sea snakes are attracted to motion in the water. They may approach a diver or a swimmer, but they

Common in shallow waters from Broome, WA, to the central coast of Queensland, the elegant sea snake Hydrophis elegans *has also been caught in Sydney Harbour. It often travels far upstream in tidal rivers. The average length is 1.75 metres but a specimen of 2.1 metres has been recorded in Queensland.*

SEA SNAKE BITES

First aid and medical procedures for dealing with the bites of sea snakes are just the same as those for ordinary snakebite (page 360). But because bites are often not felt, the companions of a victim may have to cope with the surprise onset of symptoms such as nausea and dizziness, and the possibility of sudden collapse on a boat or in the water. And doctors have to watch for a small complication — the delicate fangs of sea snakes may break off, remaining embedded in a wound.

The Commonwealth Serum Laboratories make an antivenom, prepared from the blood of horses that are injected with venom from both the tiger snake and the beaked sea snake *Enhydrina schistosa*. It works against the venom of all dangerous sea snakes. Where it is not available, tiger snake antivenom can be used.

usually keep their distance once they perceive the size of a human interloper. They are every bit as shy as their land-based relatives, and able to move more freely to avoid a confrontation.

Divers occasionally report having had snakes coiling round their limbs or spearguns. Even so, there is very little evidence of aggression. Researchers trapping snakes under water have handled hundreds of them without a bite.

No record of a killing

Sea snakes are dreaded in Southeast Asian waters, where they are said to have taken the lives of many unprotected fishing folk. But there is no record of an Australian fatality. Apart from fishing accidents, the chance of trouble here seemed to be non-existent until 1979, when a toddler was bitten and gravely poisoned near Yeppoon, Qld.

Venoms of most sea snakes are highly toxic and fast-acting, for if the fish that are preyed on do not succumb quickly they can escape. But even the biggest types, sometimes 2 metres long, have relatively small venom glands and their outputs are low.

The fangs are usually short, too, so some attempted bites are mere scratches. Stokes' sea snake *Astrotia stokesii*, the species responsible for the 1979 attack, is an exception. Its fangs can

Pelamis platurus, *the pelagic or yellow-bellied sea snake (left), is an ocean wanderer that enters inshore waters and is sometimes beached in storms. It is the only species found in the south throughout the year. It averages 70 cm in length and can reach 1 metre.*

The beaked sea snake Enhydrina schistosa (right) *is considered to be the world's most dangerous, accounting for many deaths in Southeast Asia. It is found between Darwin, NT, and Rockhampton, Qld, and hunts catfish in shallow estuaries and tidal creeks. It normally grows to 1.2 metres long.*

grow to more than 6 mm and pierce a wetsuit, and its venom yield compares with that of a tiger snake.

Modified for a life under water

Sea snakes are air breathers, closely allied with the elapid family of land snakes. They almost certainly evolved from elapids — perhaps in the Australian region. The evolution must have been fairly recent, for while they are common around the Indian and western Pacific Oceans, the modern positions of the African and American continents bar them from the Atlantic.

Their most obvious adaptation for marine life is a flattened, oar-shaped tail. The nostrils point upward rather than forward, and have flaps that close when a snake submerges. A more surprising modification has taken place internally: the right lung of a sea snake, extending all the way down to its anal region, ends in a non-functioning part without a blood supply. This is a reservoir of air for diving. A sea snake can also take in about a fifth of the oxygen it needs from

The olive-brown sea snake Aipysurus laevis *frequents tropical reefs, often making close inspections of divers. It is among the biggest sea snakes, sometimes growing to more than 2 metres.*

Hardwick's sea snake Lapemis hardwickii *is among the most abundant species in tropical waters and a likely hazard in prawn trawling. Its average length is 1 metre.*

the water through its skin, while excluding salts. Other excess salt from its diet of fish is excreted through a gland in its mouth.

Chasing prey, some species can dive to 100 metres and stay submerged for nearly two hours. In short bursts they have a swimming speed of up to about 4 km/h.

A skin may last only a fortnight

All true sea snakes bear their young live, in litters generally much smaller than those of land snakes. Although they have predators, including cannibals of their own kind, the toll taken is not as heavy or consistent as that exacted by birds on the land species.

The main complication of their comparatively easy marine existence is the accumulation of growths such as barnacles, slowing them down and interfering with their intake of oxygen. This is solved by the frequent sloughing of skins — as often as every two weeks.

Sea snakes can stay alive out of water. A few species hunt on mud flats. But they are virtually helpless if washed ashore because they lack the wide belly scales that land snakes use to lever themselves along the ground. □

THE ODD ONE OUT

One kind of dangerous marine snake lays eggs — and clambers ashore to do it. Sea kraits of the *Laticauda* group retain the wide belly scales of a land snake and can even climb cliffs to deposit their eggs in caves or rock crevices. Scientists do not regard them as true sea snakes but as an intermediate type. They adapt their breathing rhythm to suit either fresh air or an underwater environment. Two similarly marked species, occasionally netted in Australian waters, are strays from western Pacific breeding areas.

The black-banded sea krait Laticauda laticaudata *(left) averages about 80 cm in length. Its close relative the yellow-lipped or banded sea krait* Laticauda semifasciata *(above) is bigger, averaging 1 metre — though one caught in the Philippines is said to have measured 3.6 metres.*

Shoreline surprise

A paddling toddler was bitten in only a few centimetres of water.

Australia's worst known attack by a sea snake happened at the very edge of the water on a sandy beach. The snake, which presumably was trodden on, was of a species that is more usually encountered around reefs and in open water, to depths of about 20 metres.

The victim was a two-year-old girl, paddling at a beach near Yeppoon, Qld, in October 1979. Her mother heard her scream and saw a snake coiled around the child's ankle. It swam away but was caught and killed by teenagers. Identified as Stokes' sea snake *Astrotia stokesii*, it was 1.6 metres long.

Puncture wounds and jagged scratches covered most of the instep and side of the little girl's foot. Fingers clamped around her calf, her mother carried her to a nearby ambulance station. On the way to hospital she released her grip and symptoms of severe poisoning set in within 30 seconds. Four minutes later the child was unconscious, and soon after reaching hospital she was unable to breathe.

Treated with antivenom, the girl regained consciousness 4½ hours later but required artificial breathing aid for a further day. During her recovery she had muscle spasms and periods of hallucination, doctors at Rockhamptom Base Hospital reported in the *Medical Journal of Australia*. Her walking was hampered at times for a month after her discharge.□

'A snake was coiled around the child's ankle.'

Stokes' sea snake Astrotia stokesii *grows to an average length of 1.2 metres — but the one responsible for the Yeppoon attack was a record 1.6 metres. The species is found from North-West Cape to southern Queensland, venturing occasionally as far south as Tasmania.*

A CRUEL FLUKE OF CHEMISTRY

Spiders are everywhere. More than 2000 species in Australia are spread from treetops to caves and from seashores to deserts. Some even hunt in rivers. And who needs reminding how readily they have taken to sharing human shelter?

All species are carnivorous, and with scarcely an exception they are armed with venoms to subdue their prey. But their natural victims are tiny and the output of venom is minuscule — compared, for example, with snakes. It is a strange accident that certain spider venoms contain substances powerful enough to kill or disfigure people. Few of us are harmed. But for the spiders the consequences are disastrous. Whether or not they pose any threat, they often provoke an unreasoning revulsion and fear that lead automatically to their destruction.

That is our loss, for spiders are the leading natural controllers of agricultural, horticultural and household insect pests. The venom properties of most of them have never been investigated, so there is a chance that some new dangers will be exposed in years to come. But the fact will remain that the menace of spiders is slight in relation to their overall numbers.

BURROWERS, BUILDERS AND VAGABOND RAIDERS

Among the first creatures to come out of the sea, some 400 million years ago, spiders have diverged so widely in form and habit that they deserve to be seen as many different animals. They vary as much as tigers and mice.

Their common characteristics, shared with scorpions, ticks and mites, class them as arachnids — not insects. Their bodies are formed in only two parts, a fused head-and-thorax and an abdomen, encased by an outer skeleton that moults as they grow. The forward part carries four pairs of jointed legs.

Two organs of touch, called palps, extend in front. Sometimes they are so long that they look like extra legs. Above the mouth is a heavier pair of appendages, the chelicerae. Tipped with fangs, these are used for grasping, piercing or crushing prey. In nearly all spiders they are fed by venom glands.

The abdomen contains the spider's heart, procreative organs, main digestive system and devices for taking oxygen into its circulating body fluid. At the rear are silk glands with ducts leading to tiny protuberances — the spinnerets.

What sets different spiders apart

What a spider does with its silk makes a major distinction in the way it lives. But there are also important structural differences. The chief ones, all pointing to varied styles of existence, are in the size and arrangement of the eight eyes — a few species have only six — in the breathing apparatus and in the operation of the prey-killing chelicerae.

Scientists classify Australian spiders in two main groups — the mygalomorphs ('formed like mice') and the araneomorphs ('formed like spiders'). The division is made on anatomical grounds, but it corresponds closely with obvious differences in their habits.

Trapdoor spiders and funnelwebs make up the 200 or so species of mygalomorphs. Their chelicerae strike downwards in parallel, oblig-

Above: The fang-tipped chelicerae of 'true' spiders, such as this female huntsman Olios calligaster, close on their prey laterally in the manner of pincers. Primitive mygalomorphs — the trapdoor and funnelweb spiders — strike downwards.

Left: A few spiders have only six eyes but most, like this wolf spider Lycosa, have eight. Their size and arrangement on the head help identify different types.

ing the spiders to raise the front part of their bodies to attack. And they breathe through two pairs of lung 'books' — sets of delicate sheets that work like the gills of fish. These are visible as pale patches under the abdomen.

Mygalomorphs use their silk to build permanent homes, usually in or on the ground but sometimes in tree holes. They dig burrows and line them, often fashioning protective lids, or they make lairs in densely spun funnels. Prey is taken as it comes nearby.

The hunters and the weavers
Araneomorphs, more highly evolved, are sometimes called 'true spiders'. They have only one pair of lung books, the others having been modified as tubes that allow a faster oxygen intake and a more energetic life. And their chelicerae work laterally, like pincers.

Among the araneomorphs are two distinct kinds — weavers that construct sticky webs and wait for their prey to trap itself, and hunting spiders that roam at large. If the hunters build at all, it is often only a covering for temporary refuge. Their main use of silk is for quick aerial escapes from danger. Among the hunters, jumping spiders are distinguished by two enlarged front eyes, giving binocular forward vision to judge pouncing distances.

Both kinds of araneomorph, as well as the mygalomorphs, include some spiders known or presumed to be dangerous. So it is comforting to make one further distinction. None of the orb weavers — those that build the classical circular or part-circular web with radiating spokes and an inner spiral — need be feared.

How prey can be sucked dry
Insects are the principal prey of spiders. But many take whatever else they can overpower — frogs, lizards, mice, tiny birds occasionally, and other spiders frequently.

However firmly the prey is grasped, or bound up with silk, it is further immobilised with venom. The fangs are fed from glands in the bases of the chelicerae of mygalomorphs, or farther back in the head region of other spiders.

A spider cannot digest anything solid. It regurgitates some of its stomach contents over and into the prey to help liquefy internal organs and start the digestive process. Then, using the pumping action of muscles in its throat and a 'sucking stomach' in the front part of its body, it draws the juices out of its victim and down into the main gut in its abdomen.

The casing or skin of the victim, sucked dry, is left crushed and torn by hunting spiders and mygalomorphs in their efforts to squeeze out all of the contents.

But the weavers that truss up their feasts with silk, and have no need to grapple with them, often leave empty shells of recognisable shape.

Annuals and perennials
Most hunting and web-building spiders live for a year or so, breeding only once. But trapdoor spiders and funnelwebs spend three years or more just reaching sexual maturity. Then the

A redback's blowfly prey, already hopelessly snared, will be bound up with silk before its juices are sucked out.

Above: A barking spider Selenocosmia *dispatches a frog on the Atherton Tableland, Qld. This type includes Australia's biggest spiders.*

Right: Huntsman spiders and others that do not build snares have to grapple with their prey, and usually leave the remains crushed.

females breed season after season. Some female trapdoor spiders live for over 20 years.

A growing spider moults two to eight times in a year. For hours each time, its soft body is in danger, not only from predators but also from drying out. Hunting spiders, normally home-less, spin silken envelopes just big enough to crawl into. These refuges, containing moulted husks, are often found in and around buildings.

Sexual distinctions show up early among the weavers. Males are usually much smaller than females, and often so differently built that they are unrecognisable as the same species. Male hunting spiders and mygalomorphs, however, resemble females until after their final moult.

When a male reaches sexual maturity he develops spines and other modifications on his legs. These will help him grasp a female during mating. More extraordinarily, his palps — the pair of sensory organs in front of the legs — swell at the ends to take on a club-like shape.

From bridegroom to wedding breakfast

Mating takes place during a short season that seems to be regulated by temperature and humidity. It is usually during summer, but different types of spiders require different con-

Above: Wolf spiderlings emerging from their egg cocoon cluster on the mother's abdomen, hooking their claws onto knobbed hairs.

ditions to trigger the instinct of mature males to go in search of females. Senses of touch and smell, as well as sight, may be used in the quest.

When the male identifies a suitable female, he spins a small pad of silk and ejects sperm onto it. Then he draws the sperm into the

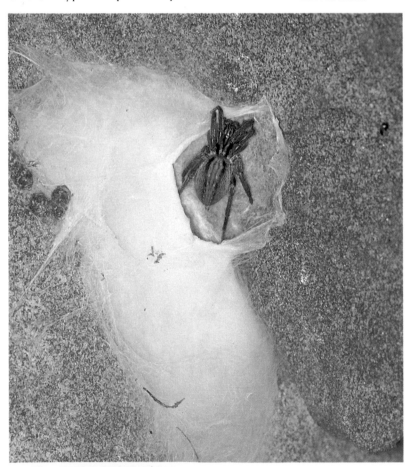

The nocturnal hunter Miturga gilva *builds a dense envelope of silk under rocks or logs, as a daytime refuge and a place to raise its young.*

AVOIDING BITES

As a general rule it is unwise to handle spiders that have bodies bigger than your little fingernail. Most bites can be avoided if simple precautions are taken:

- When gardening, or handling firewood and other materials that are left outdoors, wear strong gloves of canvas or leather.

- Before you lift rocks, branches or other objects that have lain on the ground, shift them slightly. If possible tilt them *towards* you, so that anything underneath retreats in the opposite direction.

- If a big spider appears on loose clothing, don't brush at it but shake it off, making sure that it falls clear of your footwear.

- If the spider is on your skin or on tighter clothing, don't try to trap or squash it. Allow it to walk off if you can, or brush it away from you with a fast, sweeping motion — if possible with an object other than your hand.

- To trap a spider in a house, use a glass jar. Slip a sheet of cardboard under, invert the jar and shake the spider to the bottom. If the spider is on a ceiling, dislodge it with a broom. But don't stand directly underneath — it is likely to drop straight down.

- If clothing left on a floor or furniture is to be worn again, shake it out. And make a habit of turning shoes, slippers and boots upside down and knocking them together before putting them on.

- When picnicking or camping, shake out blankets, groundsheets, sleeping bags and so on.

- NEVER blindly poke your fingers into holes in the ground or rock crannies, or under peeling bark.

swollen ends of his palps. Now ready to mate, he begins a procedure of courtship that is aimed to secure recognition by the female and to suppress — temporarily — her predatory impulse.

For weaving species the courtship often consists of tweaking a strand of the female's web with a certain rhythm. In others it could be a matter of cautious touching, or a visual display of leg-waving and dancing as elaborate as those of birds. The ritual may take hours.

When he senses that the female is submissive, the male dashes in and takes control — absurdly in those cases where the female is many times bigger. The palps are placed against an aperture in her abdomen and the sperm is ejected once more. Males use their spines, chelicerae and even silk in various ways to immobilise females during mating. When they let go, they are in mortal danger. Male spiders are often, though not invariably, killed and eaten by their mates. They are near the end of their lives anyway — no male spider survives more than one breeding season. And at least

this cannibalism ensures that the female is well fed as she retires to produce eggs.

How life starts for new generations

Eggs are nearly always laid into some kind of silken envelope or ball. This protection of the young was probably the basic purpose of spiders' silk glands, from which their various other spinning skills developed.

Females of some hunting species — notably the wolf spiders — carry egg cocoons about with them. A few guard them in nesting chambers. But weavers and most hunters simply attach envelopes, containing hundreds of eggs, to vegetation, rocks or buildings. Then they are abandoned to the weather and to predators. Mortality rates are high.

Eggs of the funnelweb and trapdoor spiders are laid in far smaller numbers. Chances of survival are greater because they receive the further protection of the mother's lair. And once hatched, the spiderlings may stay there for months, sharing her food. Female wolf spiders

An Olios *huntsman spider, discovered in her resting place under tree bark, adopts a threatening posture in defence of her egg sac. Some of this type are burrowers, some take up residence in debris and some are wanderers. Whether or not their lair is permanent, each guards an individual territory.*

SILK THAT IS STRONGER THAN STEEL

Finger-like spinnerets of a desert wolf spider.

Silk is extruded at will from the spinnerets of all spiders. Liquid as it reaches the spinnerets from the silk glands, it hardens under tension. Though always pliable, it is stronger than steel of the same thickness. Yet silk is amazingly light — a strand 0.005 mm in diameter, long enough to encircle the world, would weigh less than 170 grams.

When a spider is letting out a dragline or making a lifeline for jumping, the silk simply streams out. But when silk is used for building it is manipulated with the hind legs. In some snare builders two of the spinnerets are fused into a sieve-like plate that extrudes multiple strands of extremely fine silk. These are combed with special leg bristles to make lacy sheets. The web weavers also have different silk glands that give them the choice of dry or sticky strands.

are frequently seen with their bodies covered by newly hatched spiderlings, three or four deep. But the young do not stay long; they soon fall off or scuttle away. They are fully formed miniatures, quickly able to fend for themselves — and eager to eat one another if they are crowded together too long.

Taking flight in a gossamer cloud

Spiders are essentially solitary creatures. Clustered populations may be found where there is plenty of food, but evidence of truly social behaviour or co-operation is rare. Among web-weaving species, for which a successful hatching means instant overpopulation, the instinct to disperse is shown in a way that became the stuff of fairy tales in older times — and UFO reports more recently.

Clutching leaves or plant stems, spiderlings exude strands of silk into the air. Picked up by breezes, the strands exert a tension that plucks the tiny creatures up and takes them floating, sometimes for kilometres. Where huge numbers have hatched at around the same time, their migration may be seen as a glistening cloud of drifting gossamer.

Not only the web weavers but also some hunting species disperse in this way. Most trapdoor spiders and funnelwebs do it on foot, so they seldom go far from their birthplaces. However the dangerous mouse spider *Missulena* is able to float on gossamer, and as a result is the only mygalomorph to have populated the whole Australian continent.☐

A threatening attitude is the best means of defence for many spiders when they are caught out in the open, like this roving male funnelweb.

UNDER CONSTANT THREAT

A spider's life is one long struggle against predators and parasites.

Birds are the most visible natural enemies of spiders. They snap up migrating spiderlings as they float on gossamer, snatch weavers from their webs, pluck hunters from the ground and peck at the entrances of burrows.

Other persistent predators include bandicoots and wasps. Bandicoots are skilled at digging trapdoor spiders out of deep lairs. Wasps pursue and paralyse hunting spiders, then carry them off to store as food for their larvae.

Ground-living spiders also fall prey to frogs, lizards, cats, scorpions, centipedes and some bugs and beetles.

Under threat, most web weavers drop on a thread of silk and feign death, or scuttle into a prepared refuge. But some stay put and shake their webs violently to confuse predators. Hunters, trapdoor spiders and funnelwebs often rear up in a display of aggression.

Camouflage and mimicry

Though some weaving spiders are gaudily coloured as a warning to predators, most ground-dwellers have earthy or leafy colours that make them inconspicuous. Many huntsmen are speckled to imitate the dappled sunlight on a forest floor. One small orb weaver is said to be able to change its colour.

Spiders from several different families have markings and modified body shapes that enable them to look like ants. This not only discourages birds, but it also allows some of them to mingle with foraging ants and prey on them.

Parasites take a heavy toll of unborn spiders. Wasps, lacewings and some flies lay their eggs in or on spider cocoons. The insect larvae hatch first and feed on the spider eggs. Survivors may also be parasitised later in life, chiefly by a fungus that eventually kills them.□

Newly hatched spiderlings drifting on fine threads of gossamer await a breeze to carry them to new territories. Aerial dispersal — sometimes scattering a new generation over many kilometres — avoids overpopulation among the web-weaving species.

Funnelweb spiders

The terror of the Sydney region has far-flung relatives. Their menace is likely to grow with the spread of settlement.

Rearing up, with beads of venom already glistening at the tips of its massive fangs, a big funnelweb spider is an unnerving sight. The threat is no bluff. And it is being encountered more often. While the natural habitats of funnelwebs are destroyed by human settlement, our homes and surroundings are often more in their favour.

Uniquely Australian, the funnelweb group includes what is probably the world's deadliest spider, *Atrax robustus*. Yet it was not recognised as a cause of fatal bites until 1927. Popularly called the Sydney funnelweb, it was thought at first to be restricted to northern suburban areas.

That was merely a reflection of Sydney's growth pattern. As housing spread into other wooded uplands and people took to holiday motoring, the spider was encountered more and more widely. It turned up in coastal areas as far apart as Newcastle and Nowra, and over the Blue Mountains at Lithgow.

Before 1981, when funnelweb antivenom started its successful trials, males of *Atrax robustus* killed at least 13 people. The victims included a two-year-old who was reported to

have died in just 15 minutes. Worried attention was also focused on other funnelweb species, occurring in most of the moister parts of the east and south. Smaller types, such as those around

Above: The venom of the female Atrax robustus *deals a swift death to a lizard, but does not harm humans.*

Left: Gardening spadework exposes a burrow.

Below: Funnelwebs rear up because their fangs can only strike downwards.

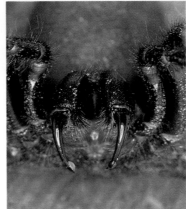

Globules of venom appear even before the strike. The fangs, up to 7 mm long, can pierce the skulls of smaller prey.

WHERE *ATRAX* SPIDERS ARE FOUND

NSW

ACT

● Sydney

Canberra ●

VIC

■ *Atrax robustus*   ■ Other *Atrax* species

Melbourne and Adelaide, seem to pose no threat. But some that favour the subtropical rainforests of NSW and southern Queensland grow even bigger than the Sydney funnelweb. They are undoubtedly capable of killing people — and in their case the females seem just as dangerous as the males.

Where funnelwebs live
The spiders make their homes in sheltered spots that are cool and consistently humid. In bushland they are most often found in gullies. Silk-topped burrows are hidden under rocks, in or under logs, among tree roots or under grass tussocks. Rainforest funnelweb species such as the big and dangerous *Hadronyche formidabilis* may climb high to take advantage of holes and clefts in trees.

Atrax robustus finds places just as suitable in the foundations of houses or in well-watered rockeries and garden shrubbery. It must always have cover for its burrow, so it is not found in open spaces such as lawns unless rocks, rubbish, building materials or other objects are left lying. Visible, funnel-like burrows in a lawn may belong to trapdoor spiders.

The web above a funnelweb's burrow is not characteristically funnel-shaped. Most often it is a purse-like tube of silk with more than one entrance, leading back to the web-lined burrow. The spider may dig this retreat in soil, to a depth of half a metre or more, or use a natural crevice in rock.

A highly distinctive feature of nearly all funnelweb burrows is a system of surface threads leading away from the entrance tube. These are triplines that alert the spider when prey is within striking distance. It waits just inside an entrance, usually at night, and rushes out when a line is disturbed. Insects, slaters and snails are most often taken. Small frogs and lizards also fall prey to funnelwebs.

Spiderlings share their mother's burrow and prey for some weeks after hatching. If food is scarce, some will fall victim to cannibalism. They go their own way in late summer or autumn, very few surviving to maturity.

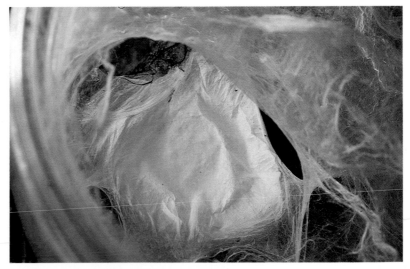

The egg cocoon, a pillow-shaped envelope of silk, is spun in spring or early summer at the bottom of the female's burrow. She lays 80-250 eggs.

167

When funnelwebs go wandering

Reaching sexual maturity, probably in the fourth summer of his life, a male funnelweb has to search for a mate. Eating nothing, he may roam widely, entering homes and sheds. If at last he finds a suitable female he gains recognition by a series of back-and-forth movements and then courts her by touch. Male funnelwebs usually survive mating and may roam again.

Breeding activity extends through summer and autumn, with a January-March peak for *Atrax robustus*. But funnelwebs of both sexes may wander at any time for other reasons. The most common is the flooding of their burrows. Excavations for buildings and swimming pools often disturb them. So do pesticides, which, although generally ineffective against funnelwebs in their burrows, excite the spiders and cause them to emerge aggressively.

ATRAX ENVENOMATION

EARLY SIGNS Pain and redness at bite site. Numbness around mouth, twitching tongue. Nausea, vomiting, stomach pain, heavy sweating, salivation, watering eyes. Laboured breathing. Mental confusion.

ACTION Keep victim still. Apply pressure/ immobilisation first aid exactly as for snakebite (page 360). Obtain urgent medical attention.

High humidity around swimming pools attracts funnelwebs. Wandering males frequently fall into in-ground pools. They can survive immersion for many hours, so in the *Atrax* zone a careful look is warranted before a pool is entered, especially at night or early in the morning. A spider should not be handled, but scooped out with a net and shaken into a jar.

Live *Atrax* funnelwebs are gratefully received at the Macquarie University school of chemistry, where they are 'milked' for vaccine research, or at the Australian Reptile Park near Gosford, where their venom is collected for the production of antivenom in Melbourne.

A funnelweb in good condition will stay healthy for some days in a closed jar with a little moist soil in the bottom. Air holes are not necessary, but it is important not to let the jar get too warm.□

Northern tree-dwelling funnelweb Hadronyche formidabilis.

BREAKING OFF RELATIONS

Recent findings have resulted in reclassification for funnelweb species.

Sweeping changes have been made to the scientific classifications of funnelweb spiders. They result from years of study during the early 1980s by Dr Michael Gray, arachnologist at the Australian Museum in Sydney.

New groupings are based mainly on structural distinctions. They take into account differences in such features as mating and genital organs, leg modifications, spinnerets and teeth.

Formerly all the funnelwebs of the southeast were classed as *Atrax*, a genus embracing 13 described species and many others still requiring study. But since 1986 this genus has been restricted to the so-called Sydney funnelweb *Atrax robustus* and only two other species, not yet officially named. One lives in coastal habitats from southern NSW to far-eastern Victoria. The other is a high-country spider found in Canberra and the Snowy Mountains.

All the others have been reclassified as various species of *Hadronyche* — a resurrection of the old name first given to funnelwebs in the 1870s. But Dr Gray expects that as a next step this group will be broken into at least two new divisions.

The puzzling power of atraxotoxin

The spider doesn't need it — and most animals are unharmed.

Human suffering from the bite of the male *Atrax robustus* is an unaccountable quirk of natural chemistry. In a complex venom mixture, one particular component does most of the damage. It can also kill monkeys and newborn mice. But even overdoses have little or no effect on other animals. Their nerve surfaces are less vulnerable, though no one knows why.

This lethal component, atraxotoxin, is not what the spider relies on to immobilise prey. If it were, the females would be disadvantaged, for they do not seem to produce it. Yet the venoms of other funnelwebs, now classified as *Hadronyche*, are of similar toxicity in both sexes. So atraxotoxin is a freak in every way.

How bite victims are affected

A bite by either sex of *Atrax robustus* is immediately painful because of the great size of the spider's fangs and the acidity of the venom. The fangs can penetrate the skulls of small vertebrates such as lizards and have been known to pierce the fingernails of children. If they are deeply embedded in a victim's flesh, the spider may be difficult to remove. In untreated cases a wheal develops at the bite site. Local pain and red discoloration may persist for hours or even days. Muscle tremors could occur in the bitten area within a few minutes.

If a male *Atrax* has delivered an effective bite and first aid is not given at once, symptoms of general poisoning can show within 10 minutes. They follow a typical pattern of envenomation but do not involve paralysis — though a violent

WHERE *HADRONYCHE* SPIDERS ARE FOUND

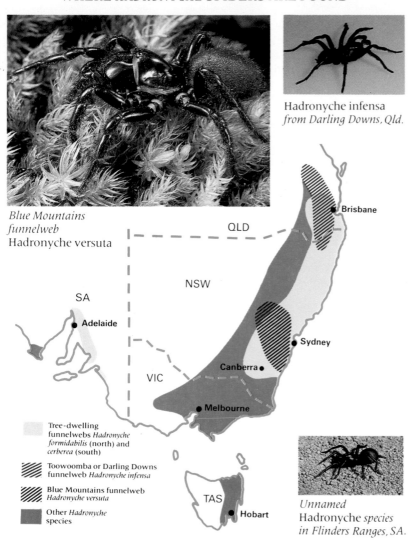

Blue Mountains funnelweb
Hadronyche versuta

Hadronyche infensa
from Darling Downs, Qld.

Tree-dwelling funnelwebs *Hadronyche formidabilis* (north) and *cerberea* (south)

Toowoomba or Darling Downs funnelweb *Hadronyche infensa*

Blue Mountains funnelweb *Hadronyche versuta*

Other *Hadronyche* species

Unnamed Hadronyche *species in Flinders Ranges, SA.*

Southern tree-dwelling funnelweb Hadronyche cerberea.

twitching of muscles can make proper care of a victim almost impossible. Doctors usually have to induce paralysis with a drug.

Mental confusion is quickly evident in severe cases. If not treated the victims eventually fall into a profound coma, during which brain damage can occur. Without antivenom they are likely to die — within hours or days, depending on their body size and previous health — from lack of oxygen or loss of blood pressure.

Bites by other funnelwebs

Symptoms nearly as drastic have been recorded in a few cases of envenomation by the northern tree-dwelling funnelweb spider *Hadronyche formidabilis*. It seems certainly capable of killing children. One child is said to have been killed many years ago in the Darling Downs area of southern Queensland, by the spider now recognised as *Hadronyche infensa*. Some other species are suspect, either through limited venom studies or simply because of their size, but the effects of their bites have not been recorded.☐

The quest for a cure

For over 20 years, every trail led to a dead end.

Venoms were a fresh field for Struan Sutherland, naval doctor turned immunologist, when he joined the Commonwealth Serum Laboratories in 1966. Leafing through old case reports of deaths from Sydney funnelweb bites, he became intrigued by the problem — and found that the CSL had given up on it.

In the late 1950s Dr Saul Wiener, the conqueror of redback spider and stonefish venoms, had turned his attention to the funnelweb. After he left the CSL in 1961 his excellent research was carried on by various other workers. But in 1964 the goal of producing an antivenom was seen as unattainable. The project was abandoned.

Sutherland suggested another try, using newer techniques. The CSL director, Dr William Lane, was opposed to a major diversion of funds but approved a small-scale exercise. Neither man dreamed that the stealthy pursuit embarked on would take 14 more years.

Funnelweb venom had posed thorny problems in research. It was hard to obtain in quantity. It had little or no effect on the small animals usually tested in laboratories. And when it was given to horses, in the same way that makes them produce antivenoms against redback spiders and snakes, nothing happened.

Sutherland's opening shot was to modify the main toxin of the venom, linking it to a carrier protein, to provoke a stronger response from immunised animals. Using rabbits, he seemed to succeed — they produced antibodies. But these failed to work against the natural venom. What was believed to be the main toxin was only a minor component.

The case of the vanishing toxin

Comparing the refined laboratory product with the natural venom, it was discovered that the true main toxin was missing. The moment that purification was attempted, it seemed to disappear. Eventually this component, christened atraxotoxin, was found to have an affinity for glass. It attached itself to apparatus and was lost, leaving a complicated but useless mixture of lesser components.

With that problem solved by using specially treated glassware and a more acid solution, test systems had to be developed to study the venom more carefully and to examine animal serum for possible antivenom activity.

Monkeys were known to be sensitive to the

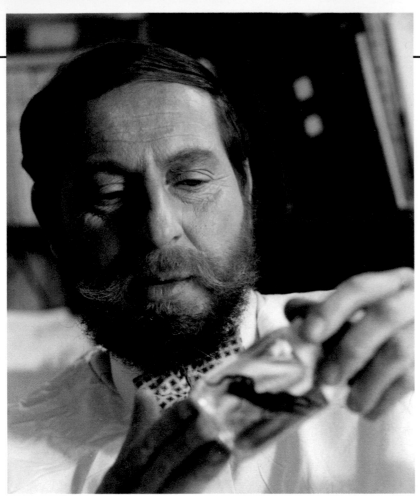

venom but they were large and expensive. Other laboratory animals — rabbits, guinea pigs, rats, mice and so on — were highly resistant. Some small, susceptible form of life had to be found if maximum use were to be made of the tiny quantities of venom that were tediously 'milked' from spiders.

All sorts of animals were tried, from slaters to flies. It turned out that mice were the ideal creatures after all — but only if they were newly born. For 24 hours they are highly vulnerable to the venom, and especially to the atraxotoxin component. After that their resistance increases rapidly, perhaps because their nerves become better insulated.

Another important test system involved the use of tiny strips of muscle with pieces of nerve still attached. Kept alive, the muscles twitched on exposure to venom. They showed that human or monkey tissue was 50 times more responsive to funnelweb venom than material from rats or mice.

By 1978 a variety of animal immunisation courses had been investigated. Weak antibodies to atraxotoxin were detected occasionally, never at levels sufficient to be of practical value. The quest for an antivenom seemed hopeless, and efforts turned to medical alternatives.

A fresh hope quickly shattered

With a growing understanding of how the venom worked, common and not so common drugs were tried against it. Experimental findings pointed to a new method of managing bite cases. Given the benefit of modern intensive care therapy, victims treated promptly were thought to have a good chance of survival.

That hope was shattered at the beginning of 1979. Christine Sturges, a healthy 31-year-old, was bitten while making a bed at her home in Helensburgh, NSW. She reached hospital just as the effects were becoming evident. Still she could not be saved. Campaigning by her husband inspired lavish public donations and the funding of various new research projects. In an unrelated venture, the CSL and the Royal Children's Hospital in Melbourne collaborated to study the effects of the venom on monkeys. Treated just like young hospital patients, they were monitored with the very latest intensive care skills. Previously unsuspected effects were discovered — some of which could be dampened down by drugs.

This work was proceeding well when in January 1980 *Atrax robustus* claimed another victim — a 2½-year-old boy bitten at Wamberal Lagoon, near Terrigal, NSW.

At the CSL it was decided that one more try for an antivenom was warranted. Even if the serum was weak, it could prove useful together with drugs. And a new material from overseas might allow its concentration.

Rabbits were dosed with the best quality male funnelweb venom. Serum from several rabbits, given to poisoned newborn mice, delayed the deaths of some of them by a few hours. This was not the answer, but something to work with.

Next the new material was employed to extract antibodies from the rabbit serum without damaging them. When a concentration was given to more baby mice, they survived high doses of venom. A bigger batch, administered to dying monkeys, brought a dramatic restoration of their normal health.

At long last, the triumph

By mid-winter 1980 Sutherland had clear experimental evidence that the concentrated rabbit product should work on humans. A race started to have enough ready by summertime, for clinical trial at a few hospitals in and around Sydney. The deadline was met — and the antivenom saved its first life within two months of its distribution.

At the end of 1985, by which time 10 critically ill bite victims had been saved, the antivenom was freed from trial restrictions. It is available to any hospital requesting it, and has been used successfully not only against bites of *Atrax robustus* but also against other funnelwebs and the mouse spider *Missulena*.□

PRESSING ON WITH VACCINE RESEARCH

An ambitious project hopes to provide the model for future research into vaccine production.

Biochemist Alison Fillery 'milking' a Sydney funnelweb Atrax . robustus *at Macquarie University, Sydney.*

Just as funnelweb antivenom was undergoing its first trials on bite victims in 1981, Sydney scientists launched an ambitious project to develop a vaccine that would give people protection in advance.

With the success of the antivenom, their continuation of costly research came in for some scathing criticism. The whole aim was described as ludicrous, because it was hard to see how enough venom could be obtained for large-scale production, or who would run and pay for vaccination programmes.

But the team of biochemists and pharmacologists, under Dr Merlin Howden, associate professor of chemistry at Macquarie University, pressed on. By 1984, they had made a vaccine and tested it successfully on monkeys. It is produced on the same principle as tetanus and diphtheria vaccines, by taking the natural toxin and modifying it to suppress its lethality.

Having gone as far as they could — without testing on humans — to confirm that vaccination would work, Howden's researchers turned their attention to making a vaccine by artificial means. They are investigating techniques that would allow synthetic mass production without any need for the natural venom. Along the way they plan to make a totally specific antivenom by cell culture. It would act directly against the lethal component rather than the whole venom, so much smaller quantities would be needed.

Well over $200 000 has been spent, and the criticism has not eased. But the project is assured of funding until the end of 1988, mainly from the National Health and Medica Research Council.

'What the council understands,' Howden says, 'and what the public ought to appreciate, is that our work isn't only about dangerous spider bites around Sydney. The methods we develop are likely to provide a model for the manufacture of better vaccines against any number of human and livestock diseases.'

More than 20 local authorities in the Sydney region also grant support to the project. They see the problem of funnelwebs literally at grassroots level, in the anxiety of parents and home gardeners and tradesmen such as plumbers, builders, landscape gardeners and pest controllers. It is hard to tell any of these people that prevention, whatever the price, is not better than cure.

Living proof

The first victim to be saved by funnelweb antivenom.

Gordon Wheatley — out of danger after two doses.

Just after midnight on 1 February 1981, intensive care staff at Sydney's Royal North Shore Hospital geared up for the first human test of funnelweb antivenom. Supplies for clinical trial had been received only a few weeks before. Now the hospital had a bite victim.

It was an anxious time for Dr Malcolm Fisher and his team, and for the victim's wife, whose consent had to be sought for the treatment. But the edgiest of all, on the other end of an interstate telephone link, was Struan Sutherland. He was about to know if his punishing, often heartbreaking 14-year pursuit of an antivenom had been worthwhile.

Gordon Wheatley, a fit 49-year-old, had been bitten at 11 pm in the lounge of his home in a northwestern suburb. A neighbourhood doctor, fortunately aware of the latest first aid procedures, bound the bitten foot with a pressure bandage and had the patient at the nearest hospital within 15 minutes.

Already Wheatley's heart was palpitating and his pulse racing. His skin sweated and rose in goosebumps and his mouth tingled. Pain from the foot reached up his leg. Within five minutes his arms were wrenched by muscle spasms. Tears and saliva streamed, and he started to vomit.

Soon Wheatley was gasping for breath, blue in the face, with a blood pressure so high that a stroke threatened. In worsening condition in spite of mechanical aid in breathing and an array of drugs, he was transferred at high speed to Royal North Shore.

After careful examination and premedication, a dose of the antivenom was slowly infused. No change was observed after 15 minutes so a second dose was given. In the hour that followed most of the symptoms were reversed, and by morning Gordon Wheatley was clearly out of all danger. Though he suffered some weakness and bouts of heavy sweating for three weeks, he made a full recovery.

A report of this and a subsequent case in the *Medical Journal of Australia*, late in 1981 after a cautious delay, gave professional recognition to the worth of the antivenom. Soon hospitals were clamouring for its wider distribution.□

The redback spider

Agony from the female's bite provides the worst experience of venom that most Australians are ever likely to suffer.

As the oldtime 'redback under the dunny seat', *Latrodectus hasselti* gained a special place in the lore of dangerous Australian creatures. Nothing else so intimately familiar was also so fearsome. Now most outdoor lavatories have gone — but redbacks bite more people than ever.

Latrodectus is common throughout the mainland and Tasmania, in all but the coldest alpine climates and the most hostile deserts. It has taken especially well to manmade structures and the material that people leave around. The prominently marked females — the only ones capable of harming us — can be found in association with most dwellings, as well as in bushland areas.

Bites by the redback are known to have killed at least a dozen Australians in the first half of this century. In the 1930s, when deaths were occurring at a rate of less than one a year, scientists estimated the mortality rate at 5 per cent — indicating that they knew of fewer than 20 bites a year.

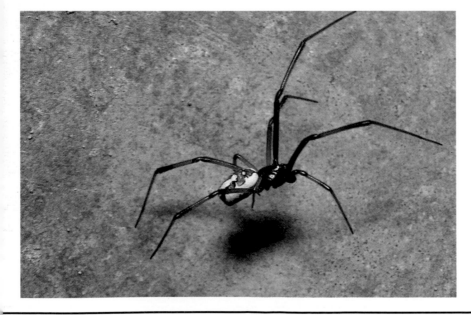

Above: Red or orange markings on the female's back may be broken into spots in front, and sometimes there are also thin white lines.

Below: The harmless male is smaller and carries more sombre markings.

Now in some years the use of antivenom is reported in more than 300 cases. The true figure could be twice as high. Redback bites are the greatest single cause of serious envenomation in this country, outnumbering all the snakebite cases and severe marine stingings put together.

Fear abates — but the pain remains

Dread of the possibly lethal effects of a bite dissolved with the introduction of an antivenom in 1956. The treatment has been totally successful because the neurotoxic component of the venom is very slow-acting. Antivenom can be administered hours or even days after a bite, and no special first aid techniques are necessary.

In the meantime, however, the victim of a severe bite suffers agonies from other components of the venom. After an initial pricking pain the bitten area usually reddens, swells and often sweats. Within about five minutes a more intense pain starts in the area and soon spreads, getting steadily worse. Some victims feel it so badly that they become irrational.

After about half an hour there may be a painful surface swelling of lymph nodes near the bite, and an abdominal ache from deeper nodes. Through a peculiar action of the venom on sensory nerves, severe pain may occur in

A SPRING-LOADED TRAP THAT PLUCKS UP PREY
Guylines and traplines surround the redback's ingenious snare.

Redback spiders build snares just above the ground or other surfaces on which prey animals walk. The snares are untidy labyrinths of strong, dry cobweb anchored by guylines above and at the sides. Stretching straight down are traplines coated with sticky globules. The whole device is called a gum-footed web.

When prey blunders into the vertical traplines it becomes stuck. Its struggles break the lines. The release of tension hoists the prey into the air, where it is helpless. Lighter animals are sometimes plucked all the way up into the web. Slaters, bugs, beetles and other spiders are the usual prey animals, but tiny lizards and even mice can be caught.

At night the redback hangs upside down in the snare web, waiting for action. It has modified spinnerets and hind-leg appendages that enable it to comb out multi-stranded swathes of fine silk. These are used to bind the prey. Then it is taken into the web, bitten, and sucked dry at the spider's leisure.

In daylight or when danger threatens, redbacks make for dark, sheltered retreats. Maturing females build tubular nesting chambers in these spots. In natural environments, in a log or under a rock, the retreat leads directly off the snare web. But in a building it could be high above, and reached by long connecting threads.

The nest and trap may be joined, or metres apart.

Where human settlement has overtaken the spiders, they may make for undisturbed spots under roof eaves, in rafters or under floorboards. They favour storage areas and outbuildings rather than permanently occupied rooms. Building materials and machinery lying about outside, or such things as boxes and cans on rubbish heaps, give them ideal cover.

Males of the species are seldom noticed. Short-lived, they are less vividly marked than their mates and they are tiny — usually growing to only about 3 mm long. Their fangs are too small and feeble to deliver an effective bite on a human.

Even the females are relatively small spiders, not often exceeding 10 mm in body length. They are aggressive only if squeezed, or if forced to defend the egg sacs in their nests. They live for a year, taking 2-3 months to reach maturity, and may lay as many as six sacs of eggs between spring and autumn. Each of these can contain up to 300 eggs, but they are prone to heavy parasitism by wasp larvae.

Widow spiders: feared worldwide
Redbacks are also found in New Guinea, the western Pacific and much of Asia, where they may have originated. They are the most colourful members of a worldwide, highly dangerous group often referred to as widow spiders.

The most notorious, the black widow spider of North America and the Caribbean, is so closely related that authorities used to class it and the redback spider as the same species, *Latrodectus mactans*. Another near relative is New Zealand's only dangerous venomous animal, the coastal *katipo* spider.□

areas far from the bite. In the worst cases nausea, vomiting and headache, often with heavy sweating, set in about an hour after the bite. Muscle weakness, tremors and a failure of co-ordination may follow, with an accelerated heart rate and raised blood pressure.

Where redbacks lurk
Redback spiders seek dry, dark places for their retreats, with space below in which to erect snares. In the bush they are commonly found in logs or stumps or under bark and rocks. But they can also be in the foliage of dense shrubs.

REDBACK SPIDER ENVENOMATION

EARLY SIGNS Redness, swelling and sweating at site of bite. Local pain spreading and increasing.

ACTION Wash bite area and try a pack of iced water to ease pain — but not ice applied directly to skin. Cover bite lightly. Obtain medical attention.

A redback's killing ground is an untidy tangle of dry web with gummed traplines stretched below.

Mouse spiders

Laboratory findings signalled a menace that had never been suspected. Six years on, they were frighteningly confirmed.

Bites by the big, sluggish trapdoor spiders known as mouse spiders have rarely caused concern. In spite of their massive biting parts and long fangs, the venom seemed to be weak. So Commonwealth Serum Laboratories staff, comparing venoms as part of their funnelweb research, were astonished by what they discovered in 1979.

Venom collected from the fang tips of a female mouse spider *Missulena occatoria* was tested on newborn mice. It proved to be even more lethal than the atraxotoxin of funnelwebs. Its possible effect on humans could only be conjectured — until a male of another mouse spider species caused the dangerous illness of a child in southern Queensland six years later.

At least six species of *Missulena* occur on the mainland, though Tasmania has none. Populations are spread all over the continent, in arid habitats as well as in bushland. Their wide distribution is owed to their ability — rare among trapdoor spiders — to disperse through the air on gossamer. Juveniles do this at about one year of age.

Mouse spiders are broadly built and short-legged, always with a very dark basic body colour. But in some species the eye is taken more by the vivid coloration of the males. In *Missulena occatoria* — often called red-headed mouse spiders — they have bright red heads and jaws and a blue sheen on their abdomen. A common mouse spider of the southeast, *Missulena bradleyi,* has a pale blue patch on the upper surface of the abdomen. Other species may show a yellowish patch

Named through a misconception

The popular name of mouse spider is thought to have derived from a belief that the burrows of

Below, left and right: Missulena occatoria *(male at left, female right) was the first mouse spider to be scientifically described, in 1805. Common in southern coastal regions, it was also the first to be tested for toxicity. Venom injected under the skin of newborn mice was even more lethal than that of the dreaded funnelweb* Atrax robustus.

this group were deep and winding, like those of mice. In fact the burrows, oval in section, go straight down to only a moderate depth. Their most distinctive feature at ground level is a lid built in two parts, separately hinged.

Female mouse spiders spend all of their long lives in their burrows. Male mouse spiders, which mature at five or six years, go roaming in search of mates in winter in warmer climates, sometimes by day. Though slow-moving, they are highly aggressive if harassed.□

Unwelcome stranger

A bite stole 12 hours of a toddler's life.

Heavy rains swept the ranges west of Brisbane late in February 1985. At night in the Gatton district, a farming couple noticed their 19-month-old daughter playing with a big, dark spider on the kitchen floor. It was captured, but not before it had bitten her finger.

After 15 minutes the child started vomiting, and in another 15 minutes she was unconscious. From Gatton Hospital she was quickly transferred to Toowoomba, where she arrived sweating heavily and breathing noisily, with a racing pulse.

The spider was tentatively identified as a mouse spider — later confirmed as a male *Missulena bradleyi*. There was no medical precedent for treating a bite of this kind, so the casualty doctor rang Dr Struan Sutherland in Melbourne. All he could suggest was that funnelweb antivenom be tried.

Dramatic improvements were observed within 15-20 minutes, but the girl did not recover consciousness. Another dose was given two hours later. Her full revival came after a total of 12 hours of unconsciousness.

The child was kept in hospital for two weeks as a precaution. She needed treatment for a blood disorder during this period, but suffered no permanent effects.□

Missulena insignis (female above, male at left) has a range from inland WA right across the southern part of the continent. Other species, adapted to particular kinds of habitat, have a more limited territory.

Below: A pale blue patch distinguishes the male Missulena bradleyi, *which was responsible for the first known serious envenomation of a human by a mouse spider.*

WHEN A DRIVER COULD BE SCARED TO DEATH

Our 'tarantula' is harmless — but alarming.

Big and hairy, the scuttling huntsman or giant crab spider *Isopeda* looks all too formidable. It has never been known to harm anyone with its bites. But there is every chance that it has been behind some unexplained traffic fatalities — those in which drivers careered off roads or into collisions for no apparent reason.

The two common species of *Isopeda*, distributed from the southwest to the tropical east, are regular visitors to homes and outbuildings. They often stake claim to the same ceiling or wall every night.

Caught out in wet weather, these wanderers may shelter under parked motor vehicles, then crawl into them. They enter through ventilation ducts and drainage or rust holes. Sometimes they are still there the next day, resting in dark recesses or behind fixtures such as sun visors. And when a vehicle moves, a spider may jump recklessly in its search for an escape route.

During the preparation of this book, an editor driving into Sydney during the morning rush-hour had one land in his lap. He professed some knowledge of spiders and should have known it was harmless.

'But I totally lost my wits,' he recalls. 'All I know is I made an unearthly noise and started flapping my legs about. The spider hopped onto my thigh. I kicked some more and he jumped to the other thigh. Another kick and he was on the door. Only then, with my heart hammering, did I recollect that I was supposed to be in control of a car...'

Only luck prevented a disastrous smash. 'After that,' he says, 'I believe the real value of knowing dangerous spiders is in being able to quell panic by telling yourself instantly when one is *not* dangerous.'

Unnecessary worries about *Isopeda* are fuelled by the tendency of some people, especially in Sydney, to give it the fearsome name of 'tarantula'. In the Americas that name is bestowed on huge, dangerous bird-eating spiders. In Europe, it properly belongs to a species of *Lycosa* wolf spider.

Skin destruction

A mysterious ingredient of some spider venoms eats away tissue. In bad cases, only surgical grafting can replace the loss.

With the lethal power of redback and funnelweb attacks countered by antivenoms, the disabling and disfiguring effects of other spider bites are emerging as a greater worry. Evidence is mounting that more than one kind can cause massive tissue destruction, along with maddening pain.

Scientists are far from certain which species are responsible. They have no idea what substances in venoms are doing the damage, or how they occur. Without heavy investment in research funding they are unlikely to come up with answers — let alone antidotes. But the problem is uncommon, and not seen officially as a threat to life. So health authorities and governments have accorded it a low priority.

The normal process of tissue death, or necrosis, starts because of a loss of the blood supply — through injury or frostbite, for example — or because of bacterial action. On the body's surface it results in gangrene, which can spread to healthy tissue.

Necrosis from animal venoms is different. It is usually self-limiting both in area and duration,

TWO AT THE TOP OF THE SUSPECT LIST

The white-tailed or white-spotted spider *Lampona cylindrata* is thought to be the likeliest cause of skin destruction in bite victims. Less than 1.5 cm long, with a cigar-shaped body, it is a wandering hunter that often takes up temporary residence in houses. *Lampona* seeks its prey of other spiders during the early part of the night, usually sheltering by day in a sac of silk that can also be used as a nesting chamber. Its favourite spot in a house is at the top of a wall. In the bush, it may be found under peeling eucalypt bark and dry logs or among foliage.

Lycosa godeffroyi, the most commonly seen of the wolf spiders, is also strongly suspected. A ground-dwelling hunter, up to 3 cm long and aggressive, its populations range across the temperate zone of Australia in all kinds of habitats.

The shallow burrows of *Lycosa* wolf spiders often have wide entrances built of silk-bonded grass or leaves, jutting from the ground. Most species are strongly marked on their backs with radiating bars or bands, earning them an old nickname of 'Union Jack spiders'.

The elongated body of the white-tailed spider Lampona cylindrata *is unmistakable.*

and not gangrenous. Often the body can replace the damage. Small blisters and sores from insect stings eventually heal. The worst wheals inflicted by jellyfish tentacles may be replaced by scar tissue. But in severe cases of necrosis from spider bites the loss is extensive and irreversible, reaching through all the layers of skin.

This phenomenon — called necrotising arachnidism — used to be known only from overseas, mainly in connection with certain species of *Loxosceles* fiddleback spiders. Its effects in Australia, which seem to be somewhat different, were not documented until the end of the 1970s.

At first the problem seemed to be concentrated in Victoria, but during the 1980s it emerged in NSW and Queensland. Outdoor workers, backyard gardeners and even people confined to their homes have been afflicted. The offending spiders are seldom seen. In many cases no bite is felt.

Symptoms have no set pattern

The speed of tissue destruction and the incidence of other symptoms vary widely, suggesting that different kinds of spiders are involved. Skin death may start with surface blistering, or with a darkening below. It can be rapid and agonising, and accompanied by drastic attacks of vomiting and diarrhoea. Or it can be gradual, with relatively little pain. In one Melbourne case it had both a fast onset and a slow, relentless continuation.

Where necrosis is extensive, the only way of

Lycosa godeffroyi is just one of the many types of wolf spider. Most occur in moister regions, from highland ridges to coastal swamps. But some species have adapted to the arid inland. In deserts, they even use pebbles as stoppers for their nests.

Wolf spiderlings cluster on their mother's back.

repairing the damage is to remove the dead tissue and scrape around it — actually enlarging the loss — and to replace it with grafts of skin from elsewhere on the victim's body. Sometimes even that is not feasible: fingers and toes have had to be amputated.

Grafting was a complete success with one of the youngest victims, a three-year-old Melbourne girl bitten while in a disused storeroom. Nearly half of the upper surface of her left foot was replaced. But before that the other effects of the bite had almost killed her. She reached hospital prostrated by dehydration from vomiting and diarrhoea. Transfusions and drugs brought her out of shock only just in time.

The spider under greatest suspicion as a cause of necrotising arachnidism is the white-tailed *Lampona cylindrata*. A smallish, night-hunting member of the sac spider group, it is found throughout Australia. It has adapted to living in houses and sometimes crawls into discarded clothing in bedrooms and bathrooms. Bites are known to cause a burning local pain, followed in some cases by blistering and ulceration.

Wolf spiders of the genus *Lycosa* have also been implicated. With species distributed all over the continent and in Tasmania, they make up the most dominant ground-hunting group. The common grey wolf spider *Lycosa godeffroyi* is a proved killer of cats. And related species in South America and Europe — including the infamous tarantula — have caused skin destruction in human victims.

One Australian sufferer of severe necrosis has confidently identified her attacker as a species of *Lycosa*. So did an elderly farmer, bitten in north-western Victoria in 1978. Skin death was not reported in his case. But after swelling at the bite site had subsided, both of his legs puffed up with fluid and serious kidney damage was discovered. In spite of treatment he failed to recover his good health and died in 1980. ☐

Once bitten...

Australia's first known victim calls in vain for research.

Joan Vivian, in blooming health and recently home with a new baby daughter, wished she were back in labour. That pain had been nothing compared with the agony that a spider bite brought.

It went on and on, in spite of all that hospital staff could do. Half out of her mind, she pleaded to have her hand amputated. Meanwhile, baffled doctors watched as skin started to rot from it.

Mrs Vivian's case was the first of necrotising arachnidism to be documented in Australia. A wheat farmer's wife in north-western Victoria, she was bitten in May 1978 while planting spring bulbs in a rockery outside her kitchen window. The spider was not seen.

About 2 hours later she noticed redness and slight soreness in her left index finger. Within 12 hours of the bite the whole arm was intensely painful, and vomiting and diarrhoea set in. She was taken to a country hospital before dawn the next day.

Soon Mrs Vivian was in shock, with a swelling travelling rapidly up her arm. When it reached her neck she was rushed to a major Melbourne hospital, where blisters were found to be breaking out all over her hand. Tests for any likely disorder proved negative.

The blisters became ulcers that ate down through layers of skin. Even four weeks after admission the hand remained grossly swollen. Operations were carried out to reduce pressure on nerves, and later to graft skin into the damaged areas. Mrs Vivian spent eight weeks in the Melbourne hospital and a further month in her local hospital.

In the meantime her husband had searched the area where she had been gardening. He found *Lycosa* wolf spiders near the back door and — unusually — inside the house. Later an expert search turned up 19 more *Lycosa*, two huntsmen and a black house spider. Three of the wolf spiders, including the biggest of all, were in the rockery where Mrs Vivian had been working.

In spite of later correctional operations — she lost count but knows she had surgery more than a dozen times in all — and intensive physiotherapy, Mrs Vivian did not recover complete use of the hand. Some actions are hampered and, if she forgets herself, painful.

Years after her case was publicised, she found that more were occurring — and still no research work had been done towards proving the cause and finding an antidote. Like Dee Starr, the victim whose story is told under the heading 'A millimetre a month', Joan Vivian has persistently bombarded state and federal politicians with requests that they press for funding. That the situation is unchanged today, she says, is 'nothing short of tragic'. ☐

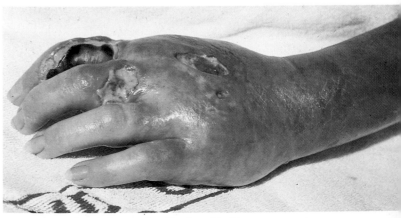

Joan Vivian's hand, four weeks after the spider bite.

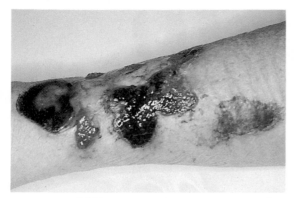

A millimetre a month

The creeping decay that no medicine can cure.

Christmas Day, 1981, brought no glad tidings for Dee Starr. The spider that bit her has blighted her life, apparently forever.

Working in her garden in a Melbourne suburb, Mrs Starr had taken the proper precaution — she was wearing gloves. But the spider, alighting on her arm, slipped into one of the gloves. Trapped there, it bit her just above the wrist.

Mrs Starr is a professional nature photographer, and she is in no doubt that the spider was a wolf spider. She captured it and kept it in a jar for some time.

Intense pain set in within a few minutes and the bite area quickly became discoloured. Bathing her wrist in an effort to ease the pain, she was amazed after about 20 minutes to see a section of skin the size of a 50 cent coin 'just wash away', leaving a deep hole.

That was the first of a mass of sores that were to consume most of the skin on the upper side of Mrs Starr's forearm. Five years later the ulcer was approaching her elbow, still progressing at a rate she gauged at 1 millimetre a month.

A grey fluid oozes out continuously, says Mrs Starr, along with blood as small vessels are destroyed. Drugs to help the blood to clot have only a brief effect.

Specialists in many branches of medicine have been consulted, as well as herbalists, naturopaths and other exponents of alternative treatments. None could come up with a way of arresting the necrosis, other than by amputation. Skin grafting cannot be considered while the rotting continues.

Mrs Starr has a theory that her skin is being consumed not by a toxin but by a digestive enzyme that a spider would use to liquefy the internal tissues of its prey, so that it can suck it dry. Somehow this enzyme is able to renew itself, she believes.

Despair and constant pain — 'like a knife going in' — have changed her nature, she says. Impatience and bad temper have cost her dearly

Above: Dee Starr returns to the spot where her ordeal began. Above left: Nearly five years later, the ulcers on her arm continue to advance.

in her private life. Still she tries to keep up with her work, although her muscular command of the arm is starting to weaken.

The ulcer is cleaned twice a day and fresh dressings are put on. To bathe or go out, Mrs Starr must swathe the arm in plastic. Water — even a sprinkle of rain — heightens the pain unbearably.

Reflecting wearily on her five-year ordeal, she says: 'If only I'd been bitten by a funnelweb or a redback... I'd have been OK in no time!'

Dee Starr makes no secret of her 'pathological loathing' for any kind of spider, or of her anger at officials who fail to see a need to fund research into necrotising arachnidism. Exhausted of all hope for herself, she has no wish to see others in the same plight.□

'If only I'd been bitten by a funnelweb or a redback'.

Ending it all

When an anguished sufferer could take no more.

A spider bite is believed to have led to one of the most tragic crimes in Victoria's history. In 1983 Dr Struan Sutherland of the Commonwealth Serum Laboratories in Melbourne was visited by a man from the Bendigo district. He had heard of Dr Sutherland's funnelweb antivenom triumph. The visitor had had toes amputated because of the spread of an irreversible skin loss. Dr Sutherland confirmed that necrotising arachnidism was the likely cause. Then the man revealed that he had to go back into hospital — this time to have the lower part of the leg amputated.

Dr Sutherland had to tell his visitor that medical science offered no alternative, and that research into the syndrome had not even begun. 'What was worst,' he recalls, 'was that no one could tell the poor fellow when his suffering was going to end, or indeed if it ever could.'

The following year, at his home, the man wrote a note explaining his depression about the illness and the effect it was having on his family. Then he smashed his wife's skull with an iron bar and turned a shotgun on himself. Seven children were orphaned.□

Other spiders to avoid/1

Evidence against these is sketchy. None deserves persecution as a killer. But their formidable biting powers demand respect.

Much remains to be learnt about Australian spiders. Many types have not been scientifically studied at all. Even among the well-known ones, few have been 'milked' of their venoms for exacting and costly investigations of their toxic properties. Spiders shown here and overleaf are known or reputed to cause serious effects with their bites. It would be most surprising if there were not others, still to reveal themselves as new areas are settled and worked.☐

Below left **SAC SPIDER** *Miturga gilva*. Various species of *Miturga* (another is shown above) are suspected of having caused painful bites that have left weeping, ulcerating sores. They are found in bushland and behind beaches all over Australia, including Tasmania. Big and aggressive, they look much like wolf spiders but have distinguishing pale stripes or lines of dots under the abdomen. Silken bags are built as semi-permanent homes, close to the ground in vegetation or under logs and stones.

Left **BARKING** or **WHISTLING SPIDER** *Selenocosmia*. This genus includes Australia's biggest spiders, growing to more than 6 cm in length. Their names come from the noise they can make by stridulation — a rasping similar to that of crickets. Related American species are called bird-eating spiders. Ours pounce on frogs and lizards. Because they kill vertebrate animals of their own size very quickly, their venom is presumed to be capable of causing severe illness in people. Of perhaps six species in Australia, one is common in dry inland areas and another in the tropics.

Above **HUNTSMAN SPIDER** *Heteropoda*. Subtropical and tropical species are said to be feared by Aborigines in northwestern Australia. Bites reportedly cause vomiting, headache and prolonged pain.

Left **HUNTSMAN SPIDER** *Olios calligaster*. Most species of *Olios* are found clambering in the foliage of shrubs. They have humped backs and a raised stance — unlike the more usual flattened huntsmen that scuttle over tree trunks and walls. Most huntsman groups are harmless, but *Olios* reputedly causes illness with its bites.

Left **BRUSH-FOOTED TRAPDOOR SPIDER** *Idiommata blackwalli*. Species of this burrowing genus, up to 3 cm long, are found across inland Australia west of the Great Divide. Only *Idiommata blackwalli*, common in sandy country around Perth, is reported to have produced significant symptoms with its bites. It can be brown, black or grey.

Other spiders to avoid/2

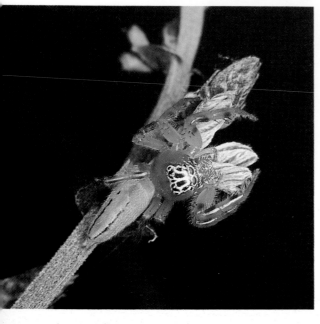

Above **JUMPING SPIDER** *Mopsus mormom*. Also known as *Mopsus penicillatus*, this tropical species has caused painful swellings. Its relatives, usually small, are found all over Australia. They are often seen on walls or plants, stalking prey and leaping on it. A brownish jumping spider causes trouble in Melbourne gardens. Bites remain sore for up to a week and may be responsible for skin damage.

Above **BLACK WISHBONE SPIDER** *Aname diversicolor*. Wishbone spiders take their name from the shape of their burrows. One of the two arms is partly closed, but available as an emergency escape route. This species, formerly known as *Dekana diversicolor*, is found from Eyre Peninsula, SA, to the west coast of WA. Its bites have been blamed for temporary symptoms of general poisoning.

Left **ADELAIDE TRAPDOOR SPIDER** *Blakistonia aurea*. Though it has not been conclusively blamed for serious bites, this species is regarded as a likely danger because of its readiness to attack and its commonness in suburban gardens and parks on the Adelaide Hills. Aggressive males also roam onto verandahs. The species is distributed across SA and into bordering districts of WA and Vic.

Left **GREY HOUSE SPIDER** *Achaearanea tepidariorum.* Distantly related to the redback, this common introduced species may be dark brown or grey. It has been responsible for painful local swellings, nausea and prolonged headaches. But these occurrences have been extremely rare, considering the spider's almost inevitable presence in most houses and other buildings.

Above **FIDDLEBACK SPIDER** *Loxosceles rufescens.* Worries about the fiddleback — named for the faint violin-shaped marking on its pale body — stem not from any known activity here but from the reputation of its relatives overseas. American members of the genus have caused tissue damage and destruction of red blood cells, leading to death from kidney failure. *Loxosceles rufescens,* which was discovered in Adelaide houses in the 1970s and then identified among earlier museum specimens, is reported from Israel to have much milder effects.

Above **SAC SPIDER** *Chiracanthium mordax.* This house-dwelling spider has an evil reputation acquired from its European relatives. In Australia its bites have been implicated in cases involving burning pain and swelling, nausea, faintness and headache. It hunts on walls, ceilings and floors by night, retreating during the day into a small silken envelope. This is usually located in the angle between a wall and the ceiling, or behind pictures or cupboards.

Above **BLACK HOUSE** or **WINDOW SPIDER** *Badumna insignis.* Dark and furry, with sometimes a bluish sheen, *Badumna* species take up permanent residence in sheltered exterior corners of almost any wooden structure. They build typical 'cobwebs' — lacy, tangled shawls. *Badumna insignis,* which prefers a dry climate, is the most common species in Adelaide and Perth. Its bites have been implicated in a few cases of fairly severe illness, involving intense pain, vomiting, dizziness and sweating. In moister regions of the east and southeast, the black house spider is more often *Badumna longiquus.* This species is smaller and presumably less likely to cause problems.

Left and above **SYDNEY BROWN TRAPDOOR SPIDER** *Misgolas rapax.* Burrows of this species, commonly found in gardens around Sydney, emerge through a wide tube of silk. In spite of the spider's name, the tube has no trapdoor lid, and can be mistaken for a funnelweb's nest. But the spider itself is much lighter in colour. Serious effects from the venom of a big specimen are supposed, but not proved. The *Misgolas* genus is distributed from Queensland to Tasmania.

Stinging caterpillars

Venomous spines and toxic hairs can inflict painful skin and eye disorders. Victims don't even have to touch some of the animals responsible.

To defend their soft bodies, the larvae of many moths and butterflies have spines or hairs that sting the mouths of predators. Often their armaments are advertised by gaudy colouring, so that birds and lizards learn not to eat caterpillars of a certain appearance.

In contact with human skin, the stingers of many caterpillars promote itching. Those of a few Australian species may cause considerable pain. Angry wheals may stay for days, and rashes may spread widely. None of these effects is permanent. Eye injuries are potentially more harmful, but fortunately less common.

The most common painful skin reactions are caused by caterpillars with spines that are fed by venom glands. In Australia the known types are moth larvae of the *Uraba* and *Doratifera* groups. Both are mainly eaters of eucalypt leaves.

Venomous spines are barbless, and do not stay in the skin. They are made with weakened tips that rupture under pressure, releasing fluids that contain histamine among other chemicals. Problems usually arise through contact with

---

CATERPILLAR ENVENOMATION

EARLY SIGNS Strong stinging sensation within seconds; painful wheal within minutes.

ACTION If pain is not relieved by calomine lotion or soothing creams available from chemists, obtain medical attention.

Below: The larva of Doratifera oxleyi, a common cup moth of the southeast, is often called a Chinese junk because of its raised parts fore and aft. The venomous bristle clumps are bright yellow.

---

live caterpillars. But moulted casings rubbed against the skin can also raise wheals.

Hairs of other troublesome caterpillars have no fluid venom supply but they contain substances that cause a form of dermatitis. On contact the barbed hairs adhere to skin. They disintegrate easily and fragments work their way in. Shed hairs have the same effect as growing ones — so no direct contact with the caterpillars is needed.

The irritant hairs of non-venomous caterpillars can cling to the skin or clothing of people brushing against trees or carrying firewood. They can be blown in the wind, or picked up in washing blown against trees. But they are most likely to be encountered in harmful quantities by sitting or lying at the bases of tree trunks.

Tender areas on bare arms and legs and around the neck are the most likely sites of rashes. The hands and feet, though they more often come into forceful contact with caterpillars, are usually too tough-skinned for the hairs to penetrate.

Right (top and bottom): Caterpillars of the gumleaf skeletoniser moth Uraba lugens *do not merely strip off the outer layers of eucalypt leaves, as their winged parents do, but chomp their way right through. Nature writer and photographer Densey Clyne has coined her own name for them — 'mad hatterpillars'. Instead of casting off their head capsules with the rest of their skins each time they moult they retain the whole succession of them, sticking up to deceive predators. The stinging hairs are short, in two rows along the back.*

NOT AS BAD AS IT LOOKS

A fearsome-looking 'weapon' cannot hurt you.

Interference with certain big caterpillars can trigger a disconcerting response. From a pocket just behind the head a bright red, Y-shaped organ suddenly emerges. Probing towards the source of the nuisance, it looks fearsome.

Called an osmeterium, this organ cannot hurt anything. But it is a weapon just the same. Its purpose is to give off an acrid odour that is highly repugnant to natural predators. Caterpillars equipped with osmeteria are species of the swallowtail butterflies *Papilio*, and not dangerous in any way.

The osmeterium of a swallowtail butterfly larva emits an odour that repels predators.

Finding out the hard way

Scientific knowledge of Australian caterpillar stings is largely the work of one man, Dr Ronald Southcott of Adelaide. In the 1960s and 70s he investigated and collated reports from many parts of the country, and subjected himself accidentally or deliberately to scores of stingings, carefully noting their effects.

The most severe skin effects are caused by the larvae of *Uraba lugens*, often called the gumleaf skeletoniser. Parent moths, small and drably grey, frequent taller eucalypts throughout the country. Eggs are laid twice a year. The two larval seasons, during which the voracious caterpillars moult many times and grow to more than 2 cm in length, run successively from January to August.

Dr Southcott's records include the case of a Sydney woman stung by one or two of these caterpillars on the leaves of a eucalypt in her garden. One sting was on her bare neck, the other below a shoulder blade — through her dress. She suffered pain and severe itching for four days, over a reddened, swollen area that covered her neck, shoulder and most of her back. The rash remained uncomfortable for a week and took a month to clear up completely.

Spectacularly decorated species of *Doratifera*, the other venomous genus, go by various popular names such as 'Chinese junk' and 'saddleback', for their shape, or 'spitfire', because of the way they suddenly erect bunches of stinging spines. Some may also be called cup moth caterpillars, from the appearance of the cocoons that they build.

The stout-bodied, reddish-brown moths are common in all coastal and inland river areas. In settled districts they often make pests of themselves by laying eggs in orchard or garden trees instead of eucalypts and brush box, the natural host plants. Again there are two generations a year, one going through its larval stage in summer and autumn, the other in spring.

Before retreating into cocoons for their transformation, the caterpillars may grow as long as 3 cm. Among the fleshy lumps along their backs and sides are some containing rosettes of spines, usually folded away. Some species have only four near the head; others have four more at the rear. If a caterpillar senses danger the spines are thrust out, looking like tiny sea anemones, and the body is arched to direct them towards the threat.

Wheals from the stings of *Doratifera* may be sore and itchy for many hours, and the marks remain visible for days. But there is no thicken-

The hairymary caterpillar Anthela nicothoe, *seen feeding on the needles of Monterey pine, weaves its blistering brown hairs — which may be 2-3 mm long — into its cocoon (right).*

Below: Irritant hairs from the caterpillar of the mistletoe browntail moth Euproctis edwardsi *often accumulate at the bases of tree trunks. But they can also blow in the wind.*

ing and hardening of the skin, which is often the case when people are stung by *Uraba lugens*.

Dermatitis from caterpillars

Larvae of tussock moths, especially the mistletoe browntail moth *Euproctis edwardsi*, cause distressing summer rashes in communities throughout the southeast. *Euproctis* infests the parasitic mistletoe that forms garlands in eucalypts and melaleucas. Its toxic spines, readily shed, collect on bark or at the foot of a tree, and are light enough to be carried by breezes.

Because the spines work their way through clothing, reactions can occur on any part of the body. Along with itching, pimply rashes, some sensitised people suffer a swelling of the face, neck and hands.

The night-flying adult moths are yellowish-

NON-VENOMOUS DERMATITIS FROM CATERPILLARS

EARLY SIGNS Itching or slight stinging sensation within minutes; pimply pink rash developing in hours or days.

ACTION Remove hairs with fine tweezers under a magnifying glass if possible, or place transparent adhesive tape gently over them and lift off. Apply soothing lotion or cream. Obtain medical attention if discomfort persists.

NOTE If eye irritation also occurs and is not relieved by bathing, obtain medical attention immediately.

brown or fawn. Caterpillars, growing as long as 4 cm, are chestnut with a prominent white stripe along the back. Very long hairs extend all around the body, but the detachable stinging spines are tiny. They emerge in golden clusters on the back, from lumps just behind and ahead of the stripe.

Another common source of skin irritation — said by some people to be worse than the effects of stinging trees or nettles — is the furry processionary caterpillar *Ochrogaster contraria*. It infests acacias. Thousands of them may be on one tree, feeding and sheltering communally.

Travelling up or down a tree, processionary caterpillars move head-to-tail. They follow the same trails, marked by strands of silk, day after day. In coastal regions their shelter is a tent-like rigging of silk at the base of a tree. Inland they build a spherical pocket among high branches, and are called bagmoths or bag-shelter moths.

When bristles get in our eyes

Hairs or spines of many caterpillars cause intense irritation if they are blown into people's eyes or transferred there from the hands. Most trouble seems to come in summer from the hairymary *Anthela nicothoe*, which feeds on acacias. It has long, stiff, barbed bristles that are pointed at both ends.

Though they can be maddeningly painful, causing a constant streaming of tears and an aversion to light, injuries are usually slight and superficial. Bristles lodge under the eyelids and scratch the surface of the cornea with each blink. But sometimes they can be embedded deep in the eye and require specialist attention. In one case, which called for repeated surgery and led to a permanent impairment of vision, a caterpillar or its moulted casing was blown into a man's eye by an electric fan.

Bathing with an eye lotion may flush out offending hairs or bristles. If severe irritation persists, medical attention should be sought.☐

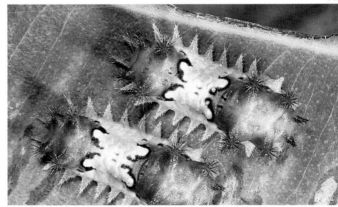

Chinese junk caterpillars, above and right, of the cup moth species Doratifera vulnerans *show different colouring at different stages of development. Caterpillars moult many times during their months of growth. Zoologists call their progressively larger forms 'instars'.*

Above and left: Processionary caterpillars Ochrogaster contraria *form head-to-tail chains when they are on the move.*

Scorpions

Abundant though seldom seen, some are as big as any in the world. We can count it a blessing that their venoms are weak.

Only two human deaths are blamed on scorpions in Australia, and even those reports are not convincingly detailed. Both victims were baby girls — one at Pemberton, WA, in 1929, the other at around the same time at Smithton, Tas. The Tasmanian child is said to have been stung twice, on successive days.

In both cases the scorpions were forest types, and among the smallest of about 30 species found in this country. It is doubtful whether either one could have exceeded 3 cm in length. But among the inland species are some that are four or five times bigger. The record goes to an unidentified species of *Urodacus* from the Flinders Ranges, SA. It measured 16 cm — as big as any scorpion in the world.

A similar type, nearly as large, occurs in the east of the state, close to where it borders NSW and Victoria. Though much of the region is arid it is not unfrequented, for it is crossed by the Murray River and the Sturt Highway and dotted with old-established river towns. Several soldiers in training near Barmera in 1942 were stung, and suffered pain and illness for a day or so. But it is apparent that these scorpions, in spite of their size, pose little if any threat to life.

Scorpions abound throughout the country, in all climatic conditions. But stingings are relatively uncommon. The animals occasionally wander into houses, just as hunting spiders may. They are mostly furtive and sedentary, however, hiding under bark, leaf litter, stones and logs, or burrowing deeply in arid regions. Stingings are never acts of aggression — simply retaliations when scorpions are accidentally squeezed by someone's hand or foot or body.

The worst penalty is pain

Burning or throbbing pain is sometimes the consequence. It may last for hours and if a hand, say, is stung the whole arm may be affected. There is generally also some swelling but further symptoms indicating a wider poisoning of the system are rare. There is no specific treatment. If common painkillers bring no relief, medical help may be needed.

A general human dread of scorpions, rooted in mythology and the religious rites of many ancient cultures, originated around the Mediterranean and in tropical America. There, some of the creatures are clearly more venomous than any Australian species and thousands of people are killed every year.

From North African experience in particular, a misconception has arisen that scorpions are characteristically animals to be found on desert

---

ADOPTING AN ATTITUDE
Earwigs mimic scorpions as a means of defence.

Earwigs have nippers, but no venom.

Harmless earwigs, common in cooler parts of the country, raise their tails like scorpions. The superficial resemblance is no mere co-incidence. It is an example of mimicry as a defence, found frequently among insects.

Earwigs are leaf-litter dwellers, along with some scorpions. They have the same predators — birds, lizards and small mammals. To take advantage of the fact that predators learn to leave scorpions alone, the earwigs have developed a similar-looking stance.

The pincers at the end of a big earwig's tail can give people a painful nip, but they are seldom able to pierce skin. And it is not true that earwigs are much given to crawling into ears — let alone that they bore into the brain and lay eggs there, as an old horror story had it. In the unlikely event that an earwig were to enter an ear and cause irritation, a simple syringing would flush it out.

Newborn scorpions, carried on the back of their mother, do not start to take on the colouring of their species until after their first moult.

sands, revelling in scorching heat. In fact all are largely nocturnal, and while some species do thrive in hot deserts, they burrow as far as 2 metres under the surface to stay cool. Many more species live in forests, woodlands and grasslands, often close to our major cities.

A scorpion differs from a spider, its fellow arachnid, in some obvious ways. Its palps, the paired appendages carried ahead of the eight legs, are enlarged into heavy, pincer-armed claws. Its body is segmented, it has no waist between thorax and abdomen, and the abdomen is extended into a narrow, flexible tail called a telson.

The telson is of course tipped with the sting, which is fed by ducts from two venom glands. It can not only be thrown forward over the scorpion's back but also swung sideways.

Scorpions feed almost exclusively on live insects and spiders. The sting is used only on prey capable of putting up a violent struggle — usually the claws are sufficient to hold a victim while it is torn apart by the biting parts at the front of the head and the juices are sucked out.

Keeping a mate at arm's length

The mating dance of scorpions is a frequent delight of nature film makers. When a sexually mature male finds a female he engages her claw pincers with his own and they begin a series of movements, back and forth and side to side, that can go on for an hour. Sometimes their tails intertwine. Eventually the male drops sperm on smooth ground and manoeuvres his partner into the precise position for it to enter her genital opening.

Scorpions are born live — another major difference from spiders — in litters of a dozen or so. The newborn, pure white, are carried on their mother's back for a few days until they undergo their first moult. Then they take on the colour of their species, which in Australia varies from golden through grey to almost black.☐

Variation between body and leg colours, as in this species of Urodacus, *is common. The huge claws are sufficient to deal with most prey — the sting, though often swung up as a threat, is rarely used.*

ONE WAY OF HANDLING THE PROBLEM

If you can't avoid a scorpion, follow these simple rules.

A scorpion causing worries inside a house or on clothing is best removed with the aid of tweezers or tongs, if they are handy. If not, and the intruder is standing still with its tail up, there is a safe way to handle it.

From behind the scorpion, place a thumb and forefinger slightly apart at each side of the tip of the tail. Close on it briskly and firmly and lift the animal up. Be prepared for it to arch its body and attempt to use its claw pincers. They cannot damage the tough skin of your fingers, so don't let go.

Take the scorpion where it will do no harm and set it down on its legs. If it is holding on with its pincers, use a twig or some other small object to distract it. Then quickly let go and step away. The scorpion won't chase you.

Centipedes

They have legs to spare — and a pair in front that work as fangs. The venom of the bigger types brings excruciating pain.

Centipedes longer than 12 cm are not commonly found in Australia. But much greater sizes are claimed in some bite reports. One attacker at Mt Mulligan, inland from Cairns, Qld, was said to be about 19 cm long. Another in inner Sydney, blamed for killing a full-grown bull terrier, reportedly measured 15 cm.

The potential danger of centipedes seems directly related to their size, for their venom is not especially toxic. Though people are said to have been killed in India and South America, the world's only documented case is that of a Filipino child, bitten on the head. But there is plenty of evidence that big centipedes are capable of inflicting enormous pain, sometimes with illness lasting for days.

Prolonged swelling and soreness of lymph nodes has featured in some Australian cases. In others there has been nausea for up to a week. The most curious recent case occurred in 1976 in Melbourne, where a fit young man lost consciousness 10 minutes after a bite on the thumb. His companions could detect no pulse or breath, but the victim revived on the way to hospital.

Doctors have no specific way of treating centipede bites, beyond the use of drugs to combat pain and inflammation. The discomfort of milder stingings may be eased by common painkillers. Dabbing on ammonia, heavily diluted to avoid blistering, is sometimes recommended. In any case it is important to guard against bacterial infection, especially tetanus.

Feet galore — but never a hundred

Centipedes form their own zoological class — separate from insects. They are called chilopoda, meaning claw-footed. Each segment of their bodies carries a pair of legs tipped with claws. The last pair, turned backward and upward, are modified for grasping and fighting, and the front two for injecting venom into prey. The number of pairs of walking legs varies between species, from 15 to 177. It is always an odd

Found in rainforest in the Kuranda Range near Cairns, Qld, this species of Ethmostigmus centipede measured 12.5 cm.

Ethmostigmus rubripes, one of which is said to have killed a four-year-old bull terrier in Sydney.

23 pairs of legs and are usually yellow to brown in colour. These are the types that reach the greatest size, and presumably present the greatest danger. In Brazil, species of the same genus easily exceed 25 cm in length, with the record standing at 30.5 cm.☐

Few worries from millipedes

Some squirt toxic fluids — but the worst effect is the smell.

Millipedes are vegetarian. They have no need of venom fangs. But to discourage predators, many species have glands along their sides from which they can squirt toxic fluids. Chemicals secreted by some big tropical types overseas are intensely irritating on people's skin and damaging in the mouth and eyes. They are said to be able to blind birds.

No serious effects of this nature have been reported in Australia. At worst the fluids from some of our millipedes give off an offensive smell. Still, if a big millipede of the far north is touched, it could be inviting trouble not to wash the hands before putting them near the eyes or mouth.

Smaller millipedes, of Portuguese origin, are common in the southeast and create a nuisance in South Australia. Populations exploding on the Eyre Peninsula and in the Mount Lofty Ranges and Adelaide Hills turn to eating crops instead of decaying plant matter, and often infest houses.

The only apparent danger of these plagues is from the germs and fungal organisms that the millipedes could bring in on their bodies. Researchers have found that the invaders are attracted at night by light. The problem is largely solved, they suggest, by drawing curtains and leaving outside lights off.

Though superficially similar, millipedes have evolved quite differently from centipedes. Their bodies are cylindrical rather than flattened. Each two segments are fused together, so to a layman's eye they appear to have two pairs of legs to a segment, rather than one. A clear distinguishing feature of many millipedes is their habit of curling up like a catherine wheel when they are alarmed.☐

Cormoscephalus is another member of the scolopendrid family, to which all our biggest and perhaps dangerous centipedes belong.

number, so the total of individual legs could be 98 or 102, but never the even hundred that inspired the animals' name.

Walking is a complicated sequence of leg movements in which only one claw in eight has purchase at any one time. The loss of a few legs is no great handicap — some species throw them off, wriggling, as a distraction when they are threatened by predators.

Centipedes are carnivorous. Most Australian species favour a diet of insects. But slugs and worms are sometimes taken, and plants may be damaged if a heavy population of centipedes is going hungry. They feed at night, taking refuge by day in dark, damp places under leaves, bark, stones or logs, or in cracks in the ground.

Hairy 'feather' centipedes of the genus *Scutigera*, small and fast-moving, frequently enter houses. Their bite is said to produce a burning sensation — likened overseas to a wasp sting — but serious effects have not been noted in Australia.

The important group of big, hairless centipedes is typified by the genus *Scolopendra*. Its members, thick-bodied and shiny, have 21 or

CENTIPEDE ENVENOMATION

ACTION Disinfect punctured area. Obtain medical attention if common painkillers bring no relief — or if victim has not had an antitetanus injection within the previous 10 years.

A cylindrical body distinguishes a millipede.

FOR THE COMMON GOOD

Insects are the most persistent order of animals that the world has known, and probably ever will know. But they suffer one severe limitation — they can never attain great size. Their breathing system can carry oxygen only over tiny distances. On any grander scale it would not work.

Certain insects overcome this drawback by compulsory co-operation. In their societies no individual is biologically complete and self-sufficient. Each can only be part of the whole. The result is a super-organism — virtually unlimited in size.

Two groups of insects have perfected this form of social integration. One is the termites. Their destructiveness is a costly nuisance, but they offer no menace to human health or safety. The other group, called hymenoptera, contains species that inflict pain and death throughout the world. These are the honey bees, the social wasps and the stinging ants.

Defence of the community is but one of the activities in which hymenoptera have no choice. Those equipped for it will sting anything that they perceive as a threat, regardless of how big it is. The fact that they endanger themselves — in the case of bees, for instance, any attack is suicidal — counts for nothing. The greater good of their society is served.

CASTE SYSTEMS REGULATED BY CHEMICALS

Countless insects use glandular secretions, peculiar to their own kind, to communicate with and influence other members of their species. These signalling chemicals, called pheromones, may dissuade predators from killing one another. Very often they work as calls to mating.

In the formal societies set up by ants and by some bees and wasps, pheromones are even more important. Conveyed usually in exchanges of food, they reinforce every aspect of co-operative behaviour. And they allow the

Workers of the native paper nest wasp species Polistes tasmaniensis *tending eggs in the cells of their bell-shaped nest. Native wasp nests are small, and hang suspended by a narrow stem. They have no covering, and cell openings are visible from below, as in this photograph.*

detection of any imbalance in the social structure — a shortage of workers, for example — and trigger the means of correcting it. So they predetermine each member's role in life, its place in the caste system.

Bees, wasps and ants are classed together as hymenoptera, meaning membranous-winged. The kinship between bees and wasps is easy enough to see. In fact zoologists regard bees simply as descendants of a kind of wasp that took to feeding its young on plant pollen and honey instead of insects and spiders.

The link with ants is less obvious, especially since we see few of them in their winged form. But there is a similarity in their narrow-waisted structure and many other anatomical details. More importantly, in relation to aggressive behaviour, there are striking parallels in the social organisation of all three groups.

Only the females sting
In all societies of hymenoptera, pheromones keep egg-laying females informed of the population balance. An automatic response determines how many eggs are fertilised. Males hatch from unfertilised eggs and become 'drones' whose only purpose is mating. Females come from fertilised eggs. But except for a few who are destined to become egg-laying 'queens', their sexual development is arrested by other chemicals that are fed to them.

These neutered females become workers — the community's food gatherers, builders and nursemaids to new generations of larvae. Their secondary reproductive glands produce venom, and what could have been egg-laying organs develop as stings. So they also take charge of collective security. In certain species of ants, some neuters with oversized heads and jaws form a special 'soldier' caste. Whatever their type the guardians of hymenoptera societies are incapable of fear. And in response to danger,

Worker honey bees Apis mellifera *attack an alien bee which has entered their hive to steal honey.*

Worker honey bees cluster around the queen (marked by the beekeeper with a yellow spot). The exchange of food and secretions between queen and workers ensures the continuity of the hive caste system.

the only defence they know is all-out attack.

The larvae of all social hymenoptera are helpless grubs, resembling fly maggots but immobile. Communal life is centred on the construction of cells into which eggs are laid, and in which the larvae are fed and raised or food for them is stockpiled. The nests and mounds that various species build are all in effect continuous nurseries, expanding as much as food sources and building materials allow.

Choices about founding new colonies, and raising other females to become egg producers, are made instinctively by the queens. After mating flights, when they take wing with retinues of drone males, they may return to enlarge an existing nursery or find a fresh site. Migrating queen bees are accompanied by workers, creating the unpredictable and often dangerous phenomenon of swarming. □

Wax combs are constructed vertically by worker bees. A chain of bees forms, and those at the top secrete the wax to form the comb. The comb will eventually reach the bottom of the frame.

These bulldog ants, Myrmecia brevinoda, *can regurgitate food to each other.*

Sting allergies

*Unless bees or wasps are swallowed,
most of us risk no more than passing pain
from stings. But for an unlucky
few, any stinging can spell death.*

For about 90 per cent of the population an occasional sting by a honey bee or a wasp, or the less common attacks of certain ants, will provoke quick pain and swelling, and then some itchiness. The effects may last for hours, but there is no risk to health.

The greatest danger to most of us is the bee or wasp that crawls unnoticed into a container of sugary liquid — typically a can of soft drink or beer. If the insect is swallowed with a mouthful of drink and stings the root of the tongue or the back of the throat, the swelling can cut off our air supply within seconds.

These are normal responses to a group of venom components, such as histamine, that are chiefly inflammatory in their effect. Venoms of various insects in the hymenoptera group also contain enzymes and acidic compounds. They are not especially toxic and on most of us they have no significant effect.

But to an unfortunate minority, the chemicals in this second group are allergens. They have the potential to create a state in which victims are primed to over-react. A person who becomes sensitised in this way is constantly at risk. A sting anywhere on the body provokes an excessive response — sometimes a life-threatening one.

Victims are unaware at first

In a potentially allergic person the first sting produces only the normal pain and local reaction. Invisibly, however, the immune system manufactures an oversupply of antibodies

Above: A close-up of the sting of a European wasp. Allergic reactions to stingings are becoming more common as the wasps spread. Left: Wasps are attracted to sugary liquid; check all food and drink left outdoors before you consume it. European and English wasps are about the same size as bees. Similarly marked native wasps are bigger.

against the allergenic chemicals. Their production may continue for years. Stung too many more times by the same kind of insect, the victim is in serious trouble.

After a second sting, perhaps months later, the pain is the same but the local reaction is likely to be much greater. The area of swelling may extend for 10 cm or more around the site of the sting. On a third or later occasion an itchy rash may appear on the body. The victim may have a struggle to breathe, because of a moderate attack of asthma or because tissue at the back of the throat swells up.

Next time such a person is stung, there may be a severe general reaction within a few minutes. Urgent and extensive medical care may be required. At worst the victim suffers a condition called anaphylaxis, meaning lack of protection. A form of clinical shock, it is one of the gravest medical emergencies.

Anaphylaxis from insect stings leads rapidly to collapse, a fall in blood pressure and possible death. Doctors are taught to deal with it by administering adrenalin immediately and ensuring that the patient's air intake and circulation are maintained.

SA has the biggest problem

Bee stings kill one or two Australians every year, on average. Thousands more allergy sufferers are under lifelong medical treatment to guard against anaphylaxis. The highest per capita allergy and mortality rates are recorded in South Australia, the leading honey-producing state. It has the greatest number of bees in relation to human population, and its rainfall patterns restrict apiaries to districts in or close to urban areas.

Many other people suffer allergy to ant and native wasp stings — often through their leisure activities, because the troublesome species are more numerous in bushland.

Cases of severe allergy to the venom of the rapidly spreading European wasp have begun to show up in Melbourne, and doctors are in no doubt that they will soon be encountered over a wider area. Some believe that in the southeast at least, these aggressive wasps could become the greatest insect menace of all.

Laboratory procedures developed in the 1970s brought significant advances in the understanding of insect allergies. The most important technique in regular use allows the measurement of the level of antibodies circulating in a patient's serum.

In the case of antibodies to bee venom, there are two kinds. One type, which can be called allergy antibodies, are present at high levels in the serum of an allergy sufferer. But if the serum of a typical bee-keeper is examined, the other side of the coin is seen. He has only small amounts of allergy antibodies, but a high level of the second kind, called blocking antibodies. These combine with bee venom without triggering any other reaction. Stung week in and week out, the bee-keeper has developed an immunity. And he hardly feels the stings.

Skin tests using minute amounts of venom, along with serum examination, allow a doctor to gauge the sensitivity of a person with a history of bad reactions to stings, and to assess the patient's likely response to treatment.

Above: Honey bees have hairy, plump bodies with less conspicuous banded markings than wasps. Some people have an allergic reaction to their stings. A severe allergy can be life-threatening.

Below: Cases of allergic reaction to stingings by bulldog ants of the Myrmecia *species, found Australia-wide, are mounting as more people venture into bushland areas.*

On the wrong track
Courses of desensitising injections have been offered since the 1920s. For over 50 years they had varied results. It appears now that the successes may have been illusory — for scientists have shown that allergy in some people fades with time. Milder reactions to stings, years after treatment, may not have represented cures

at all, but cases of natural recovery. For all those years, in fact, inadequate injections were given. Attempts at immunotherapy were based on a belief that the allergenic substances in the venoms were also contained in the insects' bodies. Early researchers were confused by another sort of allergy, to inhaled particles. So bees, wasps and ants were ground up to prepare extracts. Now it is known that venom alone is

BEE, WASP AND ANT ENVENOMATION

1. NORMAL RESPONSE

EARLY SIGNS Pain; swelling limited to site of bite.
ACTION If a bee sting, scrape the sting away without squeezing it — DON'T pull it out. Apply a bag of ice and water to relieve pain — but NOT ice directly on skin.

2. SENSITISED RESPONSE

EARLY SIGNS Wider swelling; rash; swollen eyelids, lips.
ACTION As above. Then consult a doctor with sting allergy experience or ask for referral to a specialist.

3. EXTREME RESPONSE

EARLY SIGNS Laboured, noisy breathing; general distress; collapse.
ACTION If a bee sting, remove sting urgently by scraping. For stings on limbs, apply pressure/immobilisation first aid as for snakebite (page 360). Administer emergency drugs if victim is carrying them. Obtain urgent medical attention. Monitor victim's breathing and pulse and be prepared to give mouth-to-mouth resuscitation or heart massage (page 356).

4. STING INSIDE THROAT

EARLY SIGNS As for No 3 above.
ACTION Obtain urgent medical attention. Keep victim still and assist breathing if possible.

much more effective.

The new immunotherapy is invariably successful unless patients have certain other health problems. Sufferers of life-threatening allergies are injected at regular intervals with tiny but increasing amounts of venom. They experience no more than mild local reactions. Their progress towards a useful degree of immunity can be monitored by blood tests, showing falling levels of specific antibodies and an increase in the blocking type, as well as by skin tests.

At the end of a successful course, a patient can tolerate as much venom as would be delivered by two bees, stinging simultaneously. Suitable patients are treated in only six weeks, though some take three months or more. All are advised to return for follow-up doses every few weeks to maintain their protection.

Purified bee and wasp venoms are imported, packaged for individual therapy. Venoms from the dangerous Australian ants are not available, however — nor is there a laboratory procedure to detect antibodies. Only the old-fashioned whole-body extracts, of doubtful effectiveness, are marketed for treating ant sting allergies.□

Bee-keepers are often stung despite wearing protective clothing and subduing guard bees with smoke as they open a hive.

BEE-KEEPERS SELDOM REACT TO STINGS

Apiarists acquire a natural immunity — but their families may suffer.

Any bee-keeper developing a sting allergy would quickly have to find another activity. But most apiarists acquire a natural immunity. Frequent stingings are felt as mere pin-pricks, and there is no swelling. If for some reason they do not work with their bees for a few months, the first stingings when they resume can produce marked local reactions.

Professional apiarists keep their hives as far as possible from their homes. But their families, even if never stung, can suffer other allergies. Wives who look after their husband's working clothes are especially prone, probably through inhaling particles.

Keepers themselves may become allergic to beeswax and suffer a form of dermatitis when they handle honeycombs. And they or their families often develop a food allergy to honey. Acute indigestion prevents them from ever enjoying their own product.

How we attract bees and wasps

Changing habits can reduce the risk of stingings.

Some people are clearly more attractive than others to wandering bees and wasps. When people gather outdoors it is these individuals, and seldom their companions, who seem inevitably to have one of the insects buzzing round them and alighting on their skin or hair.

No one is certain why, though it is a safe assumption that body secretions exuded in sweat or perhaps the breath have something to do with it. A particular odour may simply be appetising to the insects; possibly, some of us accidentally imitate a pheromone, one of the chemical signals by which they communicate.

People whose outdoor activities are marred in spring and summer by the attentions of bees and wasps may benefit from more frequent washing — but not with flower-scented soaps — or a dip in a swimming pool. Insect repellants are worth trying. Antiperspirant preparations, applied in the usual places, may be unhelpful in hot weather because sweating on exposed parts of the body may increase.

Other steps can be taken to avoid encouraging bees and wasps. For most of us they will at least reduce the likelihood of nuisance. For untreated allergy sufferers, they could avert a possible medical emergency.

Foods — especially jams and pungent fruits — should not be left uncovered out of doors. Nor should juices, soft drinks or beer. Wasps, though they feed their young on other insects and spiders, are drawn to sweet things just as much as bees are.

Perfumes and hair sprays are often attractive to bees and wasps. Brightly coloured clothing can lure bees, in particular, over long distances. So will flowering clover. For a household in which there is an allergy sufferer, nothing could be worse than having clover growing round a swimming pool.□

Putting an end to terror

Treatment for sting allergies brings peace of mind — but it's not for everyone.

Michael Keck was only four when he showed his first severe over-reaction to a bee sting. Immunotherapy was attempted with bee body extract — a method since discredited. Stung again a year or so later, Michael collapsed and came close to death from anaphylactic shock. His young parents despaired of being able to protect him and still let him develop as a normally active little boy.

Happily for the Keck family, immunotherapy with bee venom had just been introduced. Michael completed his initial course and had to face a test with a live bee. This deliberate sting challenge, as it is called, checks on the success of the treatment under controlled conditions. Many people have a stronger local response to natural venom than they have to the freeze-dried preparation used for injections. A team of medical workers stood ready with resuscitation equipment in case Michael should have a dangerous reaction. But even before the bee touched his skin, he fainted from fright.

'Sheer terror like that is common,' says Dr Sheryl van Nunen, the Sydney allergist who treats Michael. 'Tough men can go to pieces just at the sight of a bee near them. But afterwards, when they see how their reaction has been reduced, their whole outlook on life changes. The fear is over. Even if the sting challenge weren't medically important, the peace of mind it brings is priceless.'

Dr van Nunen — a sting allergy sufferer herself — does not recommend immunotherapy in every case. She believes the treatment should be reserved for people whose lives are shown to be at risk, or those who could be prevented from earning their living by disabling reactions to stings. If someone who suffers only a bad local reaction is given immunotherapy, there is roughly a 10 per cent chance that their sensitivity will be increased.

'As far as we know at present,' the doctor says, 'once people start on immunotherapy they will have to go on getting maintenance treatment for the rest of their lives. The doses of imported venom are costly to the health service, though not to patients, who get them as a pharmaceutical benefit. And apart from anything else, the injections are simply no fun.'

In spite of that, Michael Keck turns up cheerfully enough for his booster every six weeks. He will never forget how scared he used to be, though years have passed since his life was in danger. Now he can even make a wry joke about the swarms of bees that seem to plague his school.□

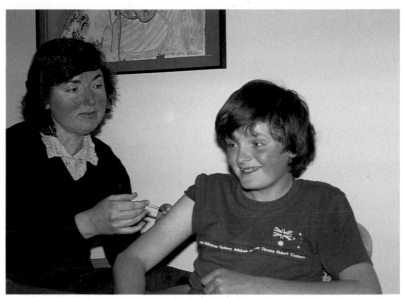

Dr Sheryl van Nunen prepares to give Michael Keck his regular booster shot of bee venom, an essential and lifelong part of immunotherapy.

HOW BEE VENOM IS COLLECTED

Supplies of bee venom to test and treat sting allergy sufferers are collected by an ingenious method that keeps the contributor bees intact. A shallow glass container, placed on the landing board in front of a hive, is covered with perforated plastic that is strung with fine wires. Bees alighting receive a slight jolt from an electric charge pulsing through the wires, and automatically retaliate. The stings discharge through the holes in the plastic, so they are not held and torn off when the bees move on into the hive.

Bitter harvest

A farmer could not reach his lifesaving pills.

Only 27, South Australian grain grower Ian McIntosh had been stung by what he called 'native' bees on seven different occasions. Each time his reaction was worse, until one attack left him unconscious for half an hour.

Courses of reliable immunotherapy were not available in the 1970s. McIntosh's doctor prescribed a supply of pills, which the farmer had to carry with him at all times. Their main purpose was to dampen the asthmatic effects of an allergic reaction.

Harvesting late in November 1975, McIntosh had the pills in the cabin of his tractor. He was stung on the back of the neck, and died before he could get the pill container open. Assuming he had not been away from the tractor when he was stung, he must have been overtaken by anaphylactic shock with terrifying speed.□

'Is he insured?'

Two later stings prove that immunotherapy works.

Rick Pointon is one of those people whom bees seem to pick out to pester. As a child, the son of a golf greenkeeper, he was always being stung. But it wasn't until he was 33, working as a radio executive in Newcastle, NSW, that he discovered he had developed a severe allergy.

Stung while gardening, he later began to itch all over. He dived into his swimming pool but pressure in his ears had an agonising effect. Back in his house, he started to feel 'woozy'. Only then did he and his wife think of the bee sting as a possible reason.

A doctor was telephoned. He told the Pointons to hurry to the nearest ambulance station. Rick was close to collapse by the time they reached it, and under treatment for the rest of the day.

When the doctor's suspicion of bee venom allergy was confirmed, Mrs Pointon asked what could be done about it. The advice she got was tactless but practical: make sure her husband was fully insured and had all his affairs in order.

That was in 1980. Rick Pointon received a course of immunotherapy with outmoded and probably ineffectual whole-body bee extract, but later he was one of the first people to be immunised with pure venom, by Dr David Sutherland at Royal Newcastle Hospital.

The bees still won't leave him alone. He has been stung twice since his treatment — but the worst effect was a headache. 'I don't even think about the allergy any more,' he says. 'It's just a matter of making a note in my diary once a month, to make sure I get my booster shots.'□

'Now it's just a note in my diary once a month.'

The honey bee

*Raising the young is everything. Selfless workers give up their lives in defence.
For drones, even the sex act is suicidal.*

Bee communities, husbanded since prehistoric times, have been taken wherever Europeans and Asians have migrated. Pollenating crop and pasture plants, or directly providing high-energy food, the honey or hive bee *Apis mellifera* has been invaluable.

It is generally a docile creature, seldom stinging unless it is trodden on or caught in clothing — or unless its nest is interfered with. But where bees are raised in hives in populated districts, alarming situations can arise if wayward swarms try to found new colonies in gardens or buildings. Workers accompanying a queen may be angered by misguided attempts to get rid of them. Removal of a swarm calls for the skills of an expert apiarist.

Feral populations, descended from escaping swarms, abound in wooded country and heathlands. People are mainly at risk through accidentally disturbing the nests. In the wild these are built in tree hollows, rock crevices and holes in banks. In daylight a steady coming and going of bees should be warning enough that a nest is nearby.

No native bees of Australia cause concern. Most are tiny and solitary. They burrow in soil or bore into plant stems. Species of *Trigona*, a small, dark type common in warmer regions, are socially organised in a way that could have spelt trouble. But they do not have stings.

A self-imposed death sentence

Honey bees in many countries cause more human deaths than all other venomous creatures put together. So it is fortunate that each worker bee can sting only once. A barb prevents the sting from being withdrawn. The bee twists

Left: honey bee pupae. Like butterflies and moths, bees develop from egg to larva to pupa to adult. Below: An adult worker honey bee collecting pollen from a flower.

its body and wrenches itself free, leaving the sting and venom sac behind. Abdominal injuries kill the insect within a day or two.

The queen alone has an unbarbed sting that she can use more than once. But like the drone males, she lacks specialised feeding parts and leg baskets for collecting nectar and pollen. Only the worker caste of sterile females can obtain food for the community. Workers also secrete wax to build the vertical combs of cells, for storing food and nursing larvae, that form the nest.

On the queen's nuptial flights, stingless males swarm in a contest to mate with her — their only purpose in life. The winner is rewarded by having his genital organs torn off, and soon dies. But his sperm is stored by the queen in a special gland and doled out as eggs are produced. Laying as many as 1000 eggs a day, each into its own cell, the queen controls the future population balance by fertilising only those that are to be female.

Any larva hatching from a fertilised egg can become a queen bee — but only if it is fed for its whole life with a special substance secreted by the workers. This 'royal jelly' is given to all larvae for three days, but after that to only a few in oversized 'queen cells'. The others, destined to be workers, are switched to a mixture of pollen and honey. When they mature and start feeding the queen her royal jelly, they seek a chemical secretion from her in return. It keeps a continued rein on their sexual development.

Expanding the empire

Sometimes, after a mating flight, an aging queen bee does not return to her nest but leads a swarm of workers to find a site for a new colony. Less often it is daughter queens that emigrate after their first mating. Where daughters are left to carry on a nest, the first action of the strongest one is to kill all her rivals.

Exchanges of food and secretions continually

When bees swarm, the mass of workers surrounding the queen settles temporarily on a convenient branch or fence post while scout bees search for a new homesite.

This swarm made a hive in the wall cavity of a weatherboard house.

WHAT TO DO IF A SWARM ARRIVES

A bee swarm in your house or yard demands expert advice and help.

Bees swarming on your property are unlikely to present any danger unless the situation is badly handled. Don't try to drive them away — they could retaliate. On the other hand, don't simply hope that they'll go away of their own accord. The longer they stay around the more likely they are to find a refuge, such as a ventilator, that could make their removal awkward.

Contact the local office of your state or territory agriculture department. From a register of voluntary bee collectors, you will be given the telephone numbers of one or two near you. Police stations in some areas may also hold a 'trouble-shooter' list. Properly equipped and clad, an expert will quickly trap the queen, bring the swarm under control, and take all the bees away.

reinforce the co-operative instinct of honey bees. One of the less obvious benefits is an unusual ability to combat extremes of temperature. In very cold weather bees cluster together to conserve body warmth, with the queen in their midst. When it is excessively hot, however, they stay well apart and agitate their wings to move air across the nursing cells — and workers bring in water to cool the nest by evaporation.☐

WHEN A BEE STINGS
Venom goes on pumping when the bee has gone.

A worker honey bee's sting consists of two barbed darts, grooved so that they can slide against each other, contained in a sheath. When the darts catch in flesh the bee can escape but the sting and the rest of the venom apparatus are torn out of the hind end of its abdomen. This wound eventually kills the bee.

The venom sac, meanwhile, goes on pulsing after the bee has escaped. Muscles controlling the movement of the darts drive them deeper into the flesh, and continue to pump venom until the sac is empty. So a bee sting should be removed as quickly as possible by scraping it away — not by pulling it out, because squeezing the sac will force in more venom.

A bee sting remains in the victim's flesh after the bee has torn free (above); valves continue to pump venom into the flesh until the sac is empty. A close-up of the sting (below) shows the barbed darts, bulb with venom, and venom sac.

The wasp invasion

*Australia's own kind are trouble enough.
Now aggressive interlopers,
revelling in a climate that's too kind,
multiply relentlessly.*

Wasp problems are nothing new in Australia. Scores of native species inflict agonising stings. Some people are stung often enough to develop dangerous allergies. But the numbers severely affected have been small, compared with those living in fear of bees. Deaths from wasp stings have been rare occurrences.

That picture seems bound to change. Wasps

of northern hemisphere origin are establishing themselves in fertile regions throughout the temperate zone of Australia. And they thrive as they never could in their homelands. In the 1980s they have defeated costly government efforts to contain them.

Wasps are much like bees in their anatomy and in most other fundamental ways. The types that are socially organised have a similar caste system. Sterile females form a working and fighting class to gather food, tend the young and defend the nest, which is governed through chemical messages from one breeding queen. Drone males are raised for the sole purpose of mating with her.

Adult wasps eat almost any protein or sugary food they can find, including the nectar that bees eat. But because the larvae in the nest are fed on insects — mostly caterpillars — and spiders, the worker wasps do not regurgitate nectar for storage in the form of honey. So they have never been sought after by humans, and made docile by thousands of generations of husbandry. Wasps in defence of a nest are fiercely aggressive — and, unlike bees, they can sting as many times as they wish.

Free from natural limits
The greater menace of wasps is counterbalanced — in normal conditions — by a natural restraint on their population growth. At the onset of winter a mated queen is supposed to go into

A European wasp Vespula germanica *emerging from its nest entrance. These rapidly-spreading introduced pests build their nests in holes in the ground, rock crevices, hollow trees or, unfortunately for us, buildings.*

hibernation and all her workers and drones are supposed to die. In spring the queen alone has to refound the colony, fetching her own food and building the cells in which to raise the first crop of workers. When there are enough to take over the care of the nest, she concentrates on breeding. There is only one such cycle a year.

This limitation is imposed on native Australian wasps, and on the so-called European and English wasps in their places of origin. But in temperate Australia the introduced species are free of climatic restraint. Instead of hibernating, queens often establish a twice-yearly breeding cycle. Where winters are mildest — along the river valleys of Sydney, for example — worker wasps survive and continue to expand the nest.

As a result, wasps that in Europe would make nests typically the size of footballs are building some in NSW, tier upon tier, that would dwarf the most hulking football player. Single colonies can comprise well over 100 000 workers.

Vespula germanica, the 'European' wasp, is also native to North Africa and temperate regions of Asia. *Vespula vulgaris*, the 'English' wasp, ranges from Europe through Asia to Canada and parts of the US. The two species are much the same in appearance and habits, but *vulgaris* is less tolerant of higher temperatures.

War may have aided their spread

Hibernating queen wasps, hidden among building materials, in packing cases or sometimes in fabrics, can easily survive transportation over long distances. Massive cargo movements during World War II, when the spread of pests was the least of anyone's worries, was the probable cause of a huge expansion of the range of both *Vespula* wasps.

Not long after the war the 'European' wasp showed itself in the Americas and southern Africa. During the 1950s it began having a severe impact on honey production in New Zealand, where it not only raided beehives in its greed for sugar but also killed the bees. Tas-

The European wasp scrapes wood (inset) and takes the rolled-up strips back to the nest. Chewed and mixed with saliva, the pulpy material is used to build the cells that house the larvae and pupae (above). The cells are then capped, but these have been opened to show pupae in different stages of development.

mania's turn came in 1959.

Public safety was not a major concern until the late 1970s, when the wasps were found to be thriving in the metropolitan areas of Melbourne, Sydney, Adelaide and Fremantle, WA. Victoria, along with Tasmania, had also had the 'English' wasp since the late 1950s. That species has not spread north of Melbourne.

The semi-arid inland belt across most of southern Australia confines *Vespula germanica* to areas near the coast. In the fertile east, scientists have their fingers crossed. They hope that the wasp's tolerance of heat will reach its limit at about the mid-north of coastal NSW. So far, no human intervention has halted its advances.

A preserved laboratory specimen of Sphecophaga vesparum.

HOPES REST ON A PARASITE

Scientists turn to biology in their war against introduced wasps.

Victorian scientists in the spring of 1986 took the first cautious steps in a testing programme that may lead to biological warfare against introduced wasps. In conditions of strict quarantine, they have been breeding yet another wasp from Europe — a parasite.

Sphecophaga vesparum, stocks of which were imported from Greece, invades nests and lays its eggs on the larvae of the social wasps. Up to 70 per cent of a new generation can be destroyed.

The parasite has been employed with some success in New Zealand. Before releasing it here, however, biologists must make absolutely sure that it will victimise nothing other than wasps. The Victorian programme is partly funded by other states, so the wasp will be available throughout the country if its use is approved.

Where a nest could start

The introduced wasps are determined scavengers of prepared foods and scraps, and favour nesting sites close to houses, picnic areas and school grounds. A founding queen could choose almost any concealed spot. Most often it is a hole dug in the ground or behind a chink in a rockery.

Compost heaps, stacked materials, tree clefts or hollows, tussock clumps and hedges are also used. Most disconcertingly, nests are sometimes built in the ceiling spaces and cavity walls of houses. Once in a ceiling, colonies seeking to expand commonly eat their way down through plywood and plasterboard.

Nests are made of a papery material. The wasps rasp at weathered wood, chew the fragments and mix them with salivary secretions. Combs are constructed horizontally — bees make theirs vertical — with the brooding cells open at the bottom. Cells are added at the edges as the colony grows. If it continues season after season, new combs are built in ascending tiers. From the outside, the covering of the nest looks like streaky grey or brown cardboard.

Wasps are not usually alarmed by movement near the nest, or by noises. But if the nest itself is even slightly disturbed, the workers swarm out to attack the cause. They sting repeatedly and in force, not only on exposed skin but also through light clothing.

A scavenging wasp could be as far as a kilometre from its nest. But if large numbers of wasps are consistently seen in one place, the chances are that their nest lies within 50 metres. While probably not obvious, its location will be traced with some patient observation. It should be left alone until proper measures have been taken for its destruction.

Native paper-nest wasps

Australian social wasps, though widespread, never form the huge communities that the introduced types may found. Species of two groups, *Polistes* and *Rhopalidia*, are usually distinguished by orange markings, rather than the lemony yellow of *Vespula*.

Nests of the native types are small, consisting of a single comb with hundreds of brooding cells in warm climates and no more than a few dozen in the south. They are fixed by a narrow stem to tree branches, rock overhangs and sometimes to the roof eaves of houses. There is no extra covering or entrance-way — the individual cell openings are visible from below.☐

Getting rid of nests

It's a householder's duty — but expert help is often needed.

Some insecticide sprays may kill or discourage wasps that occasionally fly into houses. But they do not get to the root of the problem. It can be solved only by destruction of the nest. Under pest control regulations that is the responsibility of the property owner, and it can be demanded by the local council.

If you find a nest on your property, seek advice from your council health department on the best destruction technique and a suitable chemical to use. Dusts, sprays, poisonous strips and fumigants all have their roles in various situations. Where a nest is so awkwardly placed that there is an element of danger, the council will recommend the services of a professional pest exterminator.

Removal of a native wasp nest, hanging under a roof eave, is usually a straightforward matter. A plastic bag containing an insecticide can be lifted around the nest, tied at the top, and then the stem above can be severed. Afterwards the whole nest should be burned.

All attempts at nest destruction should be made after sundown, when the wasps are inside and subdued. The nest itself should not be touched if that can be avoided. Thick clothing, gloves and a veil should be worn. If a torch is needed, use one with a red light, not white.☐

Above: Nest of native Polistes *wasp. Left: An agricultural officer searches bush for a European wasp nest.*

Bulldog ants

*Big, pugnacious and primitive, they form
an ancient link with wasps.
Life-threatening allergy to their venom
is on the increase.*

Worker ants of every kind have venomous stings or squirting glands. Most are too small to be of any consequence. But the bulldog ants, Australia's biggest, inflict stings as painful as those of bees and wasps. And there is a mounting medical problem of venom allergy. It reflects an increasing incidence of stinging, presumably because more people are building on woodland blocks or spending their spare time bushwalking and camping.

Bulldog ants of the genus *Myrmecia* are found Australia-wide. Scientists suppose that there are scores of species, but systematic studies are in their infancy. Individuals of the same species are differently coloured from place to place, while others that look exactly the same have been proved to be genetically incompatible.

Common types of *Myrmecia* are distinguished by their great size and by their prominent grasping mandibles, resembling toothed pliers. Their ferocity when a nest is threatened, and the tenacious clamping of these mandibles in attack, accounts for the popular name of bulldog. Some people may call them bull ants or bulljoes. The mandibles, though they give a painful nip, are not venomous. They hold the ant to the skin while it curls its body round to use the sting at the tip of its abdomen.

A giant type that grows to 3 cm in length and can leap nearly 20 cm is known to Queenslanders as the jumper ant. Tasmanians, on the other hand, call their smallest *Myrmecia* a jack-jumper, because of its skittish actions when

Bulldog ants of the genus Myrmecia *(above), found all over Australia, are recognisable by their great size and their huge toothed mandibles.*

The fearsome-looking mandibles of bulldog ants (inset, below) can deliver a painful nip, but they are not the creature's main means of attack. They are simply used to grasp the victim while the ant curls its body round to use the venomous sting at the tip of its abdomen (below).

excited. Bigger bulldog ants are known in Tasmania as inchmen.

Allergy to bulldog ant venom is especially prevalent in Tasmania. A recent survey showed an extraordinary incidence of over-reaction even to the first stinging by a jack-jumper ant. Some allergists, noting that Tasmania was the region of earliest infestation by introduced *Vespula* wasps, suspect a cross-reaction: people stung by wasps are pre-sensitised to the ants.

A sidetrack of insect evolution

All ants are believed to have evolved from wasps. The *Myrmecia* represent an early branch of this development. In their winged form, as mating queens and drones, they appear markedly waspish. Except for one species in New Caledonia, they are peculiar to Australia. Elsewhere their relatives are found only as fossils.

Though highly organised to defend their nests and care for the young, bulldog ants are usually solitary foragers, rarely giving trouble away from their mounds. In fact Australia has no ants of dangerous size that forage in marching teams, as do the feared driver and army ants of Africa and South America.

Even in nest defence, the bad temper of bulldog ants is sometimes overstated. The Queensland naturalist Harry Frauca made a three-year study of 'jumper' colonies near Bundaberg. He found that on cold winter days the ants were sluggish and might not attack even if provoked. If they were far underground at any time, they were unlikely to emerge.

Maximum aggression, Frauca reported in *Wildlife* magazine in 1971, was displayed only by ants close to the nest entrance on summer days. Then a disturbance of the nest would bring them charging out, to sting any alien object within a radius of about three metres.

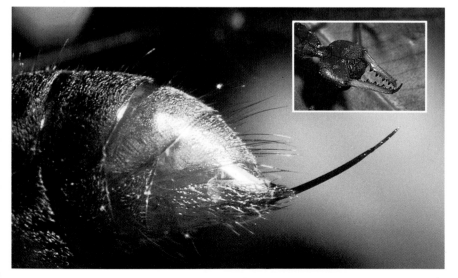

ANIMALS USING VENOM **ANTS**

Eggs that are laid for eating

Bulldog ant societies include an intermediate caste of wingless females that accompany the queens and are capable of taking over their reproductive functions. Even the true workers are not denied all sexual development — they lay eggs. But these do not produce offspring. Instead they are carried in the mandibles and fed to the larvae or shared among other adults.

The larvae also suck the juices of prey insects and spiders, brought to them by the workers. Adults take what is left, and some species are capable of regurgitating food and feeding it to the young in the manner of bees and more highly evolved ants. Liquids in the adult diet include water, plant nectar and sap, and the secretions of many other insects.

The size and formidable stinging power of bulldog ants protect them from the predators that take a heavy toll of smaller ants. Even big birds and lizards learn to leave them alone. The chief hazards for bulldog ants are the accidental blocking or flooding of their underground nests.

New colonies are established by the queens alone, after mating 'flights' that in some species amount to no more than a flopping about on the ground. A queen's wings are weak and deciduous — they drop off each season after mating and the muscle material is absorbed by the body as part of its nutrition. Males are very short-lived but workers survive for more than one season. Queens may live for many years.□

A worker of the bulldog ant species Myrmecia gulosa *laying soft eggs (above). These eggs are not used to produce offspring but are fed to the larvae (right) or, sometimes, to other adults. Queens can also lay edible eggs.*

Below left: Bulldog ants at the entrance to their nest near Alice Springs, NT. Bulldog ants are not as common in the arid outback as they are in the more temperate coastal areas. Their nests, while not always easy to spot above ground, can be up to a metre deep and contain many horizontal chambers connected by passages. Eggs and larvae are kept in the deepest, coolest part of the nest and are moved around according to weather conditions. The queen never comes to the surface.

A special fondness for lawns

Greenhead ants can cause occasional problems close to home.

Of the higher forms of Australian ants, none of which reaches the size of bulldog ants, only one type is known to endanger humans. Stings from a greenhead ant, *Rhytidoponera*, occur frequently enough to provoke allergic reactions in some people.

The ant to blame is usually described as *Rhytidoponera metallica*, though several species are probably grouped under this name. Populations are abundant in most parts of the country. Greenhead ants of this type have a particular liking for lawns as their nesting places. In subtropical regions especially, barefoot children may be stung. If they suffer unusually severe reactions medical advice should be sought and the ants' nests should be traced and destroyed.□

A worker greenhead ant Rhytidoponera metallica.

NESTS MAY GROW LIKE VOLCANOES
Bulldog ant excavations produce distinctive mounds.

Bulldog ant nests start as simple holes in the ground, up to about the diameter of a 10 cent coin. As a colony grows, material excavated from below is piled around the hole. Sometimes the mound takes a conical form resembling a live volcano in miniature; sometimes it is truncated and widely cratered, like an extinct one.

The ants normally seek well-drained high ground or slopes. They are unlikely to invade established properties on flatter land. If a nest should cause worry, however, the colony can be destroyed by the methods that are used against the underground nests of wasps.

The puzzling platypus

*Ever surprising, the world's oddest mammal comes armed with poisoned spurs
that deliver agonising, crippling stings.*

No one has found a convincing reason for the male platypus to have venomous hind-leg spurs. If the purpose is defence it is strange that the female, much smaller and weaker, is unarmed. She loses her spurs in infancy. Against prey, there seems little point. The spurs have been seen in use on frogs, but the tiny creatures that provide a normal diet are snuffled up without resistance.

If the primary function of male-only spurs is to grasp and subdue the female during copulation, then the venom is superfluous. Yet the venom glands are enlarged in the mating season. Immature males are sometimes stung to death when they challenge adults — a penalty too severe to suggest a logical means of breeding selection.

Paradoxes in the nature of the platypus have strained the credulity of zoologists ever since the 1790s. Europeans first saw the animal in the Hawkesbury River, near Sydney Town. Nowhere else on earth was there known such a combination of anatomical features: a furry body, feet that were both webbed and clawed, and a toothless muzzle that was shaped like a duck's bill.

Colonists called it a water mole. Scientists chose the name *Platypus* — flatfoot — but found that the name had already been taken by a beetle. Then they opted for *Ornithorhynchus anatinus*, meaning bird-snouted like a duck. But the first choice, more easily pronounced, stuck as a common name.

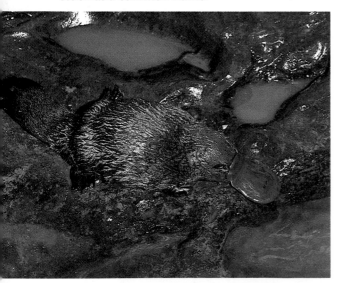

A platypus resting on a rock shelf. The long covering hairs are sleek with water, but thick underfur always stays dry.

Characteristics of a reptile

The venom apparatus itself is a striking oddity. Echidnas have something similar. Elsewhere in the world, the only mammals with what is arguably a venomous capability are some shrews whose bites inject a saliva that is toxic to small prey.

More startling was the discovery that the platypus's reproductive and excretory processes shared a single passage. This is characteristic of reptiles and birds, but it was unknown in any animal with fur. Later the echidna was found to have the same reptilian feature.

The platypus retained its status as a mammal because the female suckles her young — even though she has no teats. Glands ooze milk into grooves below two patches of fur where the young feed. But scientists remained mystified because they could find no uterus for foetal development.

Bewilderment was complete in the 1880s, when it was proved that the young hatched from eggs — laid and incubated like those of birds and most reptiles. Subsequently the platypus was found not to be truly warm-blooded. Its body temperature is as much as 10°C lower than that of marsupial and placental mammals.

In a class of their own

Zoologists had to create a special category of mammals to accommodate the platypus and echidna. The order monotremata, meaning having one opening, exits only in Australia and, in the case of a second type of echidna, New Guinea and some nearby islands. Not even fossils have been found elsewhere. The monotremes do not form any link between reptiles and the more advanced mammals, but seem to represent an earlier evolutionary sidetrack.

As recently as 1985, the platypus sprang yet another surprise. Because it is totally blind and deaf under water — the eyes and ears are sealed in a furrow of skin when it dives — it was always believed to feel for food with its muzzle. Probing the muddy depths of rivers and lakes, where sight would be of little use anyway, it did remarkably well. In one feeding session a hungry platypus may consume more than half of its own bodyweight in small crustaceans, worms, tadpoles and insect larvae.

In fact the platypus has an extra sense. It tracks electricity. Australian and West German scientists, working together at the Australian National University in Canberra, found that the

soft muzzle contains special nerve receptors, known previously only in some fishes and tadpoles. The receptors detect tiny electrical discharges produced during the muscle movements of prey animals. Food is located even under mud or stones.

Slaughtered for their fur
Its uniqueness ignored, the platypus was persecuted in the latter part of last century. It was detested as a hindrance to fishing. And although its flesh does not make good eating, its water-resistant fur was prized. To the toll taken by natural enemies — pythons, birds of prey and goannas — was added the depredation of shooters and trappers. Scores of dense, soft pelts went into making just one cloak or rug.

Fully protected now, platypuses are fairly common in undisturbed streams and freshwater lakes from northern Queensland to South Aus-

When a platypus dives its thickly webbed front feet provide the propulsion. The hind feet, though also webbed, are tucked back against the tail. The platypus closes its eyes, ears and nostrils when submerged.

tralia and Tasmania. They thrive in alpine conditions as well as on warm coastal margins. Feeding animals are occasionally seen — or more often heard splashing — at dawn or dusk. Daytime foragers before they were persecuted, they have become virtually nocturnal.

Platypuses grow biggest in the south, where some males reach a length of 60 cm — a third of it in the tail — and weigh up to 2 kg. Females do not exceed 40 cm. Weaned at four months, the animals reach maturity at two years. Some in captivity have lived for over 15 years.

Each excavates its own burrow, climbing steeply up from an entrance just below the waterline to a dry sleeping chamber. The main tunnel is usually 5-10 metres long, but some leading to female nesting chambers have been found to extend for more than 35 metres, not counting numerous dead-end branches.

A burrowing platypus rolls back the webs on its forefeet, bringing powerful claws into play. The tunnel is made barely big enough to squeeze through, so that water is forced out of the animal's fur when it returns from a feeding expedition.

Doomed by fish traps

Using only its forefeet for propulsion, a platypus can dive deeply but does not stay under for longer than five minutes. A frightened animal may seem to disappear indefinitely, but it will have emerged far away, under the cover of reeds or tree roots. If something prevents platypuses from resurfacing — and fish traps often do — they drown.

Food taken during each dive is held in cheek pouches and eaten when the platypus comes up for air. Horny pads at the back of the mouth are used to grind up the food. Grit, also stored in the pouches, probably aids the process. Platypuses are born with small teeth, indicating that their ancestors relied on them, but these are lost within weeks.

Solitary at all other times, adults pair off as water temperatures rise after winter. Breeding starts in July in the tropics but as late as October in the south. Pairs mate in the water after a ceremony of swimming in tight circles, chasing each other's tails.

The mated female excavates a new chamber far back in her burrow and pads it with leaves or grass, carried in with her folded-over tail. Before laying eggs she seals off the burrow with at least one plug of earth, again using the tail to tamp it behind her.

Usually two eggs — sometimes one or three — are laid. Smaller than pigeons' eggs, they are glossy white and leathery. The mother clasps them against her belly with her tail and incubates them for 10-12 days. Fasting, she sleeps most of the time.

Hatchlings do not feed for two weeks. In the 14 weeks of suckling that follow, the mother has to go out frequently to feed voraciously. Each time she destroys and replaces whatever protective plugs she has made. But when her milk supply wanes she leaves the tunnel open. The juveniles make their own way out and take to the water as readily as ducklings. They can forage for themselves almost immediately. □

AN ARMAMENT NO LONGER NEEDED

The echidna's venom spurs are relics of its past.

Hind-leg spurs and venom glands in the echidna — similar to those of the platypus but much smaller — have apparently ceased to have any useful function. They certainly present no danger to humans. Nor do the spines give much to worry about, unless they are pressed against the skin with considerable force — when trodden on with bare feet, for example. The spines are not venomous, but there could be a high risk of tetanus or other bacterial infection from deep puncture wounds.

The venom apparatus is a leftover from a time when the echidna's ancestors led lives very like those of platypuses. Echidnas also retain partly webbed forefeet, although they are strictly land animals now. A striking feature of their adaptation is their ability to dig with all four feet at once — virtually sinking straight into the ground when frightened — instead of burrowing head-first.

Like the platypus, the echidna created headaches for 19th-century zoologists and there was confusion over its naming. *Echidna*, the first name given, meant spined — but it had already been bestowed on a fish. Then scientists chose *Tachyglossus* — swift-tongued. Echidna hung on as one of the common names.

The other common name, spiny anteater, is misleading in most parts of Australia. Where termites are available they and not ants form the diet of the echidna, which burrows into their nests and scoops them up with its darting, sticky tongue.

Ants are eaten in the south. It seems likely that a decline in echidna populations in settled areas, through the restriction of their habitats and a high mortality on roads, has allowed ants to increase. This may partly account for the increasing incidence of ant sting allergy.

An echidna of the Tasmanian subspecies Tachyglossus aculeatus setosus.

Anglers are most at risk

If a platypus is hooked, caution should come before mercy.

Properly wary of humans, platypuses are rarely met at close quarters in the wild. But now and then one is hooked on a fishing line or tangled in a net. If it is a well-grown male and a clumsy attempt is made to free it, the fisherman could suffer severely.

A harmless female may be distinguished by red-tinged patches of belly fur. If these are not obvious and the body of the platypus is any longer than a man's hand, it should be treated with extreme caution — especially in springtime. If it has to be handled, the only safe way is to pick it up by the tail from above.

Before trying to get a grip anywhere else, in order to dislodge or cut a fish hook, the animal should be wrapped tightly in a blanket or some other thick fabric. If that is not possible it is wiser to cut the line and, though it may seem heartless, let the platypus take its chances back in the water with the hook still in.

Supple and muscular, the platypus stings by clasping its target with its hind legs. Spurs are erected at right-angles from the inner side of each ankle and driven in as the legs are clamped together. The power of a full-grown male is such that a man stung on a hand or arm cannot prise the animal off with his free hand. The platypus goes on stinging and in the breeding season, when its venom glands are enlarged and the ducts overloaded, the volume of venom injected is copious.

At Emu Creek, in the high ranges north of Toowoomba, Qld, it took the strength of two companions to detach a platypus from a night angler's wrist. His forearm swelled up within 10 minutes and next day the whole arm was so bloated that he could not bend it. Pain kept him sleepless for three more nights. And that stinging, reported by naturalist David Fleay, was at Easter — out of the breeding season.

Symptoms keep coming back

No deaths from platypus stings have been recorded. Their greatest threat to life may be indirect — a temporarily handicapped victim, alone in a remote area, could be exposed to other risks. The chief medical worry is the recurring disability of victims. Many apparently recover when they are rested, only to find that the pain and swelling return when they try to resume work. In his wildlife column in the Brisbane *Courier Mail,* David Fleay told of a man whose stung arm remained 'useless' 10 years later.

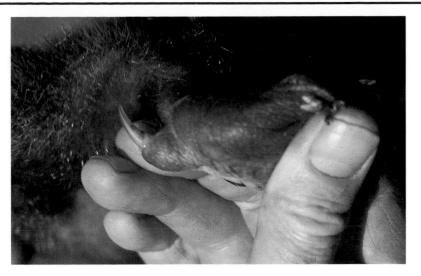

Each of the male platypus's two spurs normally lies flat against a hind leg, in a fleshy sheath just above the ankle. Curved and hollow, the spurs may be up to 15 mm long. Venom glands housed under the thigh muscles have ducts leading down the legs to the bases of the spurs. In the breeding season, when the glands are enlarged and venom output is greatest, the ducts themselves are dilated at the bottom to hold an extra supply.

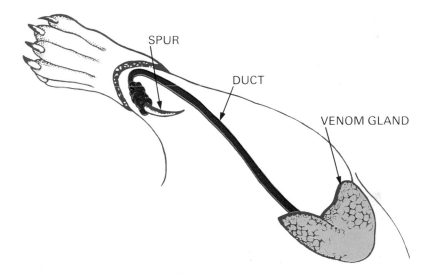

SPUR

DUCT

VENOM GLAND

PLATYPUS ENVENOMATION

No specific first aid measures are recommended for platypus stings. Any restriction applied to prevent the spread of venom may only increase pain in the stung area.

ACTION Keep victim rested with affected limb as still as possible, but without binding it up. Obtain urgent medical attention.

Fleay, the pioneer of platypus breeding in captivity, has probably been stung more than any other person. He calls the experience 'dreadful,' with the affected limb quickly swelling to three or four times its normal size and agony continuing for a week or more.

A seasoned outback Queenslander, Mr Alfred Hill, put the effects in an unusual perspective in 1982. Stung while fishing near Texas, close to the NSW border, he was in hospital for two weeks. The pain was worse than a redback spider's bite, he said, and the infection that followed was like having tetanus. Hill had suffered both.

Medical treatment concentrates on pain relief and the prevention of bacterial infection of the spur wounds. Rest of an affected limb is essential. There is no antivenom available.☐

Inset, left: At rest, the pale blue circles of the southern blue-ringed octopus Hapalochlaena maculosa *are not especially noticeable. When the octopus is excited, however, the rings become vividly iridescent (above).*

Blue-ringed octopus

Deadliest of its kind, it is so tiny that a toddler can pick one up. A warning change of colour makes it all the more appealing.

Stranded in rock pools after big tides almost anywhere on the mainland or Tasmanian coasts, the blue-ringed or banded octopus is a common sight. If not seen, it is tiny enough to be hiding in a shell, a can or a bottle. Compared with the monsters we may imagine octopuses to be, it is like a pretty toy. But it is the most lethal octopus in the world.

Fortunately, aggression is the last thing in the mind of this dainty creature. It does no harm unless it is extraordinarily provoked by being taken out of the water. Left in contact with the skin, it may make an incision with its parrot-like beak. Usually the bite is not even felt. But it is followed by the forcing of venomous saliva into the wound.

All octopuses, big and small, secrete salivary compounds to subdue and digest their prey of crabs, other crustaceans and shellfish. Elsewhere in the world, in the rare event of octopuses biting people, the result may be some swelling, soreness or numbness. Noteworthy effects have been recorded in medical literature on fewer than a dozen occasions.

But the venom of Australia's blue-ringed octopus includes a component found in no other creature of this kind. It is tetrodotoxin (TTX) — identical to the paralysing poison in the tissues of pufferfishes. Part of its purpose seems to be to dissuade predatory fishes from eating an otherwise helpless animal. And to remind those predators of its hidden menace, the octopus has developed a visual warning system.

Flashing is the danger signal

At peace, the blue-ringed octopus may not live up to its name. It is a dull yellow-ochre in colour, or sometimes greyish, with darker patches on its body and bands on its arms. Small streaks or circles of pale blue mark these darker areas, but they are not prominent. In poor light the octopus could pass for a juvenile of some other bigger but harmless type.

Its skin contains unusual pigment cells called chromatophores. Commanded by nervous impulses, these produce a sudden colour change if the animal is harassed or senses danger from a bigger creature approaching. The darker shades deepen and the blue markings become vividly iridescent. They fade gradually if the threat passes, but the bright coloration can return just as suddenly in a new emergency.

Many people find amusement in taunting the blue-ringed octopuses they come across in rock pools, to see the colours change. But children should be cautioned at the earliest possible age against picking them up. Regrettably, the spectacular display that acts as a deterrent to fish only increases our fascination.

The suckers on the octopus's tentacles are harmless.

Alike in all but size

Two species of blue-ringed octopus occur in Australian waters. They differ in size but otherwise are indistinguishable without scientific examination. *Hapalochlaena maculosa*, numerous around southern shores especially in summer, does not exceed 12 cm from the top of its body to the tips of its trailing arms. *Hapalochlaena lunulata*, common in the north, can grow to about 20 cm.

They migrate in warm currents and the limits of their distribution are not clear. The species probably overlap extensively on the east coast. Venom studies and behavioural observations so far have all been carried out on the southern species, but there is no reason to suppose that the findings do not apply to both types. Each has killed a man in similar circumstances.

A mollusc without any shell, the blue-ringed octopus has the structure typical of a cephalopod — literally a 'headfoot'. Its brain, eyes and mouth are clustered over the junction of its eight arms, which are equipped with harmless, cup-like sucker pads. The stomach and other body organs are carried above.

Much of the bulbous upper part is occupied by a cavity where sea water reaches the octopus's breathing gill and excretory duct. The cavity's outlet is a funnel that can be pointed sideways as well as down. When the animal wants to move it contracts muscles around the cavity, squirting out water and giving it a form of guided jet propulsion.

Venom glands as big as the brain

Two ducts pass right through the brain. One carries food up to the crop and stomach. The other brings venom down to the mouth from a pair of salivary glands. These are of surprising size — each as big as the brain.

When a blue-ringed octopus is starving, it makes a direct attack on its prey. It grasps a crab with its arms and pierces the shell with its beak. But if the octopus has been feeding regularly, it adopts a less strenuous approach. Hovering over the crab of its choice, it squirts venom into the

ANATOMY OF THE BLUE-RINGED OCTOPUS

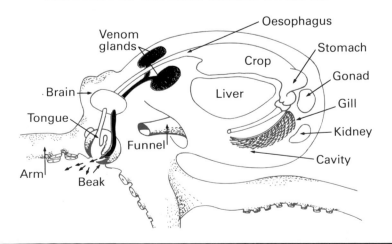

water and glides away for a few minutes while paralysis sets in. Then it settles over the helpless crab and levers the shell open. Using a rasping tongue along with its powerful beak, the octopus consumes the flesh in long strips.

Most octopuses squirt ink to confuse pursuing predators, but in *Hapalochlaena* the ink sac functions only in infancy. Another important difference is found in the mature female's care of her eggs. As far as is known all other octopuses lay small eggs in thousands and attach them to rocks or plants. The blue-ringed octopus in captivity lays only 100-200 eggs but they are far bigger, and they stay fastened to the mother's arms or to the web between the arm bases until they hatch.

In other octopuses, mating is performed literally at arm's length. The male uses the tip of one arm to transfer sperm from his body cavity into the female's. But blue-ringed pairs grapple in a close embrace. Eggs are laid about a month later and take two more months to incubate.

Embryos on the alert

Hatchlings are self-sufficient in two days, defending themselves with ink and feeding on sandhoppers. Though juveniles do not display their warning rings until they are about six weeks old, the colour-changing chromatophores seem to develop very early, at the embryo stage. If the translucent eggs are disturbed, they darken immediately.

Blue-ringed octopuses breed only once, the females dying soon after their eggs are hatched. Their lives span only about seven months, four of which are spent in reaching maturity.□

Enough to kill ten men

Victims survive if they are helped to go on breathing.

Awareness of the danger of blue-ringed octopuses is recent. Until the 1960s they had not been known to kill people. In bite cases since then, the experience of victims has varied widely. Some are only mildly affected. Others, profoundly paralysed, could not have survived without the most advanced medical techniques.

The variation may simply be due to different amounts of venom injected. Whether seasonal or other variable factors in the biology of the octopus could be involved, scientists do not know. What is known is that the amount of venom in the oversized salivary glands of a mature specimen is sufficient to inflict paralysis and eventual death on at least ten men.

Far less venom is delivered in a single bite, of course. Experiments suggest that a half-grown octopus, while it holds ample venom in its glands, is unlikely to be able to kill. And the

juveniles that sometimes occur in swarms can probably do us no harm at all. On principle, however, young children should be discouraged from handling them.

TTX, the lethal component of the venom, blocks the passage of electric impulses along our nerves and prevents muscular action. The effects of a severe envenomation are very like those of pufferfish poisoning. Victims remain conscious as long as they are breathing. Those who are not fully paralysed sometimes suffer disorientation and other signs of brain disturbance.

What can follow a bite

Victims are generally unaware of the bite. It may be noticed as a blanched area about the size of a one-cent coin, with a tiny fleck of blood. It swells in about 15 minutes and starts to darken.

Before then, an envenomed person feels abnormal sensations around the mouth, spreading to the rest of the face and the neck. Breathing becomes difficult and vomiting usually occurs. Vision is distorted. As paralysis sets in — within minutes of the bite in severe cases — speech and swallowing become impossible. If the breathing muscles are involved and the victim is not helped, death from respiratory failure will follow.

Any otherwise healthy victim can be saved if the right first aid is given promptly. Breathing must be maintained until the patient can be put on an artificial ventilator and given other life-supporting treatment. The crisis is generally over within about 12 hours. After perhaps a further day of weakness and poor co-ordination, recovery should be complete.

No antivenom to octopus bites has been developed. Chemical studies have been limited to the products of whole glands taken from dissected animals — not the pure venom they would have injected. Because mortality from bites is apparently very low, research suffers from a lack of official interest in funding.

Behind some puzzling deaths?

But many scientists cannot believe that there were no deaths until 1954 — when the culprit, now known to be *Hapalochlaena*, was wrongly identified. They speculate that before then, and perhaps since, the blue-ringed octopus was responsible for some disappearances and unexplained drownings.

Lone anglers with years of experience on the rock platforms of the southeast, for example, are occasionally swept away by waves that they would normally have avoided. Some may have been paralysed, or already dead, through handling octopuses for use as bait. Where a body is recovered after a day or so, an autopsy would reveal no trace of TTX.□

OCTOPUS ENVENOMATION

EARLY SIGNS Tingling numbness around mouth spreading to tongue, rest of face and neck. Stiffness, poor co-ordination, slurred speech. Nausea, vomiting.

ACTION Locate bite and if possible apply pressure/immobilisation technique as for snakebite (page 360). Rest victim on side and send for medical assistance. Prepare to give mouth-to-mouth resuscitation if breathing should fail. Reassure victim, who is fully conscious while breathing and can hear.

Midwinter death

Dizziness set in while the octopus was still biting.

Blue-ringed octopuses are more often seen in summer. But lower water temperatures seem to be no bar to activity — or their venomous potency. The first known killing by the southern species occurred in June 1967, three years after the menace of its tropical cousin was proved.

Three army recruits, inducted only one day before, were exploring rocks off Camp Cove, just inside the entrance to Sydney Harbour. A 23-year-old found an octopus and placed it on the back of his hand to show his companions. It had been there for about 10 minutes when he began to feel dizzy — and discovered that he could not get the creature off. One of his friends had to remove it.

Soon the victim was unable to swallow or breathe through his mouth. His breathing failed altogether and he lost consciousness as his companions were carrying him to their base nearby. Attempts to resuscitate him at the camp, in an ambulance and at a hospital were all to no avail. The young man was declared dead about 90 minutes after picking up the octopus. At an autopsy, no laceration of his skin could be found.

This case led to an official call for the Commonwealth Serum Laboratories to investigate the venomous capabilities of *Hapalochlaena maculosa*. Experimental work by Drs Struan Sutherland and W. R. Lane quickly confirmed its lethal potential.☐

Left: The northern blue-ringed octopus Hapalochlaena lunulata *is slightly bigger than the southern species.*

A case of mistaken identity

The first known killing was blamed on the wrong octopus.

Ordinary seaman Kirke Dyson-Holland, 21, made the most of his RAN posting to Darwin, NT. He became an enthusiastic member of the Arafura Skindivers' Club. Off duty in September 1954 he went spearfishing with fellow member John Baylis off East Point, now the site of the city's War Museum.

Leaving the water for a rest, Baylis noticed a blue-ringed octopus about 20 cm long. Both men were familiar with such creatures and had handled them. Baylis picked this one up and allowed it to crawl over his arms and shoulders. Then he tossed it to Dyson-Holland. It crawled from his shoulder to his back, paused there, then dropped off into the water.

As they came ashore Dyson-Holland remarked on a dryness in his mouth and a difficulty in swallowing. He was unaware of a bite but Baylis noticed a trickle of blood from a tiny puncture on his back. Soon Dyson-Holland was staggering about and vomiting. When he collapsed he was carried to a car and driven to Darwin Hospital. During a journey of less than 7 km he stopped breathing and on arrival his skin was blue. Emergency treatment failed to save him. In a post-mortem examination, not even the bite mark could be found.

Baylis later caught an octopus that he said was exactly the same as his friend's killer. This specimen was wrongly identified at first as *Octopus rugosa*, a widespread species known overseas for having a bite with effects like a bee sting. In 1964, however, it was re-examined and found to be *Hapalochlaena lunulata*.☐

'He stopped breathing and on arrival his skin was blue.'

Cones: uniquely armed

They ram poisoned harpoons into passing fish — or the hands of careless collectors.
Fortunately, the most lethal kind is rare.

CONE SHELLFISH ENVENOMATION

EARLY SIGNS Numbness spreading from mouth. Slurred speech. Blurred or double vision.

ACTION Apply pressure/immobilisation technique as for snakebite (page 360). Rest and reassure victim and send for medical assistance. Prepare to give mouth-to-mouth resuscitation if breathing should fail.

MINOR STINGS See a doctor in case wound is contaminated.

Nothing else in nature compares with the guided missile system of the cone shellfish. Only humans, with the help of tools, have developed anything like it. A humble marine snail combines arts akin to blowpipe hunting and whale harpooning to deliver venom with deadly efficiency.

Hundreds of species of the genus *Conus*, including about 70 in Australian inshore waters, manufacture and store continuous supplies of disposable darts. Up to 1 cm long in bigger species, these are hollow and bony. They carry barbs in formations that vary according to the diet of the species. Most cones prey on marine worms or other molluscs. But the most dangerous types also spear fish.

A hungry cone keeps one dart at the ready in its snout, which can be extended for a distance as long as the shell. When the slow-moving gastropod comes upon prey, or senses it passing close by, it floods the dart with venom and propels it forward so that it projects from the snout, though the base remains tightly gripped. Then the snout is rammed violently against the victim and the dart is driven in.

If more venom is needed to subdue large prey, it is pumped through the dart. Then the barbed weapon is used like a harpoon, to draw the paralysed victim back into the snout. This widens vastly so that the mouth cavity can engulf and part-digest animals almost as big as

the shellfish itself. The dart, eaten or discarded, is replaced in time for the cone's next meal.

The exact means by which darts make their way to the snout, through the throat from the sac where they are manufactured, remains a mystery to scientists.

Cones are active hunters at night in tidal shallows, generally on or around reefs. By day they wait in sand or coral rubble, or under stones. Often they bury themselves, leaving only a siphon poking up. This draws in water for pumping over their breathing gill. While buried, the cone seems to be able to sense prey from chemical traces in the water. Emerging, it uses eyes that are mounted on tentacles.

Thousands of dollars for a shell

Intriguingly varied in their patterns, cone shells are highly prized by collectors. One of them, *Conus gloriamaris*, is the rarest sea shell in the world. Specimens have been found in the New Guinea region, but not in Australian waters. A *C. gloriamaris* shell in good condition fetches thousands of dollars.

Shell patterning is often not fully revealed unless a horny outer growth is removed. But scraping it off, if the shell is held in bare hands and the animal is still alive, could be a most dangerous procedure. The armed snout can reach to any part of the shell.

Cones occur on most coasts and island fringes. Those found in temperate regions are usually small and not thought to be dangerous — though they do sting if handled. Species that present a known or suspected threat to life seem to be restricted to the warmest waters, but their range is not well defined. Some could crop up nearly as far south as Sydney or Perth.

The geographer cone *Conus geographus*, growing to as much as 13 cm in length and weighing nearly 2 kg, shows a pattern that may suggest an antique map. It is a killer overseas, and the

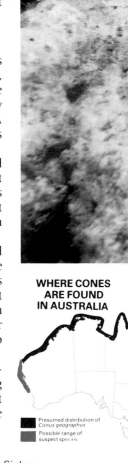

WHERE CONES ARE FOUND IN AUSTRALIA

■ Presumed distribution of *Conus geographus*
■ Possible range of suspect species

VENOM APPARATUS OF CONE SHELLFISH

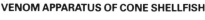
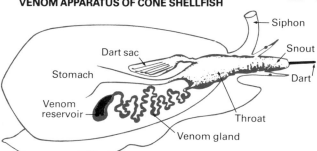

Siphon
Dart sac
Snout
Stomach
Dart
Venom reservoir
Throat
Venom gland

Marbled cone shell Conus marmoreus. *Grows to 10 cm.*

The geographer cone shell Conus geographus *is the only proven killer in Australian waters. It grows to 13 cm.*

only proven killer here. Its population is presumed to be spread from North West Cape across the north coast and down the Great Barrier Reef, but it is very rarely found.

Conus textile, the cloth of gold or textile cone, has been implicated in deaths overseas — perhaps by mistake. It does not hunt fish but preys on other gastropods, so it does not need a particularly potent, fast-acting venom. Experiments here do not suggest that it is highly toxic to humans.

Laboratory investigations have confirmed the toxicity of several other fish-eating species. But venom outputs and properties vary widely. The venom of the magician cone *Conus magus* was shown to be even more lethal — to mice — than that of the geographer cone. Fortunately this species is tiny, with a correspondingly meagre supply of venom. Perhaps the most dangerous type, after the geographer cone, is the tulip cone *Conus tulipa*. Too little is known

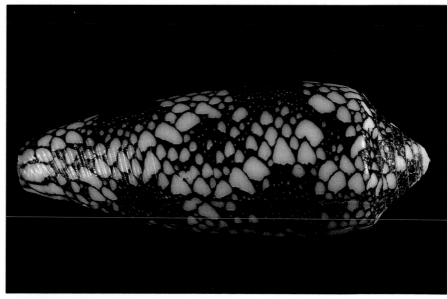

Court cone shell Conus aulicus. *Grows to 15 cm.*

Conus striatus grows to 10 cm.

Conus tulipa grows to 7.5 cm.

Conus omaria grows to 7.5 cm.

Conus textile grows to 10 cm.

about cones, however, to say that some are safe to handle and others are not. A sting that does not cause paralysis or other life-threatening effects could still produce painful local symptoms and prolonged weakness that could mar a seaside holiday.

A simple rule, at least on tropical and subtropical coasts, is to leave all cones alone.☐

Striated cone shell Conus striatus.

Tulip cone shell Conus tulipa.

Pearled cone shell Conus omaria.

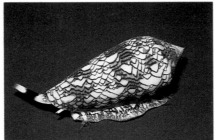

Textile cone shell Conus textile.

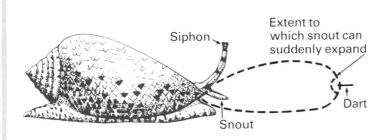

NO SAFE PLACE TO TOUCH IT

Live cones should never be picked up with bare hands.

Because a cone's stinging snout emerges at the narrow end of the shell, it is sometimes suggested that the shell can be handled safely at the other end. That is not true. Suddenly extended, the snouts of most species can reach back to any part of the shell.

Live cones should never be handled without tongs or thick, strong gloves. They can sting through clothing, so a collected specimen should never be placed in a pocket. Cones found among the coral structures of the Great Barrier Reef, where people are most likely to see them, are protected under marine park laws. Without a permit, issued only for scientific purposes, they cannot be collected at all.

A day trip to doom

Island visitors did not realise their companion was dying.

A launch cruise to Hayman Island, in the Whitsunday group off Bowen, Qld, was a rare treat in 1935. For a 27-year-old, out for the day with his family and friends, it was a splendid chance to add to his shell collection. And on Hayman's coral reef he made an enviable find — a big geographer cone.

Gripping the shell to scrape off the outer growth and expose the pattern, he felt a prickling in the palm of his hand. According to his mother, he complained of numbness in the hand almost immediately and within 10 minutes was stiff about the lips. But he was in no pain. Half an hour after the sting his legs were paralysed, and in a further half-hour he began to drift into unconsciousness.

The young man continued to breathe normally. No great concern was shown about his condition, the launch master remarked later, because his companions said he had recovered from similar 'turns' twice before.

A doctor did not see the victim, back on the mainland, until the last few minutes of his life. Nearly five hours had elapsed since the sting, and he had been in a deep coma for at least an hour. Immediately before his death his pulse became weak and rapid, his breathing shallow and slow. A coroner's report noted his perfect physical condition — he had been in training for football. Death was ascribed to heart failure due to toxemia.

That was Australia's first known fatality from a cone sting, and when this book was prepared it remained the only one. The case excited wide publicity, but should not have come as any great surprise. Killings by cones on islands not far north of Australia had been noted since 1705, and Aboriginal warnings reported.

In the Hayman Island case, paralysis does not seem to have involved the breathing muscles until the end. In other cases reported from the Pacific Islands, respiratory failure has occurred rapidly. The possible need for resuscitation should be anticipated if there are strong early signs of envenomation.

Medical management of severe cases follows the same lines as for blue-ringed octopus bites and pufferfish poisoning. With breathing maintained, artificially if necessary, the paralysis passes in a few hours and full recovery can be expected. No antivenom has been developed.

Less severe envenomations sometimes cause intense local pain, followed by widespread itching. Victims may be weakened and easily fatigued for days or even weeks, and the pain could return on contact with salt water.☐

The crown of thorns

Starfish swarming on parts of the Great Barrier Reef spell disaster for coral — and pose a threat to pleasure seekers, too.

The creeping death of coral communities, metre by metre and night after night, appals anyone who values the unique treasury of the Great Barrier Reef. The killer, the voracious crown of thorns starfish *Acanthaster planci*, is a public enemy. To the unwary reef visitor, it also offers a private agony.

Each of the starfish's hundreds of needle-sharp, brittle-tipped spines is encased in a sheath containing venom cells. A slight touch can cause a puncture that brings enormous pain. It starts within seconds and can last for hours. The stung area swells and stiffens. There is a high risk of infection, turning the wound into a suppurating sore.

The shock of a multiple stinging, through swimming into a crown of thorns or treading on one bare-footed, could cause the victim to faint in a dangerous situation. And heavy envenomation is likely to lead to repeated vomiting, so persistent in some cases that dehydration sets in.

There is no antidote to a sting. Immersing the injured part in hot water may ease the pain. Some divers, on the other hand, find that they suffer less if they stay in the sea as long as they can. In any event it is advisable to seek medical aid as soon as possible to remove any embedded spines and cleanse the wounds.

While a particular reef may not be infested by the crown of thorns, it is normal for a few of them to be about. For this and many other reasons, strong footwear should always be worn on coral reefs. People going swimming or diving should wear protective clothing and be familiar with the appearance of the starfish — in subdued underwater light among spiky coral, it can easily go unnoticed.

Many sea stars graze on coral, but no others have the appetite of the crown of thorns. It can grow to 60 cm in diameter — though 30 cm is more usual — and eats about half its diameter of reef surface every night. It turns itself inside out, spreading its stomach over a chosen patch, and its digestive enzymes liquefy all the living tissues of the coral polyps. Only bleached skeletons are left when the starfish is finished.

Acanthaster planci breeds once a year, in midsummer. A female may spawn up to 20 million eggs. But they are unfertilised — released into the water on the off-chance that they will encounter male sperm. That chance is increased by the simultaneous spawning of groups of adults. If for some reason the groups become larger, the rate of fertilisation leaps astro-

Resembling an iron-spiked cudgel, the crown of thorns starfish can grow to 60 cm in diameter. Each of its hundreds of spines is venom-tipped, posing a threat to unwary swimmers.

CROWN OF THORNS ENVENOMATION

ACTION Remove spines that can be gripped, using a straight pull. DON'T jerk them sideways or squeeze the flesh. Wash the wound and cover it lightly. Keep the affected limb still but DON'T bind it. Obtain medical attention.

nomically and a population explosion follows.

In the 1960s, soon after the first freakish infestations of northern parts of the Barrier Reef were detected, a count was taken on the tiny reef of Green Island, off Cairns. Nearly 6000 crowns of thorns were found. Grazed with such intensity, nine-tenths of a reef's living coral cover can disappear in a year. The starfish leave only a few species — mainly stinging corals.

The problem appeared to have solved itself later in the 1960s, only to re-emerge in other areas, and more damagingly, a decade later. Scientists are in wide disagreement over the ecological processes that may lead to these explosions, and over whether denuded reefs truly recover.

Government agencies propound an official view that we are seeing a natural phenomenon that occurs irregularly but is probably age-old. Costly schemes to exterminate the crown of thorns would therefore be wasteful. But suspicions remain that modern human activities somehow triggered the crisis on the reef, and only human efforts can resolve it.□

Black sea urchin
Diadema setosum.

Sea urchin
Asthenosoma
intermedium.

Flower urchin
Toxopneustes
pileolus.

Sea urchins

The most threatening kind, not noticed in our waters before 1960, is armed with venomous stingers that look like flowers.

Punctures from the spines of almost any sort of sea urchin are potentially troublesome. Risks of infection are high, especially if spines break off deep in the wounds. And some species seem to have venoms in or on their spines. The immediate swelling and the intensity and duration of the pain that is suffered are out of all proportion to the tissue damage inflicted.

Illness from spine toxins is generally not severe, however. It does not rank with the complications of having a foreign body in a wound. But some types of sea urchins have an extra way of protecting themselves from predatory fish. Beneath their spines they carry small appendages called pedicillariae. These are hooked stingers, fed by venom glands.

In the most dangerous sea urchin of this kind, the flower urchin *Toxopneustes pileolus*, the densely packed pedicillariae resemble waving, three-petalled blossoms. Their stings can cause collapse from shock, hours of muscular paralysis, and respiratory failure. The animal is dreaded in Japanese waters, where it is said to have killed divers.

Flower urchins were not recorded in Australia until 1960, when some were discovered in Brisbane's Moreton Bay. Subsequently the species was identified in Sydney Harbour and on the Great Barrier Reef. No cases of severe envenomation have been documented here.

A question of where it stings
Japanese experience of the flower urchin was puzzling to scientists who had handled it in the Pacific Islands without ill effects. They thought the wrong species was being blamed. Now it is believed that the stinging hooks are not strong enough to pierce hardened skin. Envenomation comes from contact with more tender areas.

A Japanese authority, caught on the side of a finger by seven or eight pedicillariae, likened the immediate pain to a jellyfish stinging. Soon he was dizzy, paralysed in the lips, tongue and eyelids, and struggling to breathe. His limbs went limp. The pain and most other effects disappeared in an hour, but his facial paralysis lasted for six hours.

Another sea urchin with venomous

pedicillariae is *Tripneustes gratilla*, widespread along the east and west coasts at least as far south as Sydney and Rottnest Island, off Perth. It has tiny, globular stingers. This species caused hours of intense pain to a Red Sea researcher when it brushed skin near his elbow. But it is commonly handled by skindivers here and is not known to have caused serious problems.

Beware the black types

Slow-moving sea urchins feed at night on plants or small, sedentary animals on shallow sea beds. Usually more or less pumpkin-shaped, they are closely related to starfishes and are just as prolific and varied. Of the types that produce significant envenomation effects with their spines alone, the worst are black sea urchins, *Diadema*. Two barely distinguishable species abound near tropical shores and occasionally turn up far to the south.

Their main spines are strong and enormously long — up to 30 cm has been claimed for some. In addition they have shorter spines that are needle-sharp and fragile. It is virtually impossible to touch these without their penetrating the skin and breaking off. Burning pain is felt at once, followed by a throbbing. Wounded areas are quickly inflamed, their redness often turning purplish-blue as dark pigment spreads from the spine fragments. Victims may suffer nausea and general weakness.

Embedded in flesh, the spine material of most sea urchins dissolves within a few days. This is not the case with *Diadema*. Fragments can remain for months, becoming encrusted and causing chronic inflammation and discomfort. Sometimes they work their way out in a new place, damaging more tissue.

Punctures from black sea urchins require prompt medical attention. Deeply embedded spine fragments often have to be removed surgically — they can be detected by X-ray — and the injured area may have to be immobilised for some days. Antibiotics could be needed to halt infections.☐

SEA URCHIN STINGING

I. NORMAL SPINE WOUNDS

ACTION Try hot — not scalding — water to ease pain. Carefully pull out any spines that are easily removed. See a doctor for removal of embedded fragments and cleansing of wounds.

2. SEVERE STINGING CAUSING COLLAPSE

EARLY SIGNS Numbness spreading from mouth; slurred speech. General weakness.

ACTION Send for medical assistance. Rest victim with head lower than body and wounded area kept still but not bound. Prepare to give mouth-to-mouth resuscitation if breathing should fail.

Two examples of the stinging sponge Neofibularia mordens. *Effects are not felt for an hour or more after contact.*

THE TOUCH-ME-NOT SPONGES

Their skeletons are like poisoned splinters of glass.

Sponges are among the most primitive of animals. They are merely loose gatherings of single cells, without musculature or anything that could be called a brain — not even a nervous system. Feeding is virtually accidental. As water passes through channels in the structure of a sponge, organic particles adhere and are absorbed.

For all their simplicity, sponges can be immensely successful, in fresh water as well as marine environments. In the beds of seasonally flowing rivers in Central Australia, they have adapted to withstand fierce heat and desiccation. From deep ocean anchorages they may grow metres wide and metres high.

Sponge communities hold together and grow because some cells secrete a material that forms a lattice-like framework. In the types that are harvested, usually from very warm waters, it is a flexible fibre. Stripped of the sponge cells by boiling, this skeletal fibre is sold for bathroom use.

In other sponges the skeletal material may be limy. But it is frequently silica-based and formed in brittle, glassy splinters. Handling or treading on a sponge of any size with this sort of skeleton is likely to result in tiny skin wounds, usually not felt at the time. An irritating, often painful and sometimes incapacitating form of dermatitis may follow.

Gastro-intestinal symptoms produced in experimental animals appear to confirm that toxins are involved. Signs of general poisoning are rarely noted in humans, but the local effects can be bad enough. The worst cases require medical treatment for pain relief, and days of bed rest. There is no specific treatment for the skin disorder, which may take up to a month to resolve fully.

Perhaps the most formidable sponge in our nearshore waters is the dark blue or purplish *Neofibularia mordens*. Fairly common in St Vincent Gulf, SA, it reaches 50 cm in height. Attention was drawn to it in 1960, when a scuba diver brought one up from Willunga Reef, south of Adelaide. He and seven companions all suffered to some degree from touching it.

Arranging the sponge for photography, a man in the boat party contracted lesions that took days to appear but put him off work for two weeks. A woman who rested her bare feet on the sponge called the delayed effects 'absolute hell'. Her pain and itching increased day by day. They returned even a month later, every time her feet became warm.

Without specialised knowledge it is impossible to tell which sponges may contain glassy splinters. None should be touched — not even one that has been washed up on a beach. A dead sponge can remain dangerous until the cells decompose and the skeleton breaks down.

The fire worm

*Hidden under boulders or broken coral in
the intertidal zone, it has clumps
of bristles that can pierce leather gloves.*

Careless handling of a chunk of coral on a
tropical reef flat, or of shoreline boulders
anywhere down the east coast into NSW, is
inviting trouble. A slim worm, up to 20 cm long
and salmon-pink, may be crawling on the
underside. Contact with its bristles quickly
produces burning pain and itching — with
worse to follow if the victim is unlucky.

The fire worm *Eurythoe complanata* is the
most venomous member of a family of segmented marine worms called polychaetes,
meaning many-bristled. Though built much
like earthworms they are carnivores, with
highly developed senses to detect their prey and
biting parts to devour it. And to defend themselves they have stinging bristles protruding
from fleshy appendages on each segment. Most
have just a few bristles. But the fire worm
sprouts them in luxuriously bushy clumps.

These sharp, slender bristles, made of a
material like limestone, penetrate skin with the
greatest of ease. Even gloves may offer no
protection. A pioneer of marine biology on the
Great Barrier Reef, C. T. Roughley, remarked in
a *National Geographic* article in 1940 that the
bristles of a Heron Island fire worm had readily

Fire worms Eurythoe
complanata *are
found worldwide at
the edge of tropical and
temperate seas. As well
as a hazard to people
turning over stones
and coral rubble, they
are the bane of anglers
cleaning the fish that
feed on them —
notably the yellow
sweetlip of the Barrier
Reef region.*

pierced a thick leather gauntlet, poisoning his
index finger and leaving it partially numbed for
six weeks.

Palpitations and fainting

The initial pain and itching of a fire worm
stinging may continue for up to a week. In the
meantime a rash of white, pinhead-sized
pimples appears. The area soon swells and
reddens, and if fingers are affected they may
become uselessly stiff for a day or so. Lymph
nodes may be painful and in some cases chest
pains, an accelerated pulse, palpitations and
fainting have been reported. Most symptoms
disappear within ten days — unless the wounds
become infected — but numbness can persist
for a month or more.

Bristles broken off in the skin are tiny and
almost transparent — virtually impossible to
extract with tweezers or forceps. A better way of
removing them is by gently laying adhesive
tape over the wounded area, then lifting it off.
Medical attention is advisable in case general
illness or a secondary infection should develop.

Related bristle worms from deeper waters,
including the 'sea mice' *Chloeia flava* and
Aphrodite australis, sting in the same way. But
they are smaller animals and more sparsely
bristled, so are less likely to give trouble. The
huge reef worm *Eunice aphroditois*, from the
same family, causes more problems by biting. It
is dealt with in Part 1 of this book.□

BRISTLE WORM STINGING

ACTION Lay adhesive tape gently over wounded area and lift
out bristle fragments. Wash area with antiseptic or salt
water. Try vinegar, diluted ammonia or cooling lotions to
relieve pain. Seek medical attention in case wounds are
contaminated.

Bohadschia argus.

Pentacta anceps.

Stichopus variegatus.

A QUESTION MARK ON SEA CUCUMBERS

Excretions blamed for blistering and blindness.

Last century it was claimed that some sea
cucumbers — the *bêche-de-mer* prized by
Chinese gourmets — excreted a fluid that
could cause blistering of the skin. If carried to
the eyes, it was said to be capable of inflicting
blindness. No such cases have been recorded,
and scientists more recently have disputed
the claim.

Sea cucumbers, or holothurians, are related
to starfishes and sea urchins but have no
defensive spines. They do secrete a toxin that
is poisonous to fish. It has been shown
experimentally to have a potential
neurotoxic effect on mammals.

Once commonly harvested on the Great
Barrier Reef, the sought-after species were
not sent to Asian markets without preliminary boiling, for conversion into the form
known as trepang. While there is no evidence of direct danger — short of eating sea
cucumbers raw — it would be prudent to
wash the hands after touching one.

The stayput stonefish

A master of disguise, the idlest predator of the underwater world plays a waiting game. And in defence, its armament is fearsome.

All but buried in mud or sand and perfectly camouflaged, looking like an encrusted boulder or a chunk of old coral, the stonefish is content to be motionless for hours on end. It cannot be detected by other sea creatures — let alone by people entering the warm shallows where it may be feeding. Even sting victims sometimes fail to find the cause of their agony.

Shrimps and smaller fish virtually make a present of themselves to the indolent stonefish, by swimming within lunging distance of its upturned mouth. Prey is sucked in with lightning speed — the action of opening and shutting the mouth has been timed photographically at just 0.015 seconds.

A life of such ease is not without problems. The stonefish is at risk from bottom-feeding sharks and rays that may chance on it. But its defence is formidable. If something big disturbs the water or the sea bed close by, the fish still does not budge. It simply erects a row of 13 spines that lie along its back fin. Anything meeting the spines with enough force to be deeply pierced is injected with toxins. Their potency makes this the most venomous fish in the world.

Two kinds — equally dangerous

Too elusive for scientists to be certain of its range, the stonefish has been found in scattered nearshore locations between the Houtman Abrolhos Islands, off Geraldton, WA, and far northern NSW, as well as along the Great Barrier Reef. There are at least two species, similarly armed and equally menacing.

The reef stonefish *Synanceia verrucosa*, which likes to lurk among coral rubble, seems to be fairly rare. It can be distinguished by the separate bony protrusions on which its eyes are carried. The estuarine species *Synanceia trachynis*, with both eyes sharing a single ridge, is thought to be relatively common on rocky and muddy bottoms. The most frequent encounters with it have been in Moreton Bay, Brisbane.

At home in brackish tidal inlets as well as in sea water, stonefish may conceal themselves at depths as great as 40 metres or anywhere up to the low tide zone. Nestled in damp mud or sand, they are able to tolerate hours of part-

STONEFISH ENVENOMATION

EARLY SIGN Pain increasing markedly soon after stinging.

ACTION Leave water immediately. Try to ease pain with hot — not scalding — water if available. Do NOT attempt to arrest the spread of venom. If pain continues to increase, send for medical aid. Even after a minor stinging, see a doctor in case treatment is needed for contamination.

COLLAPSE OF VICTIM Help to safe place, rest with head lower than body and send for medical aid. Monitor breathing and pulse and be prepared to use resuscitation techniques (page 356).

The reef stonefish Synanceia verrucosa, *above, can be distinguished by its two eye ridges. The more common estuarine species — pictured overleaf — has a single ridge.*

exposure to the air during extremely low tides.

Instead of scales, a stonefish is covered with warty protuberances. These exude bitter-tasting and slightly toxic secretions, along with a sticky slime to which mud and marine organisms adhere, aiding the animal's disguise.

Huge side fins, used for scooping out hiding places, make the stonefish an ungainly swimmer. Heavy-bodied and stumpy-tailed, it has none of the streamlined grace of most other fishes. But swimming skills are little needed by a creature that makes such a success of immobility. Giants about 50 cm long have been found, though a more usual size is 20-25 cm.

Smaller doesn't mean safer

The venom apparatus is efficiently developed even in tiny stonefish. If fact there is a greater chance of envenomation from one of modest size than from one that is full-grown, because shorter spines do not have to penetrate so far before the mechanism is forced into action. The critical depth is about 1 cm in stonefish of

WHERE
STONEFISH ARE
FOUND IN
AUSTRALIA

WHERE
STONEFISH ARE
FOUND IN
AUSTRALIA

average size, but not much more than 0.5 cm in one that is only half as long.

When a stonefish is alarmed, the three longest spines at the front are raised almost vertically. Ten other stinging spines are angled backwards. There are three further spines under the tail and two pointing out from the belly, but these are not venomous.

Some people have trodden heavily on big, healthy stonefish and escaped all symptoms of envenomation. Though their wounds were painful, they were lucky enough to be impaled by spines that had been used not long before and exhausted of venom.

On the other hand an apparently dead stonefish, stranded above the waterline, may be alive and highly dangerous. Even a truly dead one, though it cannot erect its spines, should not be handled carelessly or kicked. The venom remains toxic for many days.□

Wear shoes — and go gently

Impulsive actions can incur a huge cost in pain.

Foot punctures account for more than two-thirds of serious stonefish envenomations. Wearing shoes removes most of the risk, as long as people tread gently. But if they go running — or worse still jumping — into the water, the fish's spines can penetrate sandshoes, thong

WHEN A STONEFISH DELIVERS ITS VENOM, THE VICTIM DOES ALL THE WORK

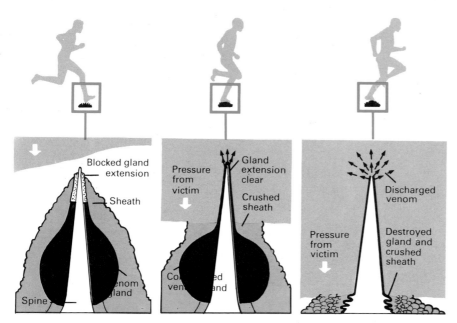

As pressure increases, venom seals break and a spine is bared.

A stonefish's 13 stinging spines are encased in thick, warty sheaths. Pairs of bulbous venom glands towards the base of each spine have narrow extensions leading up along grooves that reach almost to the sharp tip.

The gland extensions are normally sealed off with fibrous material. But if something pushes down on an erect spine with sufficient force to make it penetrate deeply, the sheath is stripped back and the blockage is removed. At the same time the glands are tightly compressed.

With explosive force the venom shoots up the spine grooves and into the wound. The glands are emptied in a single burst and destroyed. Much of their cellular material enters the wound, along with the liquid venom. The glands regenerate and are re-charged with venom in a few weeks.

The stinging spines of other fish work on the same principles. They may use muscles to direct their spines towards an enemy, but no other effort is involved. The spines function only when injured by an outside force.

An estuarine stonefish Synanceia trachynis, left, sprawls over a plucky researcher's hand. It can be picked up this way, by scooping into the sand underneath it – but extreme care must be taken that it does not slip and strike a leg or foot.

Right: A more prudent attitude is shown by this diver, filming an estuarine stonefish Synanceia trachynis from a safe distance.

sandals or snorkelling flippers. Dangerous hand wounds are sometimes inflicted after stonefish are accidentally netted or hooked in fishing. But too often they come about through people trying to pick up or turn over what they take to be rocks or coral on the sea bed.

Children in particular should be cautioned about the stonefish and its remarkable skill in camouflage. While the chance of a stinging is too slight to overstress the issue and make a child frightened of northern waters, some of a young person's natural exuberance and curiosity may have to be curbed.

Stonefish have killed many Pacific and Indian Ocean islanders. Their lethal potential figures prominently in the lore of northern Aboriginal tribes. But no deaths have been recorded in Australia during the period of European settlement. The likelihood has been made more remote by the development of an antivenom in 1959.

Nevertheless, dozens of people go through agonising ordeals every year. And along with the local effects of a stinging, severe envenomation carries the risk of collapse — which could have dire consequences if the victim is alone in the water or on a boat.

Victims may become irrational

Penetration by a stonefish spine — or several spines in the worst cases — brings instantaneous pain which rapidly sharpens if venom has been injected. It may increase in severity for ten minutes or more, and then continue unabated for many hours. Conscious victims can become delirious, or so frenzied that they strike or bite people trying to help them.

Swelling is rapid in the punctured area, and it often spreads up the limb. Muscle weakness in the limb may lead to temporary paralysis. Symptoms of shock may develop — including fainting. The pain is likely to spread to lymph

THE UNLOVELY LOOK-ALIKES

Two fish that may be mistaken for stonefish are not regarded as dangerous.

The banded frogfish *Halophyrne diemensis*, common in tropical waters, can be mistaken for a stonefish. It has much the same build and some similar habits. In fact its alternative common name is the bastard stonefish. If seen it is likely to be on the move, however, and it has only two back spines. These may be mildly venomous but the fish is not regarded as dangerous. Nor is a related fish found in NSW waters, the brown or dubious frogfish *Batrachomeus dubius*.

Brown or dubious frogfish Batrachomeus dubius.

Banded frogfish Halophyrne diemensis.

glands. Ulceration of the wound may follow, perhaps lasting for months. Gangrene can occur in untreated cases.

First aid efforts, once a victim is out of the water, can be directed only at relieving the pain and guarding against the possible effects of shock. Bathing the wounded area in hot water may ease the pain in mild cases. But powerful opiate drugs could be needed if the envenomation is severe, and only a doctor can administer them. In any case it is advisable to seek treatment because there is a high risk that the wound will be contaminated.

No attempt should be made to arrest the progress of the venom with a pressure bandage or tourniquet. That would only increase both the pain and the likelihood of tissue damage at the wound site.☐

Half-buried in sand, a stonefish is perfectly disguised.

GURNARD SCORPIONFISH *Neosebastes pandus*. Also called gurnard perch. Rocky reef shallows and seagrass meadows, southern WA, SA, Vic, Tas.

The scorpionfish family

Close relatives of the stonefish, scorpionfish can inflict similarly severe pain, with a chance of collapse from shock.

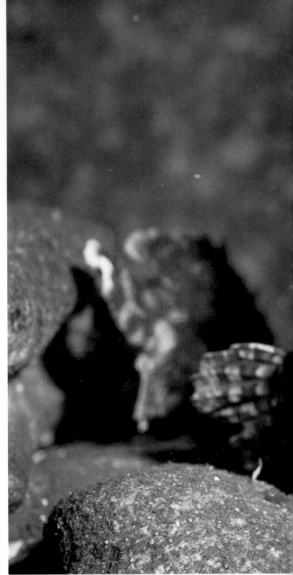

Generally more mobile than their stonefish cousins, the other members of the widespread scorpaenid family are less likely to be trodden on. But they are far more likely than stonefish to turn up in a fishing catch. No estuarine or nearshore waters around the coasts, islands and reefs of Australia, including Tasmania, are free of them. At least two or three kinds may be common in any location.

Though the venom glands of the scorpionfishes are not as highly developed as those of stonefish, their back spines and sometimes other spines can inflict maddening pain and symptoms of shock. This poses a particular menace to boat and rock fishermen, who if they collapsed could fall into the sea and drown.

Catches should be closely inspected before they are touched, and inexperienced people are advised not to handle anything resembling a scorpionfish without extreme care and heavy gloves. They should remember, too, that the spines remain venomous for days after a fish is dead, even if it is kept in cold storage.

Some, especially the firefishes and other so-called 'butterfly cods', shown overleaf, are highly attractive and are prized as aquarium fish. But many people, unaware of the danger, have been severely stung in their own homes, and even while working in pet shops.

Heat destabilises scorpionfish venoms and helps them disperse through the bloodstream, reducing pain. First aid measures and the medical management of severe stingings are the same as those employed to help stonefish victims — but there are no antivenoms.☐

Right **DEVIL SCORPIONFISH** *Scorpaenopsis diabolus*. Mainland and offshore reef shallows, Qld, northern NSW.

SCORPIONFISH ENVENOMATION

EARLY SIGN Pain increasing markedly soon after stinging.

ACTION As for stonefish envenomation (page 221). Do NOT attempt to arrest spread of venom.

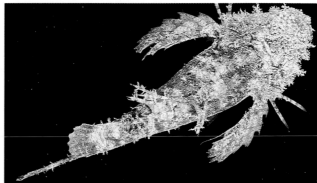

GOBLINFISH *Glyptauchen panduratus*. Also called saddlehead. Rocky reef shallows, subtropical WA, SA, Vic, Tas, NSW.

CARDINAL SCORPIONFISH *Scorpaena cardinalis* (above) and **CHAINED SCORPIONFISH** *Scorpaena ergastulorum* (right). Both commonly known as red rock cod, also prickly heat, mouth almighty. Rocky reef shallows, SA, Vic, Tas, NSW, Qld.

BEARDED GHOUL *Inimicus caledonicus* (left) and *Inimicus sinensis* (above). Also called demon stinger. Mainly offshore but sometimes rivers and estuaries, northern WA, NT, Qld.

FIREFISH *Pterois antennata* (above) and *Pterois volitans* (below). Also called butterfly cod, lionfish, zebrafish, turkeyfish, fire cod, coral cod. Mainland and offshore reefs in tropical and subtropical waters.

BULLROUT *Notesthes robusta*. Also called 'croaky' because

Above **THREE-SPINED SCORPIONFISH** *Taenianotus triacanthus*. Coral or rocky reefs, Qld, NSW.

Below Swollen hand 3 days after firefish stinging.

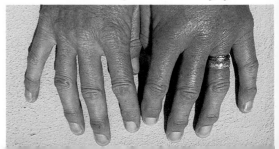

FALSE STONEFISH *Scorpaenopsis gibbosa*. Tropical reef shallows.

BUTTERFLY SCORPIONFISH *Dendrochirus zebra*
(above) and *Dendrochirus brachypterus* (below). Also called
butterfly cod. Mainland and offshore reefs in tropical and
subtropical waters.

of a noise it makes. Rivers and muddy estuaries, east coast.

FORTESCUE *Centropogon australis* (above) and *Centropogon
marmoratus* (below). Also called waspfish. Estuaries from southern
Qld to southern NSW.

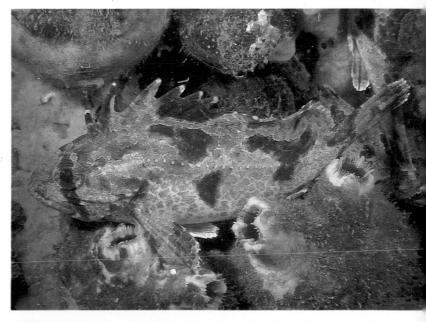

COBBLER *Gymnapistes marmoratus*. Also called soldier fish. Mainland and reef
shallows, WA, SA, Tas, Vic and NSW.

227

Catfish: hellish handfuls

They punish careless captors with a powerful twitch of their fins. Venomous spines spring out in three directions.

Catfish owe their name to slender, fleshy sense organs, sticking out around their mouths like whiskers. Called barbels, these organs are harmless. To inexperienced people trying their hand at fishing they serve as warning signs, identifying a big family of potentially dangerous stingers.

In most fishes with venomous spines, the defence mechanism is entirely passive. It works only if external force is applied. But a catfish exerts its own force. If it is interfered with it stiffens its fins and pulls on them in such a way that three spines spring out violently — one at right angles from the back fin, and others from each of the side fins.

The only safe way to hold a catfish is by the tail. Attempts to grasp it anywhere around the body invite a severe stinging. Hours of intense pain may follow, and in the worst cases there is a risk of collapse from shock. Wounds are easily contaminated by bacteria. Spines are barbed in many species, increasing tissue damage.

More than 30 species of catfish are found in and around mainland Australia and Tasmania. Bottom-feeders, active mainly at night, they are

Above: Striped catfish Plotosus lineatus *commonly swim in tightly packed spherical formations. This is the species blamed for fatal stingings overseas.*

Below: The blue catfish Arius graeffei *frequents fresh and saltwater shallows in all tropical and subtropical waters.*

commonly netted in estuaries. But their other habitats include rivers, freshwater lakes, tidal lagoons, mudflats and rocky or coral reefs, to depths of up to 40 metres.

No deaths from catfish stingings have been recorded in Australia. Fatalities overseas have been blamed on the vividly striped *Plotosus lineatus*. It is abundant here, often moving in dense schools, and it can grow to a length of a metre. While this may be our most dangerous species, all catfish are presumed to be venomous and any of significant size should be treated with great caution.□

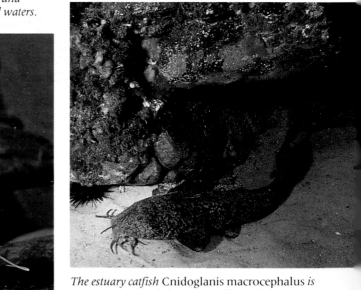

The estuary catfish Cnidoglanis macrocephalus *is common over sandy bottoms in all mainland states.*

CATFISH ENVENOMATION

EARLY SIGN Pain increasing markedly soon after stinging.

ACTION As for stonefish envenomation (page 221). Do NOT attempt to restrict blood circulation.

Other venomous fishes

*All can cause prolonged and sometimes shocking pain with their defensive spines,
as well as leaving troublesome wounds.*

None of the fishes shown here, or their many relatives, is known to have directly caused human deaths in Australia. But the pain that they can inflict with their spines is sometimes so overwhelming that there may be a risk of collapse in a hazardous situation.

Other fishes, though scientists have found no evidence of their having a venom mechanism, can produce pain out of proportion to the injuries that they inflict. The lesson for inexpert fishing enthusiasts and divers is not to handle or kick *any* unfamiliar fish, or to let one leap about in a boat.

Little or nothing is known about the composition of many fish venoms. But they all seem to share some important characteristics. They are destabilised by heat, and their effects are mainly local — they can safely be allowed to disperse in the bloodstream. So the emergency advice given for stonefish envenomation on page 221 applies to all stings from fish spines.□

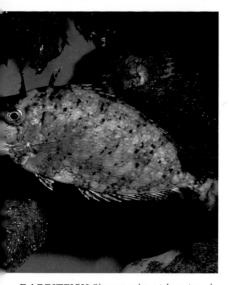

PORT JACKSON SHARK *Heterodontus portusjacksoni*. Also called doggie, oyster crusher. Single venomous spine at front of each of its two fins — sharpest in younger sharks. Bottom-feeder at depths to 200 metres but breeds in shallows, all non-tropical waters except Tas.

RABBITFISH *Siganus spinus* (above) and *Siganus doliatus* (below). Also called spinefoot. First back spine most dangerous, but other back spines and pelvic and anal spines also venomous. At least 10 species in rivers, estuaries and around reefs, all regions except Vic, Tas.

OLD WIFE *Enoplosus armatus*. Also called zebrafish. Back spines cause envenomation unless lacerations bleed freely. Common in harbours, on seagrass meadows or around reefs, all regions.

RATFISH *Hydrolagus lemures*. Also called elephantfish. Single venomous back spine. Sometimes trawled from deeper non-tropical waters, WA to NSW and Tas.

FLATHEADS *Platycephalus fuscus* (above) and *Thysanophrys cirronasus* (right). These and many other flathead species have dangerous head spines. Bottom-dwellers in bays, estuaries and offshore waters in all regions.

THE AUTOMATIC AMBUSH

In every sea, at any depth or temperature, certain animals catch and consume swimming prey without even knowing that it exists. Grappling and killing devices are triggered automatically, in response to a victim's touch. The action is no more conscious than that of an insect-eating plant.

These unwitting predators are known as coelenterates or cnidarians. They are the simplest animals to have definite organs and muscles and nerves. Jellyfishes, sea anemones and corals are the most prominent types. Others are easily mistaken for seaweeds.

Lacking a brain, a heart, a circulatory system or even an intestine, a coelenterate is essentially a cup of digestive cells generally armed with contractible tentacles. The most advanced can swim, and even evade danger, instead of merely drifting or staying anchored to the sea bed. But no coelenterate has any other deliberate choice of action.

Primitive though they are, the coelenterates readily dispose of prey animals that are much more highly evolved. Their attacking mechanisms are remarkably efficient. And the action of the venoms some of them produce can be terrifyingly fast.

ARMED CAPSULES THAT FIRE AT A TOUCH

Tentacles of coelenterate animals are covered with myriad capsules, called nematocysts. They take many forms, but each has some sort of triggering mechanism — a projecting bristle, perhaps, or a chemical receptor — that responds to the slightest stimulus.

When the trigger is activated a tube emerges forcefully, pushing aside a lid if the capsule has one. Gaining its energy from a rapid pressure change, the tube turns itself inside out, like the collapsed finger of a rubber glove blown suddenly into its proper shape. A slender thread that lay coiled at the tail of the tube is fired out.

Nematocyst threads in some coelenterates simply act as snares, lassoing minute animals or restraining them with spines or sticky substances. But most are designed to penetrate the bodies of prey and inject venom. Often these injector threads carry barbs. The threads hold the victim harpooned while paralysis sets in.

Fortunately the venomous threads of many of these animals are too small and weak to pierce human skin, particularly where it is hardened. We can run our hands over many feeding reef corals and most sea anemones, for example, and the worst sensation may be a sliminess. Some coelenterates that do succeed in stinging people on more tender areas of skin have a negligible effect because the animals themselves are tiny.

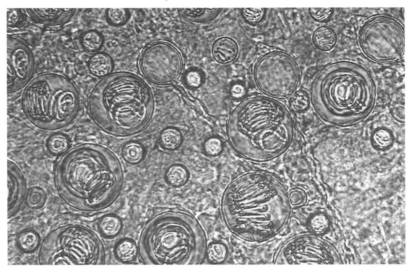

Above: Nematocysts of a Portuguese man o'war under a microscope. Some capsules have discharged; others remain ready to fire.

Right: How a typical nematocyst discharges, when it is triggered by the touch of other living organisms. An internal tube turns itself inside out, injecting a tube through which the capsule's load of venom is forced.

Left: A magnified cross-section through a chironex box jellyfish tentacle shows stinging capsules tightly packed around the blue core.

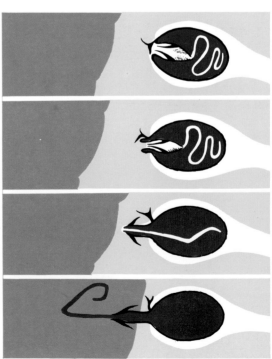

230

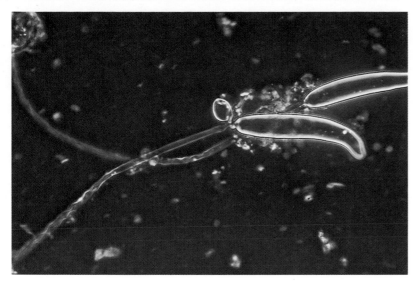

The extra menace of jellyfishes

At the other end of the scale of coelenterate danger are big jellyfishes with long, trailing tentacles. Extensive areas of our bodies can come into contact with thousands of nematocysts on a single animal. Several jellyfish species inflict immense pain and severe tissue damage. The chironex box jellyfish can bring rapid death to widely-stung people.

Since the diet of dangerous jellyfishes consists mainly of shrimps, prawns and sometimes small fish — all easily speared and subdued — the potency of their venoms may seem excessive. But having no choice in what may blunder into their tentacles, and no way of restraining themselves from the act of capture, they often have to defend themselves from what they unwittingly attack.

A fish of moderate size, small enough to be seized, may be too big to eat. But it still has to be paralysed quickly. Otherwise its struggles, engaging more and more tentacles, could wrench them off and leave the jellyfish with no means of feeding.

Human victims, in that sense, are like even bigger fish that a jellyfish has no chance of holding. People have ample strength to break away, snapping the tentacles. The jellyfish may survive and regenerate the lost parts. But a victim's plight does not end. The broken-off tentacles continue to cling and the stinging can increase as movements trigger the firing of more nematocysts.

How sting effects can vary

The severity of a coelenterate stinging depends not only on the potency of the venom but also on how many nematocysts are discharged into the skin. Two people stung by the same sort of jellyfish can experience very different effects. One may have a fleeting contact with part of a single tentacle; the other may be wrapped around by many.

In jellyfish of the same species, the one with the biggest bell usually has the longest tentacles — and in the case of the deadly chironex, the greatest number of tentacles. In theory it follows that the biggest-belled jellyfish has the most nematocysts. But in practice there is a factor of sheer luck. If we happen to encounter part of a tentacle that has been used recently in feeding or defence, many of the stinging capsules may be exhausted and harmless.

Shed individually or on tentacle fragments, nematocysts keep their stinging power long after detachment from their owners. In high concentrations — in containers in which jellyfish have been kept, for example — they can produce a phenomenon known as 'hot' water. People suffer stings with no animal present.

In tidal lakes and estuaries, detached nematocysts are thought to be the cause of some irritating skin eruptions suffered by swimmers. Their complaint can be distinguished from the usual 'bather's itch' — caused by worm larvae — because the stinging is felt immediately.□

Long injector threads lead from discharged nematocysts of a chironex box jelly. Not all are venomous — some are used simply to cling to prey animals.

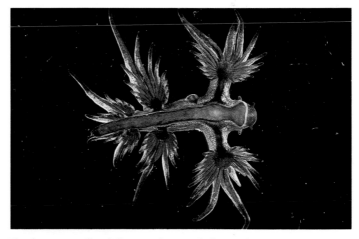

The glaucus or sea lizard Glaucus atlanticus *feeds on coelenterates and uses their stinging capsules for its own defence.*

A SECOND-HAND SYSTEM OF DEFENCE

Nematocysts are exclusive to the coelenterates. No other animals have developed anything of the kind. But one sort of creature, entirely unrelated, converts nematocysts to its own use.

Nudibranchs or sea slugs of the aeolid group are specialist feeders on coelenterates. Somehow they can nibble off and swallow the nematocysts without triggering their discharge. In the glaucus or sea lizard especially, the still-potent capsules work their way out through the tissues and emerge on the surface of the animal's own tentacles, giving it a borrowed defence against predators.

A brush with a glaucus, often found floating on ocean surfaces, can incur a stinging — more or less painful according to what it has been eating. But because it seldom grows to more than 3 cm in length, serious consequences are most unlikely.

231

Chironex: the unseen threat

All but invisible as it cruises in calm tropical shallows, this jellyfish is the world's most dangerous marine stinger.

Below: A Chironex fleckeri *in mangrove flat shallows.*

One species of jellyfish, *Chironex fleckeri*, is blamed for the deaths of about 60 people — mostly children — in our tropical inshore waters since the 1880s. That exceeds the combined toll taken in the same region by sharks and crocodiles. Yet for most of the period the scientific establishment refused to believe that such a lethal creature could exist and not be seen. When broken tentacles were found clinging to victims, medical authorities were usually content to accept that the culprit must be the Portuguese man o'war or bluebottle, *Physalia*. The fact that *Physalia* was well known in the south and all over the world, and had never killed anyone, was glossed over by assuming that the tropical victims had heart conditions or allergies. Old hands in the north were not deceived — especially not those who had picked up Aboriginal lore. They knew that other stingers lurked in the shallows in summer. But in a push to expand tropical settlement and tourism, their talk of mysterious 'sea wasps' was not fashionable.

Proof that the real killer was a previously undescribed cubomedusa, or box jelly, emerged late in the 1950s. The nickname of sea wasp unfortunately persisted. It led some people, misunderstanding local advice, to plunge into dangerous waters content that there were no insects about. Others panicked when they were bitten by March flies.

Brisbane's false alarm

Alternative names offered included 'fire medusa', 'cubo' and the Aboriginal *indringga*. Plain 'box jellyfish' has gained the widest currency in recent times. It is unhelpful, because *Chironex* is just one of many box jellies. When another type was found in Moreton Bay in 1984, sketchy media reports led some Brisbane people to believe that the lethal tropical kind had advanced south.

Dr Robert Endean, associate professor of zoology at the University of Queensland and a leader in research on the creature, has proposed that the generic name, since it is exclusive to the one species, be encouraged in popular use. The editors of this book are happy to lend support.

The chironex has been found at many points between Gladstone, just south of the Tropic of Capricorn on the Queensland coast, and Broome, WA. But its distribution is probably not continuous over that range.

Apparently restricted to inshore waters, the chironex has not been found around Great Barrier Reef structures except in the far north where they lie very close to the mainland. But it frequents Melville and Bathurst Islands, near Darwin, and the islands of the Gulf of Carpentaria.

One rival known abroad

The world's only other fatal jellyfish stingings have been recorded sparsely in the Philippines, Malaysia, southern India and off islands east of New Guinea. A similar looking box jelly,

Chiropsalmus, is blamed for the deaths in the Philippines.

No other jellyfishes are believed to be capable of directly causing the death of healthy people. Of species in Australian waters, the chironex has by far the greatest total length of tentacles, and therefore the most stinging capsules. And its venom is many times more toxic than that of any other of our jellyfish.

Early ignorance of such an important animal is partly excusable. The chironex inhabits regions that were relatively little frequented by Europeans until World War II. And the circumstances of a bad stinging are not conducive to methodical investigation. When a child reels out of the water, draped with worm-like tentacles and screaming in agony, the last thing in anyone's mind is to rush in and try to net the rest of the jellyfish.

Even to determined researchers, the chironex is elusive. It is the most advanced jellyfish known, with the best-developed nervous system. It reacts to water disturbances, and sensors around its bell detect big objects by changes in light intensity. Propelling itself in the manner of an octopus, a chironex can flee more quickly than a person can wade. But its tentacles trail behind — sometimes as much as three metres away from the bell.

A shadow on the bottom

Pale blue or milky but semi-translucent, the bell of a chironex is extremely difficult to see in the water. Trained people watch for a purple tinge that shows in some of the tentacles and around their bases. In clear shallows under strong sunlight, a faint moving shadow on a sandy bottom may give a clue. But in muddied waters after storms, when the animal is most active and the sky may still be overcast, there is scarcely any chance of seeing it — or of the chironex sensing and avoiding its potential human victim.

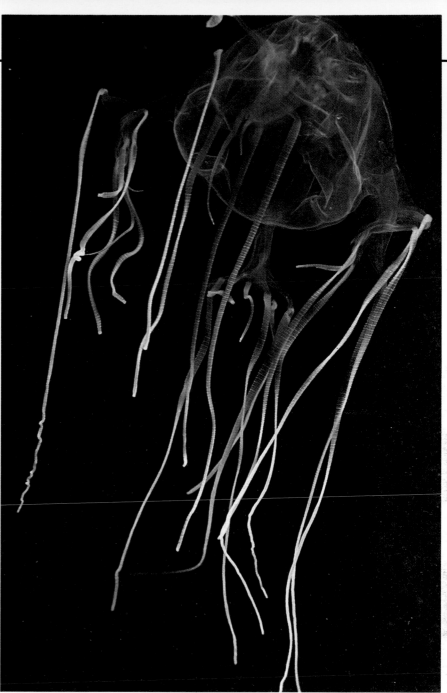

Contracted, the tentacles of a chironex reach only a quarter of their feeding length.

The box-shaped bell may measure as much as 20 cm across and 30 cm deep — bigger than a man's head. At four points around the open end, hand-like appendages protrude. As a chironex matures, these sprout more and more tentacles. A big specimen may have up to 15 in each bunch, varying in size but giving it a total tentacle length approaching 100 metres.

The tentacles are formed like ribbons, crossed with folds and ridges and coated with nematocysts. Not all of these are stinging capsules: other kinds fire hooks or exude adhesive substances to help the tentacles cling, or act as chemical sensors so that the jellyfish does not waste its venom on inanimate objects.

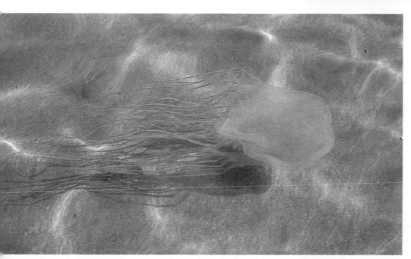

A purplish tinge near the tentacle bases may be all that is seen of a chironex.

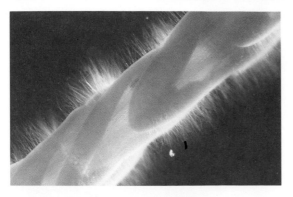

Left: The myriad stinging cells on a section of chironex tentacle, placed in methylated spirits, all fire out their injector threads. Meths or other alcohols, once suggested as treatments for stingings, only make them worse.

A thousand stings from a pinhead

Shaped like slender cigars, the stinging nematocysts are arranged in bands with their tips just reaching the tentacle surface. They are astonishingly tiny and densely packed: more than 1000 venom-injecting threads can be fired from an area about the size of a pinhead.

Injector threads may exceed 0.2 mm in length. A chironex with a bell width of 15 cm or more has threads that can penetrate the skin on most parts of adult human bodies. Small children, because their skins are thinner and softer, are at risk from half-grown jellyfish.

Estimates of the total number of stinging nematocysts carried by chironexes range up to thousands of millions. But no one is stung by all of them. More to the point is the length of tentacle contact required to receive a potentially lethal dose of venom.

Adhesion of 6-7 metres of tentacles meant probable death to untreated victims, according to the pioneering investigator, the late Dr John Barnes. His observations were based on cases in which many of the victims were children. Robert Endean's calculation of the length of uncontracted tentacles needed to kill an adult weighing 70 kg is 17.5 metres. Other authorities put the life-threatening length for adults as low as 3 metres.

Right: A chironex in one of its early polyp phases, fixed to a rock and eating a microscopic shrimp.

Below: Starting its life as a free-swimming medusa, the young chironex has still to develop most of its tentacles.

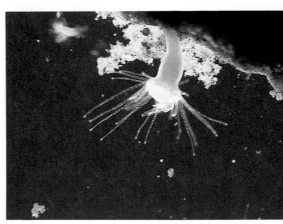

Riding a storm from below

Like any other jellyfish the chironex is easily damaged in boisterous seas. Where there is surf along the tropical Queensland coast, people sometimes suffer minor stingings from tentacle fragments left in the water.

The chironex is a survivor, however. It uses its superior swimming ability to dodge the worst effects of storms, withdrawing to deeper water and resting on the bottom. When conditions are calmer it returns to the shallows to feast on prawns, shrimps and small fish flushed out from estuaries.

Over most of its range, the chironex disappears during the dry months of winter and spring. Juveniles and some surviving adults turn up in early summer. Until the 1980s it was generally assumed that the species also sought a deep-water retreat for spawning. But infantile polyp forms, as well as juveniles, have recently been found upstream in tidal creeks.

Wiping out the chironex will probably never be feasible, however much some people would like to see it happen. The leading predator of the adult jellyfish, the hawksbill turtle, suffered at the hands of 'tortoiseshell' jewellery and soup manufacturers; its numbers may be permanently depleted. But the involvement of upstream habitats in the life cycle of the chironex raises hopes for its eventual control near popular beaches.□

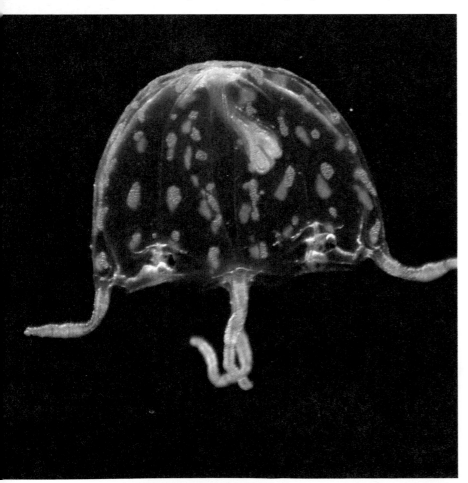

When a blackout is a blessing

A maddened victim's struggles will make matters worse.

Old-time stories of 'sea wasp' stingings often paint a startling picture of death within seconds. No envenomation could take effect that quickly. If any victims died while still in the water it was because they collapsed, and rescue efforts were slow. And in the old days, people knew nothing of proper resuscitation techniques.

In fact there are always a few minutes in which the most severely stung chironex victim can be saved. And provided that people on hand have a thorough knowledge of first aid, the victim's rapid loss of consciousness may be the best thing that could happen.

Overwhelmed by mounting waves of pain, people who are severely stung usually tear at the clinging tentacles and thrash about, increasing the stinging and the rate of absorption of venom into the bloodstream. They may become irrational and need forcible restraint.

The potentially lethal component of the venom was thought for many years to be a cardiotoxin, acting directly on the heart muscle. Recent findings indicate that it may be a neurotoxin with a more general effect.

This component works with unusual speed — sometimes in less than five minutes — because of the dispersed way in which it is injected. But its effects seem to last for only 20-30 minutes. Aided by resuscitation techniques if necessary during that period, a victim should pull through.

Antivenom prevents scarring

Supplies of antivenom are issued to hospitals, private doctors, ambulance stations and some lifesaving clubs in areas where stingings may be expected. Where first aid is effective, antivenom may not always be needed to save a victim. But it relieves pain and breathing difficulty and counteracts the effect of a second component of the venom — a skin destroyer.

Wheals like whiplash marks, up to 0.5 cm wide and often showing a cross-barred pattern, are raised immediately wherever a tentacle makes contact. About six hours later, in a major envenomation, these start to blister. Skin death may begin within a day. If the lesions go untreated there may be deep, long-lasting ulcers and permanent scarring.

A third component of chironex venom has the power to damage red blood cells. It does not seem to be of any medical importance during the emergency phase, but it may contribute to the later skin destruction.☐

MARKS ARE SLOW TO FADE

Antivenom injections after a chironex stinging are not only essential for life-threatening stingings, but also combat tissue destruction and reduce subsequent scarring. The wheals, however, take months to fade. The photographs below show, clockwise from top left: a young female victim's thigh after 3½ hours, 24 hours, ten days, and two months. They were taken at Townsville General Hospital, Qld.

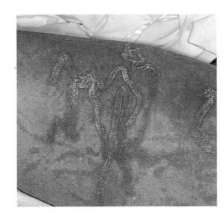

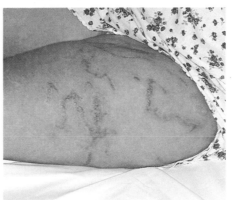
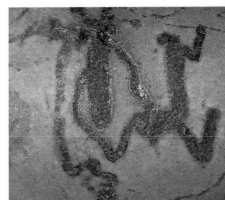

A swimmer holds the immature chironex that was capable of stinging only slightly on the calf and foot.

CHIRONEX ENVENOMATION

EARLY SIGNS Immediate pain, increasing; uncontrollable screaming. Whitish strings adhering to skin, causing red, purple or brown wheals.

ACTION 1. CONSCIOUS VICTIM Assist victim from water, if possible not touching tentacles. Lie victim in safe place with head lower than body and restrain from moving. Douse tentacles with vinegar for a minimum of 30 seconds. Send for medical attention, or help from nearest lifesaving club or ambulance station. Do not leave victim. If victim loses consciousness, monitor breathing and pulse and prepare to use resuscitation techniques (page 356). Shelter a recovering victim.

2. UNCONSCIOUS VICTIM Grip victim by the arms and drag from water onto beach, if possible not touching tentacles. Lie victim in safe, preferably sheltered place and begin resuscitation techniques immediately (page 356). Douse tentacles with vinegar for a minimum of 30 seconds. Send for medical attention, or help from nearest lifesaving club or ambulance station.

NOTE Do NOT rub or wipe off tentacles or attempt any other method of removing them unless unfired stings have been deactivated with vinegar. Do NOT use methylated spirit or other alcohol as a substitute. The advisability of applying a pressure bandage is debatable. A tourniquet should NEVER be used.

ENEMIES ON THE HOME FRONT
Invasion trainees didn't know what had hit them.

Australian governments showed no concern about fatal jellyfish stingings in tropical waters until Japan entered World War II and ports in far northern Queensland and the Northern Territory became training and staging bases for a counter-attack. Hundreds of soldiers and sailors were put out of action and a few were killed.

A young Dr Ronald Southcott, now Adelaide-based and one of Australia's most eminent zoologists, was among the first to tackle the problem. He was one of two medical officers attached to a beach assault group, stationed near Cairns in the summer of 1943-44. They soon found that they were dealing with two distinct types of stingings — one with an immediately drastic impact, the other with a delayed action.

They enlisted the local knowledge of Dr Hugo Flecker, a Cairns general practitioner who had started Australia's first register of injuries from tropical animals and plants. When the tides of war swept them on, Southcott and his colleague left Flecker with the problem, better defined but unsolved.

With no progress made even ten years later, Flecker pressed for police help in seeking the jellyfish responsible next time a severe stinging occurred in the district. The opportunity came tragically at Cardwell in 1955, and several specimens of an unknown type of box jelly were netted. They were sent to Southcott, who formally described the new genus and named it *Chironex fleckeri* in his old friend's honour.

Many tests and experiments were to follow before it was proved that the chironex could and did kill. It was also necessary to try to establish how widely it was distributed. Important in this investigation, mainly undertaken by Southcott, was a canvassing of the knowledge of northern Aboriginal tribes. Drawings of the chironex were widely recognised. And among his mementoes is a fine

An Aboriginal bark painting from Melville Island, NT (left) helped trace the distribution of the deadly chironex. The jellyfish was named by Dr Ronald Southcott (below) for his wartime friend Dr Hugo Flecker (bottom).

bark painting of the animal.

Still unexplained were the other, delayed-action stingings suffered by the wartime trainees. Flecker had found them to be common among Aborigines to the north of Cairns, and called them 'Irukandji' stingings after the tribe. He died in 1957, unsuccessful in his search for the cause.

Another Cairns doctor, John Handyside Barnes, took on the task. With no previous experience in such work, and at a heavy price in accidental stingings and loss of income, he became a tireless field researcher.

His triumph came in 1961. Southcott did the honours again — the Irukandji mystery stinger became *Carukia barnesi*. Among Barnes's later achievements was the invention of a method of 'milking' the deadly chironex for venom studies, and eventually for the production of antivenom.

Eulogies were heaped on Barnes by the medical profession when he died in 1985, as they had been on Flecker. The two small-town doctors, egged on by Southcott, performed a most notable service for science and public safety.

The mystery solved
And the answer came as no surprise to local fishermen.

The horrible death in January 1955 of a five-year-old boy at Cardwell, Qld, led directly to the identification of the chironex as the killer jellyfish of the tropics. At the urging of Dr Hugo Flecker of Cairns, the district inspector of police insisted that his sergeant at Cardwell spare no effort to find tangible evidence.

Tentacle fragments, rubbed off the boy's legs and thighs by his mother, were recovered from the beach. Next day the sergeant took out a boat and netted three different types of jellyfish —

one of them the box jelly that Flecker had been looking for. Oddly, it caused no raised eyebrows around the town. The sergeant told an inquest that jellies of this type frequented the beach for three months every summer, and were 'regarded by local fishermen as being extremely dangerous'. □

Footnote: The coroner clung to the customary official view that no healthy person could be killed by a jellyfish. Death was put down to 'allergic shock' from the stinging.

If you can't stay out, cover up

Chironex stingings can be avoided — mainly by heeding warnings.

Most well-frequented beaches in the Northern Territory and tropical Queensland have big signs near carparks or entry paths, pointing out the chironex menace and giving various information. The warnings are meant with deadly seriousness.

Lifesaving club members often drag the shallows of their beaches with fine nets — not to clear them of stingers, for that is impossible, but simply to see if any are about. They arrange for special alerts to be broadcast on local radio stations. Tune in frequently if you plan to swim. These warnings may come outside the recognised 'stinger season'.

Parents should realise that at a time of chironex risk children must be kept *entirely out of the sea*. Paddling in a few centimetres of water is no less dangerous than swimming, and vigorous play in the shallows is worst of all.

If you must enter the water when stingers may be about — getting in or out of a boat, for example — wear a T-shirt and cover your legs fully. Light wetsuits called 'stinger suits', giving all-over protection, are sold in coastal towns. But they are not cheap. Before the suits were developed, principally for professional use, men unashamedly borrowed pantyhose from their wives or girlfriends. They are still a good idea.

Vinegar stops further stinging

Driving along a tropical coast, keep a flagon of vinegar in your car. You may never need it, but someone else could. When a stinging occurs, not all the venom capsules on the clinging tentacles fire immediately. A 30-second drenching with vinegar is the best way of making sure that they never do.

Never try to wash, scrape, rub or pick off

Local councils all along the tropical coastline are making their stinger warnings bigger and bolder, taking the view that it is better to have a scared tourist than a dead one. But the Darwin sign (top left) understates the danger period. Stingings have occurred there in June, July and September.

tentacles, and never put methylated spirit or other forms of alcohol on them — in spite of what some books and magazines say. Any of these actions is likely to increase a stinging.

At unfrequented northern beaches, don't be tempted to swim in summer or autumn unless your body and legs are fully covered, or unless you have at least two companions prepared to watch from the shore, with vinegar on hand, experience at resuscitation and knowledge of where to find medical help quickly. Better still, go inland and find a stream in hilly country. The water will be far more refreshing.□

'Irukandji' stingings: wait for the worst

*Some jellyfish stingings are deceptively mild at first. But a delayed reaction brings
agonising cramps, sickness and collapse.*

People too easily shrug off some jellyfish stingings in northern waters. The pain, seldom great, is limited to a small area which reddens and develops goose pimples. There are no alarming wheals. And the animal involved — in the unlikely event that it is seen — is of unimpressive size.

But the delayed effect, after perhaps 20-30 minutes, can be drastic. Victims commonly collapse with severe pain and cramps in their abdomen and limbs, crippling backache, nausea, vomiting and fits of coughing. Helpers unaware of the earlier incident have rushed victims to hospital fearing food poisoning, spinal injury, rupture of an internal organ or, in the case of divers, the 'bends'.

Hospital care is usually desirable in any case, simply to spare people hours of agony. And it may provide a safeguard for someone with a heart condition. On some occasions, a sharp rise in blood pressure occurs about two hours after the stinging. But no Australian deaths have been attributed to this type of envenomation. All of the symptoms eventually resolve themselves. The only danger for healthy people comes from ignoring the initial signs. Then

---

IRUKANDJI ENVENOMATION

EARLY SIGNS Moderate pain in red, pimpled area the size of a card.

ACTION Warn other people, leave water and rest in safe place. Douse affected area with vinegar. Obtain medical attention — but do NOT drive.

Carukia barnesi eluded hunters for three decades, until one swam straight across the face of Dr John Barnes in 1961. Even then he spotted only its stringy white tentacles. A second specimen was detected because a tiny fish moved oddly — caught on a tentacle of an invisible jelly.

illness could overwhelm them in hazardous circumstances — in deep water, for example, or while driving a car or a power boat. Even the strongest people can be quickly prostrated.

Jellyfish stingings with a delayed effect are called 'Irukandji' stingings, after the Aboriginal people of the coast north of Cairns, Qld, where the problem was first studied. Similar stingings have been noted at many points between Mackay, farther south, and Onslow, WA.

More than one culprit

The creatures to blame seem to form a closely related group, showing microscopic variations in their tentacle structure. So far only one species, *Carukia barnesi*, is accorded official recognition in scientific literature.

Carukia is a box jelly of the carybdeid family, having only one tentacle at each corner instead of bunches of them. Considering the trouble it can cause, it looks surprisingly puny. The bell, a rounded oblong rather than a square, rarely measures more than 2 cm on the longer sides. The longest tentacle is unlikely to reach more than 60 cm.

But unlike the chironex, *Carukia* carries stinging nematocysts on its bell as well as on the tentacles. And judging by the roughly oblong pattern of most recorded skin reactions, serious envenomations probably result from contact with the bells.

Irukandji stingers seem to arrive in swarms, infesting northern beaches only for brief periods. In Queensland at least, it may be that they come in only on certain currents forced by northerly winds. In any case their occurrence is normally within the chironex 'stinger season', so the same precautions will cover both risks.

An undeserved reputation

A multi-tentacled box jelly, of the same chirodropid family as the chironex, is occasionally found in northern Queensland waters. From early this century it was identified as *Chiropsalmus quadrigatus* — and much feared, for that is the type blamed for fatal stingings in the Philippines. Now it appears that it is not *Chiropsalmus* at all, but a distinct type waiting for recognition in its own right.

It looks much like the chironex, though it is smaller and has only about half as many tentacles in each bunch. The composition of the venom is similar, too, but a mature animal's stinging power is about 100 times less. ☐

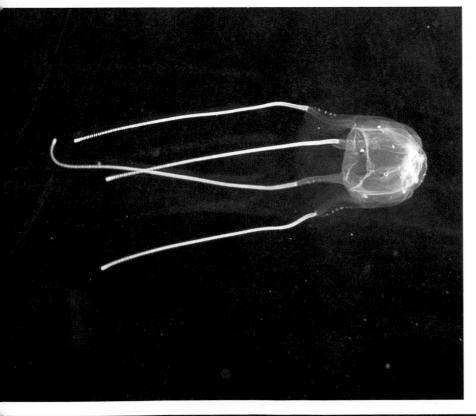

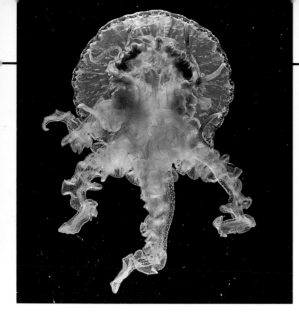

Left **LITTLE MAUVE STINGER** *Pelagia noctiluca*. All coastal waters but uncommon in south. 4-12 cm across bell. Effects: Irregular wheals resembling hives, severe local pain, coughing. Blamed for serious allergic reactions overseas.

Other jellies to avoid

Big or small, they cannot do any lasting harm. But don't take a chance on their ruining your family's day at the beach.

These are the true jellyfishes most likely to inflict significant pain, to the particular distress of young children, at coastal towns and holiday resorts. None of them causes trouble as widely or as often as the Portuguese man o'war (overleaf), which is not a jellyfish at all. The wheals their stings raise may look frightening but the skin effects are not permanent. Vinegar limits the stinging and a pack of ice and water usually eases the pain. Children should be instructed not to handle dead jellies found ashore — their stings can be just as bad as those from a living animal.☐

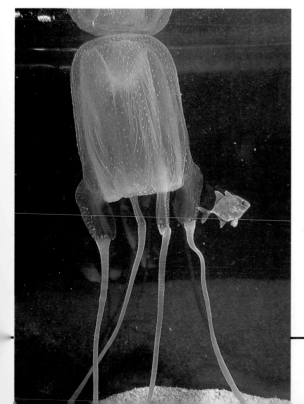

Above
LION'S MANE or **HAIR JELLY** *Cyanea capillata*. All coastal waters, more common in tropics. 20-40 cm across bell. Effects: White zigzag wheals turning red, moderate pain, itching.

Left **MORBAKKA** or **FIRE JELLY.** Tropics, south to Brisbane in summer. 10-15 cm across bell. Formerly classified as *Tamoya*, status now uncertain. Effects: Broad wheals and severe local pain, muscular aches, mental confusion, prostration.

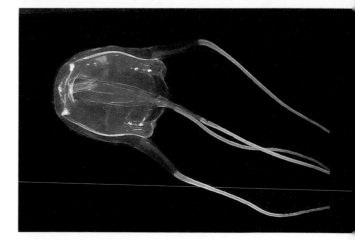

JIMBLE *Carybdea rastoni*. All coastal waters, common in southern ports in summer. 1-3 cm across bell. Effects: Thin straight-line wheals, sharp local pain.

Hydrozoan hoaxers

A mock jellyfish, a fake seaweed and a false coral share close kinship.
Each is a colony formed by smaller creatures.

Its gas-filled float acting as a sail, a Portuguese man o'war drifts at an angle of 45° to the wind. About half of a swarm shear to the left, the others to the right, so that not all can be driven ashore in a storm.

The Portuguese man o'war *Physalia* — or to most people in southeastern Australia, the bluebottle — is universally the best-known venomous marine stinger of all. What is less widely recognised is that it is not a jellyfish, but one of a related group of coelenterates, the siphonophores.

This wind-driven ocean drifter is a colony of individual organisms, born from one egg but developed and specialised to perform different tasks. One fills itself with gas and creates an unsinkable float. Others look after breeding, and the remainder stinging and feeding. Among the nearest kin of these primitive animals, sharing the class called hydrozoa, are other colonial organisms that also have perfected disguises — some to look like marine plants, some exactly like reef corals. They too are formidable stingers, not to be trifled with.

The feeding polyps of *Physalia* are assembled to form the longest dangling tentacle, which can trail as much as ten metres when extended in its 'fishing' mode. If something organic is touched, the tentacle contracts and the stinging capsules are drawn together in rows of buttons shaped like the halves of coffee beans.

On human skin the venom injected from thousands of nematocysts immediately causes a sharp pain that spreads and turns into a strong ache, lasting for up to two hours. Red lines and scattered pimples appear rapidly. The wheals often carry the pattern of the nematocysts, like necklaces of beans. Blisters may form in the worst cases, but usually all skin effects fade within a day. Vinegar is the only recommended treatment to limit the stinging.

Symptoms of general envenomation, such as headaches, vomiting or abdominal pains, are most uncommon. But dread of the local effects alone is enough to drive swimmers from the water when an inshore wind brings in swarms of the stingers. In the case of major Sydney surfing beaches, tens of thousands of people have had to flee on especially bad days.

Stinging 'coral' and fire 'weed'

Brushing tender parts of the body against several species of *Millepora*, the so-called fire coral, can result in local effects comparable with those of a Portuguese man o'war stinging, and sometimes worse. Cases of prolonged vomiting and prostration have been recorded.

Colonies of *Millepora* have a shared skeletal surface of limestone pitted with tiny pores through which the individual polyps poke their stinging tentacles. They give the bleached, branched structure a tinge of colouring, usually a bright green. The colonies can be found not only in association with real reef coral but also in subtropical waters.

Fern-like stinging hydroids, called fire weeds, occupy much the same range and have similar effects. But the worst of them, *Lytocarpus philippinus*, is limited to the northernmost parts of the coast. Said to be capable of causing cramps, vomiting and collapse, it is feared by many divers.□

Sea anemones: not for teasing

The response from some types can be a burning, blistering sting.

Poking a big sea anemone to see it close up is a sport that children adore, and a temptation that many adults cannot resist. It is too much to ask that human nature should change. But for safety's sake, the only way to do it is with a twig — never with a finger.

Nearly all sea anemones are harmless. Their stinging capsules and venom-injecting threads are too small to hurt us. In every region, however, from tropical reefs to the rocky coasts of Bass Strait, there are exceptions. Some live in deep-water obscurity, but enough are found in the intertidal zone of paddling and pool exploration to warrant great caution.

A severe stinging brings instant, burning pain. It quickly gets worse, spreading up the affected limb to the nearest lymph nodes, and it may persist for hours. The stung area reddens and swells. Blisters form and ulceration may follow. Meanwhile there is a possibility of stomach pain, vomiting, cramps and fever. In extreme cases breathing distress, delirium and shock have been reported. Victims of a major sea anemone envenomation need the same emergency care and medical attention as people suffering from one of the more dangerous jellyfish stingings. The difference is that jellyfish encounters are nearly always unavoidable accidents. Anemone stingings usually result from conscious interference.□

Stinging anemone Actinodendron plumosum.

Left: The stinging hydroid Lytocarpus philippinus *is fairly common at moderate depths off the coast of far northern Queensland.*

Below: Fire 'coral' Millepora tenera *at a depth of 5 metres off Lizard Island, Qld. Some species occur well to the south, beyond the range of true, reef-building corals.*

REVENGE OF THE CORALS

Stings are feeble, but cuts are a curse.

Reef coral polyps are generally too tiny for their stinging threads to pierce our skin. Itchy rashes are raised occasionally, with pain that can be distressing for small children. But the effects are short-lived. Significant symptoms of general envenomation are unheard-of.

The real trouble comes from violent collision. If we slip or swim or dive into a coral formation with enough force to scrape or cut our skin, the wound is unusually severe. It is quickly sore and angry-looking, and shows a reluctance to heal. Along with limestone fragments and micro-organisms it probably contains injected nematocyst threads, discharging deep into the flesh.

Serious ulceration can develop from neglected coral cuts, along with a variety of infections. A wound should be washed in warm fresh water at the first opportunity, and medical attention should be sought if it remains unduly sore or inflamed.

Note: Vinegar, invaluable in limiting the stingings of all jellyfishes and allied animals, works on superficial coral rashes. But in the case of cuts it is no substitute for thorough washing and prompt medical attention.

Common toadfish *Torquigener hamiltoni*.

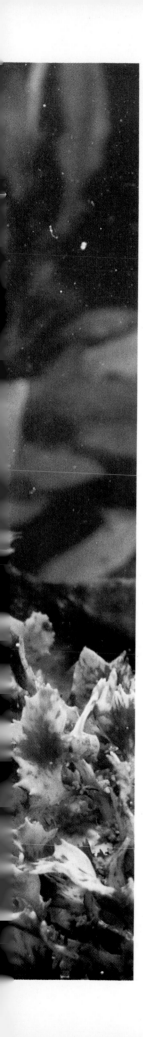

Part 4

ANIMALS THAT ARE POISONOUS TO EAT

No creature asks to be eaten. The risks are of our own choosing — as long as we know about them.

EVERYDAY HAZARDS 244
Fish is one of the tastiest and most nutritious of foods. But you should always exercise caution and common sense before making a meal of an unknown fish. Some species can make you ill; some can kill.

WHEN SURVIVAL IS THE KEY 254
It's unlikely that anyone would sit down to a meal of sea lion, snake or toad in the comfort and safety of their own home. But if you're caught in the wild without food, at sea or in the bush, knowledge of which animal is or isn't poisonous to eat could be a critical factor in your survival.

MANMADE HAZARDS 258
Chemicals used in agriculture and industry can work their way up the food chain and turn otherwise harmless animals — even those bred specially for the table — into potentially hazardous eating. Strict and vigilant monitoring of animal husbandry practices and of the disposal of industrial waste is the only way to avoid these hazards.

Pufferfish peril

Beware a comical creature that blows itself up to bluff enemies. A meal may be tasty – but it can kill in 20 minutes.

Small boys love teasing stranded pufferfish. The persecuted animals suck in air or water and inflate their bodies to absurd, balloon-like shapes. It is an instinctive response to danger, designed to confuse predators.

Older people too, if they are new to fishing, may find delight in puffers – or toadoes, as most of them are called in Australia. They are among the easiest fish to catch with a simple handline and hook. A few toadoes may be all that a novice angler succeeds in bringing in. And it would be natural to try cooking the catch.

Such a meal, if much of it were eaten, could be the last. Toadoes and their relatives – porcupine fish, cowfish, boxfish, tobies and ocean sunfish – contain a lethal toxin. Less than a tenth of a gram is enough to kill anyone. And it acts more quickly than snake venom. A Sydney victim in 1821 was said to have lasted only 20 minutes. In an overseas case, the victim died in 17 minutes.

The danger is virtually worldwide in warm and temperate waters, and hundreds of fish species are implicated. Australia has more than 30 kinds in coastal waters, estuaries and even tidal creeks. There are more of them in the tropical zone, especially around the Great Barrier Reef. But most recorded poisonings have been in the far southeast because more Europeans have lived and fished there.

Recognising a dangerous species

Fish containing the toxin vary widely in body form and colour. What they have in common is a lack of true, separated scales. They are covered instead by a defensive plating that may be abrasively bumpy – something like a shark's skin – or spiny.

The main group, the pufferfish or toadoes, are shaped much like avocadoes. They are slow swimmers and are usually seen alone. Sometimes they wriggle into wet sand when the tide recedes. Their teeth are unmistakable. On both jaws they are fused together with one cleft in the middle. So a toadfish appears to have just four wide, beak-like teeth. These teeth work well in dismembering crabs, a favourite prey. Some of the bigger species of toadoes, snapping indiscriminately, have earned a reputation for attacking human fingers and toes. In 1979 a little girl lost two toes to a pufferfish while paddling at Shute Harbour in Queensland.

Pufferfish range from less than 5 cm in length to more than 75 cm for the silver-cheeked toadfish *Gastrophysus scleratus*. This giant is found off Queensland and in most warm coastal waters from Africa to Tahiti. Captain Cook tasted one in New Caledonia during his second Pacific exploration in 1774. He had cause to be doubly aggrieved by the sharp illness he suffered – the fish was not a gift, but had been paid for with trade goods.

Europeans in Cook's time seem to have been

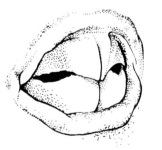

The beak-like teeth of the pufferfish or toado are fused into bony plates on both jaws, each divided by a cleft in the middle.

The flesh of the three-bar porcupine fish Dicotylichthys punctulatus *is extremely poisonous.*

Crystals more precious than gold

Pufferfish poison has been studied longest and most intensively in Japan, where gourmets have a perverse passion for eating such fish. When first extracted from eggs in 1910 the poisonous substance was called tetrodotoxin after the main family of puffers, the tetraodontidae – 'four-tooths'. Many scientists now call it TTX for short. A Japanese corporation produces it in pure, crystalline form from pufferfish ovaries and sells it to laboratories for about $50 000 a gram. It is used in research and

---

MENACE ON THE MENU

Affluent Japanese pay high prices for the privilege of eating pufferfish. They call it *fugu*, and sometimes eat course after course of it served in different ways. Restaurants allowed to offer *fugu* must have licensed chefs who have spent years learning how to prepare it safely. It takes the skill of a surgeon.

Part of the art is to allow just a trace of poison to remain, so that customers can enjoy a tingling of their lips and tongue and the thrill of flirting with death. Less skilled preparation of *fugu* in unlicensed establishments leads to about 50 fatalities a year.

---

as a painkiller for terminally ill patients.

Though it defied chemical analysis for decades because of a highly unusual molecular structure, TTX has turned out to be curiously widespread in nature. Toxins that seem identical have been isolated in some Central American frogs, in a Californian newt, in the blue-ringed octopus, in an unrelated goby fish and recently in an Australian crab.

Levels of TTX fluctuate in pufferfish. They are highest just before and during the breeding phase and some TTX goes into the eggs, perhaps giving them a better chance of surviving against predators. The heaviest concentrations are in the ovaries of females and in the liver and gut of both sexes. But some is in the skin. Only professionally trained people have the skill to prepare a fish for eating without spreading a

surprisingly ignorant of the dangers of pufferfish. In other cultures, from ancient Egypt to China, pufferfish had been notorious for thousands of years. Hebrew dietary laws almost certainly intended a prohibition of them. In the Old Testament book of *Deuteronomy*, any fish without scales was forbidden as 'unclean'.

A RECIPE FOR ZOMBIES

TTX is a possible key to Haiti's mysterious zombies. These 'living dead', buried after succumbing to sudden illness in the prime of life, reappear in other villages as the witless slaves of voodoo witchdoctors. The witchdoctors may have mastered the use of a pufferfish extract, probably diluted in a potion. Paralysed and scarcely breathing, their 'recruits' would appear dead. But they could be exhumed soon after

burial, revived, and given hallucinogenic plant drugs to keep them in a dreamlike, obedient state.

Pufferfish victims have probably been buried alive in Japan. Last century there were several reports of people recovering just in time, as they were about to be interred or cremated. Some had an anguishing recall of everything that had been said in their hearing.

COLOURFUL CRABS WITH TOXIC FLESH

At least six reef crabs found in Australia are known to have toxic flesh. Records do not show that anyone has been harmed, but some researchers believe that crab poisoning may account for unexplained deaths in the past, especially among shipwrecked mariners. These crabs or their close relatives have certainly caused many deaths in the Philippines and the Pacific Islands.

All the potentially dangerous crabs are of the xanthid family. They are smaller than the crabs favoured by Australians for eating, and they are gaudily coloured with spots or mottling. Most seem to be limited to the Great Barrier Reef region, but two at least are in Moreton Bay, Brisbane, and the biggest, *Zosymus aeneus*, has been reported from northern NSW.

A Japanese study on *Zosymus aeneus* concluded that the poison in it was saxitoxin – the cause of paralytic shellfish poisoning. A similar toxin was isolated from *Atergatis floridus* , another Queensland crab. But recent Australian research has indicated that the Japanese finding was mistaken.

A postgraduate team led by Dr Robert Endean, associate professor of zoology at the University of Queensland, purified an extract from the *Atergatis* of Moreton Bay and conducted standard tests on laboratory mice. They did not react as they would have to saxitoxin. In its neurological action and its lethal effect, the main component of the crab poison was indistinguishable from the TTX – tetrodotoxin – of pufferfish.

Zosymus aeneus

poisonous dose through the flesh, and there is no method of cooking that will lower the potency of the toxin. But it is partly soluble in water, so a starving person with nothing else to eat could perhaps reduce the risk by shredding the flesh and soaking it for hours, repeatedly changing the water.

What TTX does to the human system

TTX is a neurotoxin, blocking some of the electrical impulses that pass through nerve endings. Normal responses of feeling and movement fail. Muscle activity and the heart's pumping are reduced, causing blood pressure to fall. Victims cannot move, speak, see or swallow – though they may remain conscious and lucid. The main life-threatening effect is breathing failure.

In a serious poisoning, major symptoms develop within 10-45 minutes of eating the fish. The sooner they start, the worse the case. Paralysing effects set in so quickly that vomiting and diarrhoea – early signs of most other poisonings – are uncommon. And all the TTX stays in the system, making the effects even more devastating. The death rate in Japan for untreated cases is about 60 per cent.

For all the scientific attention given to TTX, no antidote has been found. Doctors' priorities are to flush the toxin from a victim's stomach and to keep him breathing. Hospitals use mechanical ventilators and administer drugs that can relieve some of the symptoms. Since a patient in the apparent state of a vegetable may hear and understand everything going on, reassurance and discretion in what is said in the patient's hearing are important aspects of first aid and medical care.□

Paradise lost for two young lovers

The holiday haven that became a grave.

Together a year and planning to marry, the two young Tasmanians had an idyllic vision of their mainland holiday: to find an empty beach and live off the land – or sea. Perhaps Eden, NSW, had a special attraction in its name.

Phillip Cartledge, a Hobart psychiatric nurse, was 23 and his girlfriend Jocelyn Jones was 19. It was August 1965, and the beaches were indeed deserted. Driving north out of Eden, they turned off Princes Highway down a track to Long Beach – now part of Ben Boyd National Park. The couple parked in scrub behind the beach, set up camp and began exploring their private world. Phillip had fishing lines and

Jocelyn her frying pan, ready for the first catch.

Three days later, Phillip Cartledge staggered to the verge of the highway. A passing motorist rushed him to a doctor in Eden. Jocelyn had collapsed, Phillip told the doctor. She was back down the beach track. They had been starving, he explained, but that morning they had managed to get together a meal of tough-skinned fish, 15-20 cm long with spots on the back.

Soon the young man could no longer speak, and in a few minutes he was dead. So was Jocelyn, found by police on the lonely track. Tests on their stomach contents confirmed the doctor's guess – toadfish poisoning.□

Deep in the south, a grisly gift

The pufferfish danger lurks even in our southernmost waters.

Some species of pufferfish are abundant around Tasmania, where in summer they are common in warm estuaries and tidal creeks. This sad story happened in 1950 on the Huon estuary, south of Hobart.

An 11-year-old boy, watching anglers beside a creek near his home at Castle Forbes Bay, was given five fish to take home. His mother cooked four – the cat stole the other. The boy's father tried some and disliked it, so the cat got that as well. After eating two small fish, the boy went out to ride his bike.

He returned 40 minutes later, complaining of numbness in his hands, legs and feet. Soon he could not move them and was unable to swallow. He was taken to a doctor but died about two hours after eating the fish.

His father was sure it was 'mountain trout' that they had eaten, but pufferfish poisoning was suspected – especially when the cat became paralysed and died. Hobart police detonated an explosive in the creek and recovered a quantity of the toadfish *Spheroides liosomus*. It matched the appearance of what the boy had eaten.

After warnings were issued in the district, it was discovered that another family had fed similar fish to their three cats. All had died – along with a hen that pecked at their vomit.☐

> *'He was taken to a doctor but died about two hours after eating the fish.'*

Toadfish Spheroides pleurogramma

Long-nosed boxfish Rhynchostracion nasus

Pet's death went unheeded

The early warning sign that was ignored.

Pufferfish given to pets can have dire effects, though sometimes the speed with which animals vomit saves their lives. Animal illness sets in so quickly that it can serve as an alarm to a pet's owners. For the Lang family of Five Dock, NSW, the sudden death of a magpie should have been warning enough.

Early in 1972 the Langs were enjoying a camping holiday near the beach at Currarong, on Beecroft Peninsula 25 km southeast of Nowra. They fished from rocks and caught about two dozen small pufferfish, gutted and cleaned them, then left them soaking in sea water overnight.

Before the fish were boiled for next day's lunch, one was given to a magpie that the children were rearing after finding it injured. Almost immediately it started to reel about. Then it fell over and lay with its wings twitching. Soon it died.

Nevertheless, the family ate the rest of the fish. The portion given to Hans, at 14 the oldest of the three Lang boys, included one fish that looked different from all the others. Described later to experts, it was thought to be *Amblyrhyncotes richei*. As rain was coming, the family broke camp as soon as the meal was finished and packed for the drive to Nowra. All but Mrs Lang began to feel ill.

The two younger boys vomited quickly and suffered no lasting effects. Hans did not vomit until at least half an hour after eating. Mr Lang did not vomit at all. His arms, legs and neck went limp, and he eventually spent two days in hospital. But meanwhile Hans lay in the car, fighting for breath and increasingly paralysed.

Mrs Lang took over the wheel and sped to Nowra's Shoalhaven District Hospital, where the quick assessment and action of Dr J.E. Spivey is credited with saving Hans's life. At first sight the boy appeared dead. Air was pumped into his lungs through a tube in his throat and he was sent on by ambulance to Wollongong Hospital, where a wider range of life-support systems was available.

In the ambulance, Hans started breathing again and regained consciousness. Although he couldn't move a muscle he could hear. A full account of his case, prepared for the *Australian Medical Journal* by doctors who treated him, includes many of Hans's own recollections.

'I remember ... being in the ambulance receiving air and

Stars and stripes toadfish Arothron hispidus

breathing,' Hans wrote. 'I could hear them talking but I couldn't move or anything . . . They were laughing and chattering and they even played with the hooter.'

At Wollongong Hans remained totally paralysed, although many of his bodily systems seemed fairly normal. He was fed intravenously and treated with drugs, but his failure to respond pointed to brain damage from lack of oxygen. Again he was transferred, this time to a specialist respiratory care unit at Sydney's Prince Henry Hospital. Hearing his attendants, Hans was not cheered: 'One of the men . . . asked the nurse how I was and she said I still looked worse.'

Prince Henry staff, however, were aware that a victim of TTX paralysis could be fully conscious. 'I heard nurses later on and they were trying to talk to me,' wrote Hans. 'They opened my eyelids every now and then and I found out I could see.' Regaining control of his eyelids was the first sign that Hans was on the mend, four days after his poisoning. He eventually made a full recovery.□

The brown-spotted rock-cod Epinephelus tauvina – *also called the giant estuary cod or greasy rock-cod – grows to as long as 2.5 m. Commonly found around coral and rocky reefs, it can give divers a fright if they come across it unexpectedly. But these giants do not attack human beings. Their size, however, means that their edibility should be treated with caution.*

The ciguatera riddle

Little fish are sweet, an old saying goes. Among fish that feed near coral, they are also much safer to eat.

Science has a long way to go before all the mysteries of ciguatera fish poisoning are unravelled. Researchers are not sure they have got to the heart of its origins in the marine food chain. They do not fully understand the nature of the toxin, or know whether it undergoes any changes when passing from one animal to another. And they are far from confident of ever finding a way of protecting people from it. As far as the general public is concerned – and the tropical tourist industry, for many a holiday has been ruined by poisoning – the questions are what fish may be implicated, where, and when. The fishing industry, too, is anxious to know. Outbreaks of poisoning and marketing bans put its revenues in jeopardy.

But there are no simple answers.

CIGUATERA POISONING

EARLY SIGNS Weakness, dizziness, dull aches in limbs and head. Prickling of mouth and hands, then numbness. Muscle pains or cramps. Reversal of temperature sense.

ACTION Induce vomiting with fingers down throat if victim fully conscious. (Urge others who ate the fish to do the same.) If victim unconscious or severely distressed in breathing, give mouth-to-mouth resuscitation. Obtain medical aid.

The scourge of mariners

Symptoms suggesting ciguatera poisoning were described in Alexander the Great's time, more than 300 years BC. The association of the illness with predatory fish was understood in China over 1000 years ago. It became a scourge of Europeans as maritime exploration flourished in the 16th century: heavy reliance on a tropical fish diet played havoc with many crews, especially in the West Indies. The name ciguatera was coined in Cuba from *cigua*, the Spanish Antilles name for a small turban shellfish that causes digestive and nervous disorders.

In Australia, fish capable of causing ciguatera can be caught at least as far south as latitude 30° – beyond Grafton, NSW, and Geraldton, WA. Poisonings happen mainly between Hervey Bay, Qld, and Darwin, NT, but outbreaks occur in cities farther south if toxic fish from the tropics are marketed there, or taken home frozen by returning holidaymakers.

Ciguatera poisoning does not have to be reported to health authorities, so records of its incidence are patchy. Surveys of hospitals and private doctors indicate hundreds of cases each year in Queensland, but only mass poisonings – when a big fish such as a mackerel is shared among several people – attract much publicity. At Hervey Bay in 1983 an entire ship's company was laid low. They ate a barracuda.

The culprit: an ambiguous traveller

A toxin resembling that in ciguateric fish has been isolated from *Gambierdiscus toxicus*, a single-celled creature that feeds on algae colonising dead corals. The toxin is not consistently present. What triggers its production is unknown. Bacteria may be involved.

Gambierdiscus is one of the dinoflagellates – important elements of the microscopic, drifting plankton on which ocean food chains are based. They are classed as plants. Yet they are capable of independent movement, directional and rotational, using whip-like appendages. When in motion, they seem more like animals.

Their populations explode when reefs are damaged and corals die in huge numbers. That can happen naturally through the violence of waves in tropical cyclones, through silting from freakish mainland floods, or through the depredations of the crown-of-thorns sea star. But heavy damage is also caused by industrial water pollution, jetty construction and so on.

The amount of toxin in the dinoflagellates is infinitesimal. But other little creatures eat them, and they in turn are gobbled up by something else. On the toxin goes up the chain of fish-eat-fish, growing in menace at each step. Most of the toxin is held in the internal organs. In a process sometimes called bio-accumulation or bio-magnification, the amounts increase according to the size of a fish and its age – for it may take in many doses during its lifetime.

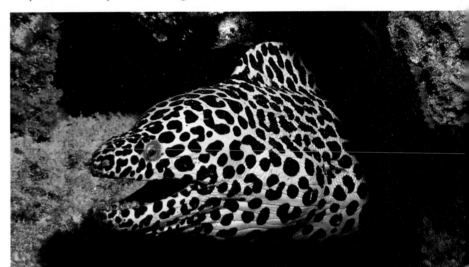

A tessellated moray Gymnothorax favagineus *photographed at Heron Island, Qld. Because of their stationary habits, lurking in lairs and eating a necessarily restricted diet, moray eels are likely to have highly toxic flesh and should never be eaten.*

HOPEFUL NEWS FROM HAWAII

Researchers in Honolulu develop a fish toxin detection kit.

Many cases of ciguatera poisoning could be avoided if fish marketing or public health authorities could run random tests on commercial catches. But in spite of worldwide efforts, no workable method has been found.

Feeding tests on experimental animals, including cats, mice and mongooses, are of some use in confirming medical opinion after poisonings have occurred. But they take days to complete. A recent proposal called for the injection of mosquitoes, apparently without success. Radioactive tracing methods produced inconsistent results.

Work in Honolulu has brought the most encouraging news so far. Pharmacologists under Professor Yoshitsugi Hokama of the University of Hawaii have developed chemical reagents that show a colour change if there is toxin in a fish sample. In 1985 their materials were sent to both coasts of the US mainland and tested on fish there. Consistent results were recorded.

Professor Hokama is confident that his method can be incorporated in a cheap, portable field kit – similar to a snake venom detection kit. If trial successes continue and commercial interests are prepared to take up the manufacture and distribution, he expects kits to be on the market before the end of the 1980s.

249

These fishermen would be well advised to think twice about eating their Great Barrier Reef catch of barracuda and mackerel: the bigger a reef fish, the more likely it is to have toxic flesh.

Which fish are poisonous?

Whether any particular fish could be seriously poisonous is a matter of sheer chance. No one can be sure where it has been, let alone which other fish it has eaten and where they have been. While higher levels of fish toxicity are almost certainly related to reef damage, the poisoning does not show up until months or years later, perhaps hundreds of kilometres from the site of the damage.

Moray eels are an exception – probably because of the stationary lives they lead, lurking in their lairs, and their necessarily restricted diet. When they are toxic, they are very toxic indeed. Moray eels should never be eaten.

RULES THAT CUT THE RISK

- Avoid the biggest specimens in a batch of reef fish.
- Discard the brain, spinal cord, intestines and internal organs.
- Eat sparingly.
- Eat no more than one meal of reef fish in a day.
- Never eat two meals from the same fish.

The only other general advice that can be given is to avoid the biggest predatory fish of tropical coastal waters. Every expert has a different list of leading 'public enemies'. About 20 species – several of them popular commercial fish – are known to have caused ciguatera poisoning in Australia. Many more are capable.

Nothing about the appearance or smell of a fish serves as a warning that it is toxic. It does not matter how fresh the fish is, or how it is prepared and cooked – except that if any of the gut or internal organs are left in, the risk of poisoning is very much higher. If a cat is nearby when experienced fishermen are cleaning their catch, they will throw it a small piece of flesh. The cat will go stiff-gaited and quickly vomit if much toxin is present in the fish.

Before the toxin was isolated from *Gambierdiscus*, quantities for research were extracted from the livers of moray eels. Findings had to be judged with caution, however, because there was no certainty that the eel toxin and the ciguateric fish toxin were the same.

That problem remains. Molecules of the dinoflagellate toxin are extremely small and have defied complete analysis of their chemical structure. They cannot be matched convincingly with traces in the remains of fish that have actually caused ciguatera. So it cannot be proved that laboratory discoveries apply to human poisoning.

It has even been conjectured that different toxins from other sources may cause ciguatera. Some American researchers think that as many as five organisms could be involved. That could explain why, again and again, promising techniques for detecting poison in samples of fish have failed. Trials in different places have produced different results.

The effects of ciguatera poisoning

The poison as it occurs in fish is obviously a neurotoxin, interfering with nerve functions, but how it works is far from clear. Effects in the gravest cases are paralysis, breathing failure and heart failure, although few poisonings are as serious as that and the acute symptoms usually subside in a few hours or days. Some of the long-term consequences, however, can be disturbing and painful.

Symptoms have been known to develop in minutes in the most serious cases. The usual delay is 2-12 hours after eating a toxic fish. The severity of the illness generally depends on the amount of fish eaten, but sometimes members of a family who eat the same amount are affected to different degrees. Those who suffer more may have accumulated the toxin from small previous doses.

The early signs are much like those of pufferfish poisoning, but ciguatera is often distinguished by a reversal of the victim's temperature sense. On the skin or in the mouth, hot feels cold and cold feels hot. Given ice cream or a chilled drink, a patient thinks he is being burnt. He feels frozen under a hot shower and may want to turn it up to scalding point. Muscles are weak and there may be headaches or chest pains, but for many victims the worst symptom is a swelling and itching of the hands and forearms, the soles of the feet and sometimes the whole skin.

Treatment concentrates on flushing the toxin from the digestive system as soon as possible, maintaining breathing and circulation and avoiding dehydration. Drugs are used to relieve inflammation and pain, and to ward off an arthritic condition that can linger for months. Muscle weakness and lethargy continue in convalescence, and some patients need anti-depressant drugs or psychiatric counselling.

Ciguatera victims are not immune to later poisonings. In fact another toxic fish will probably give them a worse illness. And they may be so sensitised that any fish – harmless to other people – brings on a recurrence of muscle spasms, weakness, stiffness or fierce itching. Eating pork or poultry that were fed on fish meal may be enough to trigger the effect. Heavy drinking, months or even years later, can do it. In the most sensitive cases, so too can mild stimulants such as tea and coffee, even pepper or chocolate.

Though the acute symptoms can be devastating and the after-effects debilitating, ciguatera poisoning seldom kills. Since 1970 it has been firmly blamed for only one Australian death and strongly suspected in two or three other cases. Our low mortality rate compared with records from Asia and the Caribbean partly reflects a wider availability of medical services..

But as fatalities among South Pacific islanders are also extremely rare in relation to the known number of poisonings, it seems that the fish of the whole region may be generally less toxic. A possible reason is that our coral reefs are less damaged than those of other regions.

If that is so, the severity of ciguatera poisoning in Australia could increase if our reefs are not conserved.☐

WHERE CIGUATERA POISONINGS OCCUR

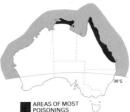

■ AREAS OF MOST POISONINGS

▨ ZONE OF POSSIBLE DANGER

BANNED FROM SALE

If you catch any of these fish in tropical waters, don't eat them.

Queensland law forbids the marketing of three kinds of fish – red bass (also known as bohar snapper), chinaman and paddletail – if they are caught in tropical waters. A spate of poisonings shortly before the ban was imposed indicated that these were the commercial fish most likely to cause ciguatera.

In the light of more recent research, it is now known that other popular fish may be just as much to blame. No two experts agree on their lists of leading suspects. They all agree, however, that the public would benefit from a more flexible approach that allowed for the prohibition of different fish in different areas, subject to frequent review.

Chinaman Symphorus nematophorus

Paddletail Lutjanus gibbus

Red bass Lutjanus bohar

And baby made eight

Ciguatera toxin can pass through the placenta and endanger a foetus.

'Seven people sat down to the feast. Eight people were poisoned.'

A plump and succulent coronation trout, sent frozen from the Great Barrier Reef, was the centrepiece of a Brisbane barbecue on Father's Day, 1981. Seven people sat down to the feast. Eight people were poisoned.

Unborn children are shielded from many diseases and poisons. The foreign substances do not pass the placenta, the spongy organ through which foetal nourishment is filtered. But ciguatera toxin can.

Four hours after the fish was eaten, everyone at the barbecue had fallen ill with symptoms of ciguatera. One woman, expecting to give birth two days later, felt strong movements from the baby interspersed with what she described as shivering. The movements continued until the next day, then gradually subsided.

After two more days a boy was delivered to her by Caesarian section. The baby had a facial palsy and needed intensive care for respiratory distress. Some muscular deficiency in his hands was also suspected.

Happily, the baby responded to hospital care and soon began a normal childhood. □

Expert became his own guinea pig

After ciguatera poisoning struck, Bryan Pratt deliberately ate big meals of fish.

'Violent spasms swept over me. Pain was intense and I couldn't speak or swallow.'

Bryan Pratt knew all there was to know about ciguatera – except how it felt. A trained microbiologist and a professional in natural sciences, Dr Pratt is also a seasoned yachtsman in tropical waters, an angler, a diver and a specialist photographer of fish. But when he found he was poisoned, 10 000 metres in the air on an interstate flight, he was helpless.

Dr Pratt is assistant secretary for land management in the ACT parks and conservation service. In 1980 he attended the World Wilderness Congress in Cairns, Qld. Delegates feasted regularly on reef fish, including black cod and coral trout, in their hotels. Many, like Dr Pratt, went on boat trips and brought back more.

After a last breakfast – again of fish – he caught an early flight to begin the long journey back to Canberra. About five hours later, on the second leg between Brisbane and Sydney, sudden illness overwhelmed him.

'A can of beer triggered it,' says Dr Pratt. 'Violent spasms swept over me. I shuddered and sweated and palpitated. Pain was intense in my arms and legs, and my head felt as if someone was ramming a welding rod into it. My mouth and hands were numb and I couldn't speak or swallow. I was lucid – I knew what was going on – but I couldn't tell anyone.'

Under hospital treatment in Canberra – delayed while airline officials debated his fitness to travel on – Dr Pratt made a substantial recovery. But it was far from complete. 'For weeks I had difficulty with my balance – even just placing my feet where I wanted. The old symptoms kept returning in spasms, without warning. I didn't dare drive. And for months I had spells of memory loss.'

Bryan Pratt did not experience the confusion of heat and cold that is pronounced in many ciguatera cases. 'I just wouldn't be sure of my taste. If I was drinking a cup of tea, say, I'd have to take a second look.

'There's a lot more to be learned about ciguatera. Why does it affect different people in different ways? Why is it a more serious illness in certain regions? Why are the same fish more dangerous in one place than in another? I find it hard to accept that it's all the fault of one organism, and that a toxin comes all the way up the food chain, unchanged. It just doesn't fit.'

Dr Pratt, who believes that allergic reactions may follow ciguatera poisoning, set about testing his own sensitivity. Once the direct after-effects had disappeared, he deliberately ate big meals of fish and other seafoods – but nothing caught north of Byron Bay, NSW. A wine connoisseur, he gave up drinking for a period so that he could compare his reactions with and without alcohol. Three years after the Cairns poisoning, a blunder put him back in hospital for two days. On the far south coast of NSW he ate dolphin fish. Later he realised that they could have migrated from northern waters.

Dr Pratt argues for wider publicity of the dangers of ciguatera. 'People anywhere should be far more aware of the risk. Even some of my angler friends don't know.

'I believe there are many more poisonings than the records show. Milder cases could go undiagnosed, especially in the south. As for more serious poisonings, I wonder a lot about disappearances at sea – people who go overboard for no apparent reason.

'Ciguatera could account for some unexplained car accidents, too. When it hit me, I lost all control of myself. If it had struck a couple of hours later, or if I'd lived in Brisbane, I could have been driving home from the airport...' □

Shellfish illness

Few countries are as safe as Australia. Our molluscs let us off very lightly. But some people do suffer from allergic reactions to shellfish.

Suspicion of shellfish goes back to the dawn of history. Ancient Hebrews were enjoined never to eat them. Coastal Amerindian tribes would rather starve than gather mussels when they noticed certain sea conditions. And stories from overseas of mass poisonings with appalling death rates still appear in the press.

Australian seas are spared the tiny marine organisms that cause the worst shellfish poisonings. And because of our relatively low population and high standards of hygiene and official food inspection, risks of serious bacterial infection are fairly remote.

SOUTHERN RED TIDES SEEM HARMLESS

A danger signal in the northern hemisphere, red tides are no cause for alarm around Australia.

Red discolourations, seen occasionally in coastal waters and lakes, are caused by microscopic plankton organisms when they have a sudden increase in population. A similar phenomenon is seen at night when waves glow with unusual phosphorescence.

In North America and many other regions, 'red tides' cause alarm and may force the closure of shellfish industries. There the organisms are likely to be highly toxic and may make the shellfish poisonous.

Red tides are not unusual in Australian waters, but in the south at least they are not known to have been caused by anything harmful to humans. Near coral reefs in the tropical zone, the 'blooming' of algae that is seen as a red tide could include the organisms that are blamed for making ciguatera toxin. But it gives no hint of where or when ciguateric fish may turn up.

Harmless red tide in Lake Macquarie, NSW

PARALYTIC SHELLFISH POISONING

Australian waters are mercifully free of this menace.

The deadly scourge that has made shellfish so feared in the northern hemisphere is paralytic shellfish poisoning (PSP). Australia seems never to have had it, though it occurs as close by as Papua New Guinea and New Zealand.

PSP comes from toxins produced by various dinoflagellates – single-celled plant-animals in the same group as the ciguatera organism. These toxins go directly into grazing molluscs and are accumulated by them. The poisons go farther up the food chain only if the shellfish are eaten by other predators. Then chemical changes may take place – which could account for some of the unexplained toxins in fish.

Most PSP deaths have been on the colder coasts of North America. Shellfish for sale are tested regularly and fisheries are closed down if the 'red tide' danger signal appears near collecting areas.

An organism thought to be closely related to the one that causes trouble in North America was identified in 1985 from the Derwent estuary near Hobart, Tas. Studies were started immediately but there is no evidence so far that it could make shellfish poisonous, or that it ever occurs in dangerous numbers.

Allergic reactions to shellfish

Most harm is done to a small minority of people who acquire an exaggerated sensitivity – an allergy. Eating molluscs such as oysters, scallops or mussels, and in rarer cases crustaceans such as crabs, prawns or crayfish, sets off a reaction like that of a hay fever sufferer to pollen. But it can be much more severe. Victims soon learn to leave the offending seafoods completely out of their diets.

Some protein in the food triggers the reaction. It does not occur the first time a shellfish is tried, unless a similar protein has been absorbed previously from something else. Instead of digesting the protein in the normal way, the systems of a few people resist it. They create antibodies to destroy the protein. These remain in the bloodstream, and the next time the protein is encountered the reaction is excessive. Exercise, heat and emotional excitement make it worse.

In rare cases shellfish allergy can lead to death from shock, or from asphyxiation because swelling blocks breathing passages. For most sufferers the worst effects are days or even weeks of disability with headaches, muscular aches and itchy skin swellings. Hospital treatment is sometimes necessary.

The 'stomach bug' syndrome

Most gastric upsets after eating shellfish in Australia are caused by bacterial infection. It rarely occurs in commercial produce – farmed oysters, in particular, flush out foreign organisms when they are kept in purification tanks after harvesting.

A few serious bacterial or viral infections that can be passed to humans through shellfish, such as typhoid, are dealt with in Part 5 of this book. Otherwise the usual symptoms of gastro-intestinal shellfish poisoning are mild.

Weakness, followed by nausea and vomiting, sets in after a long delay – 8-12 hours commonly, and sometimes as much as 36 hours.

ALLERGIC REACTIONS TO SHELLFISH

EARLY SIGNS Redness, swelling, itching spreading from head and neck. Nausea, headache.

ACTION If victim fully conscious, induce vomiting with fingers down throat. Encourage rest. If breathing is distressed give mouth-to-mouth resuscitation and obtain medical aid.

Diarrhoea is often accompanied by griping stomach pains. Often there are muscle pains, headache and a mild fever. In most cases the illness is all over in less than two days. Medical aid should be sought for people who seem abnormally distressed, but usually the best medicine is rest, with a high intake of fluids. Ordinary home remedies for stomach upsets can help.

Tasmania has a limited occurrence of diarrhoetic shellfish poisoning, which can come on in 30 minutes and does what its name suggests. It is traced to *Dinophysus* dinoflagellates in southern waters. Victims recover in three days, with or without medical aid.☐

The timebomb tuna

Within hours, a harmless substance can turn into a toxin. Cook the fish promptly or chill it – otherwise don't eat it at all.

Scombroid poisoning, a peculiarity of tuna and other big oceanic fish of the mackerel type, is rare. But changing styles of Australian dining may heighten the risk.

Virtually all of the tuna eaten in Australia used to be canned, under strict rules to ensure its purity. Now there is a growing demand for cuts from fresh whole tuna, by people interested in Asian food. In Japanese *sashimi*, for instance, slivers of raw tuna are considered essential.

No warning is given

Bacteria from the air or from contact invade any dead fish that is not refrigerated. That is the start of a normal process of rotting. Eating infected

---

SCOMBROID POISONING

EARLY SIGNS Vomiting and diarrhoea with severe headache. Flushed, puffy face.

ACTION Induce more vomiting with fingers down throat. (Anyone else who ate the same fish should do the same.) If breathing is distressed give mouth-to-mouth resuscitation. Obtain medical aid.

---

fish can result in gastroenteritis and a bout of vomiting and diarrhoea with perhaps a mild fever. Food poisoning of this sort is simply treated and seldom has any more serious effects. And there is usually a warning. The flesh of stale fish looks dull or discoloured. It may smell 'off' in preparation, and taste bad.

In tuna, bonito, mackerel or albacore, however, another process can take place. The muscle tissues of these fish naturally contain a substance called histadine, which is harmless to humans. But if certain bacteria invade the fish, they start converting the histadine into saurine, or scombrotoxin. The bacterium mainly responsible is believed to be *Proteus morgani*. Some authorities also blame other germs such as *Salmonella* and *Escherichia*, common in normal food poisoning.

Tuna can become highly toxic in 10 hours if it is left at ordinary room temperatures – sooner if it is exposed to sunshine. Yet it is not putrefied – there is no discoloration or bad odour. Sometimes a sharper taste, peppery but not unpleasant, has been noticed. The toxin can be detected by laboratory analysis of uneaten parts of the fish – an exacting procedure, used only to confirm the cause in the most serious cases of poisoning.

The effects of scombroid poisoning

Scombrotoxin closely resembles histamine, which our own bodies produce in response to tissue damage. Histamine causes local inflammation as a first step towards repairing the

Pole fishing for albacore tuna in the South Pacific.

damage. But eating a meal contaminated with scombrotoxin is like dosing the whole digestive system and bloodstream with histamine.

Nothing happens for at least 20 minutes, and sometimes for an hour. The early symptoms are nausea, vomiting and diarrhoea – like ordinary gastroenteritis – but they are accompanied by a headache and an intense throbbing of blood vessels in the head. The pulse is fast but weak, though heart palpitations may be noticed. The mouth becomes dry and the throat burns, and the victim cannot swallow.

After about two hours of illness the face is puffy and the eyes are inflamed. The skin is red and itchy and blisters may break out, especially on the face and upper trunk.

Symptoms of a cold develop, followed by a fever with shivering chills and muscular weakness. Circulation is poor – the lips go bluish – and there may be severe breathing difficulties. Someone trying to get up from a sitting or lying position could faint.

Most victims begin to recover within a day, although the illness can last much longer. Fatal effects are possible, but highly unlikely if prompt first aid is given and medical treatment is available. Antihistamine drugs, used in treating cases of allergy, work well in cases of scombroid poisoning.

Avoiding the risk

Any catch of tuna or other scombroid fish must be kept cool – out of the sun and preferably in ice. Safe commercial marketing relies on refrigeration facilities.

Cuts from a whole fish should not be eaten if there is any uncertainty about how the fish was kept, at any stage from the fishing boat to the kitchen.☐

Professional tuna fishermen hose their catch to keep it cool from the moment it is landed, and pack it in ice as soon as possible.

Preparing a yellowfin tuna for the bleeding process that makes Japanese sashimi.

EATING WILD ANIMALS *Most creatures*

In its publication *Stay Alive: a handbook on survival*, the Australian government advises that to survive in the bush you must be prepared to eat 'almost anything that swims or crawls, walks or flies'.

That word 'almost' is the basis of these next seven pages. For while most animals, if recently killed, are safe to eat – even, despite widespread belief to the contrary, venomous snakes – there are some exceptions that the bushwalker and seafarer should know about.

Cooking and preserving flesh
Although the flesh of most wild animals is not toxic to human beings, it should always be thoroughly cooked, if possible, before it is eaten. The cooking process will destroy any parasites and their eggs that may have infested the animal.

If all of the flesh is not required as food immediately, the remainder should be preserved to avoid a possibly fatal dose of poisoning from eating decayed meat. You can preserve

are edible – but there are some exceptions

meat by cutting it into strips (discarding any fat) and hanging it over a fire to dry in the smoke; by pickling it in a salt solution; or by freezing it.

Fish should be cleaned and eaten as soon as possible after being caught. Or soak them in several changes of water and dry them in the sun.

First catch your snake

Roasted over a fire or baked in ashes, the flesh of snakes is nutritious and palatable. It has been described as tasting like something between chicken and eel. Even venomous species can be eaten – cooking destroys the toxin.

If there is no fire to cook a snake, the head should be cut off well behind the jaws. An ample safety margin is provided by taking the distance between the snout and the eyes, and going twice that distance behind the eyes. The riskiest part is catching and killing the snake.

But remember that all snakes are protected native wildlife, not to be killed except in emergencies where human life is in jeopardy.☐

The giant toad

Importing this pest was an unlucky misjudgment by scientists. Trying to eat it would be sheer folly.

Bulging poison glands show clearly on a giant toad's shoulders. The spread of the species out of canefields and into dry country is aided by tree felling and dam building. Toads shelter under logs and debris during drought, breeding in water when its temperature reaches about 26°C. Adults establish themselves in the same shelters and breeding places for life. Juveniles looking for homes spread the population – sometimes by up to 30 kilometres a year.

Giant toads – our so-called cane toads – are so repulsive that it is hard to imagine anyone eating them by choice. But a famished bushwalker could be tempted. The toads sometimes trap themselves in tents and billycans, and it takes only one to make a meal.

Warty lumps on the toad's back cover poison glands. A potent toxin can be squirted from them, sending a spray as far as 1 metre and stinging the eyes of an animal trying to interfere with the toad. Some people are at risk just handling giant toads, if they have any weakness of the heart. The poison penetrates unbroken skin. A heavy dose causes a drastic acceleration of the heartbeat, at the same time making the victim struggle for breath.

Poison also pervades the toad's flesh. Much more of it is held in bloated glands reaching over the shoulders from just behind each eye. These are modified saliva glands, evolved in the same way as those of a venomous snake. They may be a defence against pecking birds.

A full intake of the toxin, from swallowing a toad, can kill snakes, cats and dogs within an hour. Humans – especially children – would be inviting nothing less than a devastating illness. Doctors would urge the hurried inducement of vomiting in conscious victims, and try certain drugs to overcome some symptoms. But they know of no specific treatment.

Growing as big and as plump as dressed chickens, giant toads are voracious feeders. They favour beetles and bees, but dine on almost anything they can catch. Their unpleasant habit of also eating animal and human faeces means that they can be carriers of parasitic worms – another good reason not to consider them as food.

An answer worse than the problem

The giant toad *Bufo marinus* originated in Central and South America. Taken to Hawaii, it did well in keeping down a beetle that damaged sugarcane crops. Queensland had the same

Red-groined toadlet Uperoleia rugosa

TOXIC FROGS

Australia has about 140 species of native frogs, with new ones being discovered nearly every year. Early European settlers called some of them toads or toadlets because of their warty skins. All the frogs seem to have poison glands in their skins, and some people have suffered hours of pain – even collapse – after passing frog slime from their hands to their eyes or mouth.

An uncertain number of Australian frogs also have enlarged toxin-producing parotid glands like a toad's. Though most of these amphibians are tiny, they are capable of killing animals much larger. It is inadvisable to eat them.

problem. So on the recommendation of government scientists, in came the toads. They were released near Cairns in 1935, and later in sugar-growing areas to the south, including NSW.

Ineffective against the cane beetles – our native frogs did far better – the toads bred at an astonishing rate and quickly became pests themselves. They infested farms, destroying honey bees, poultry and pets. What they have done to the more subtle balances of natural life, by killing insects, spiders, birds and snakes, is beyond estimation.

Female giant toads lay up to 35 000 eggs each season. Males, if mates are in short supply, develop ovaries and turn into females. The spread of the species makes the term 'cane toad' absurd. They are capable of breeding almost anywhere. Coping with semi-arid conditions, they have advanced up Cape York Peninsula, into far western Queensland to beyond Mount Isa, and across the Gulf Country into the Northern Territory.

One use has been found for the toads. They have replaced frogs in laboratory experiments and teaching. But the chance of their accidental or mischievous release is a nightmare for ecologists. Perth, Adelaide, Sydney and Darwin have had scares. In the Darwin escape, in 1974, most of the toads were rounded up. But in 1985, they were found to be breeding there.□

Seafaring turtles

Ignoring the legal protection of these vulnerable reptiles invites a penalty far worse than any court would exact.

Poisoning scandals soon after the turn of the century took turtle soup off European menus and put paid to a lucrative tropical canning industry. Ruins of an Australian factory can still be seen on a desolate coral cay near Heron Island, Qld. It was built there to exploit the compulsion of female marine turtles to return season after season to the same beach nesting places, even though they may travel thousands of kilometres in the meantime.

Apart from its shell pattern, the hawksbill turtle is easily identified by the parrot-beak shape of its top jaw. Like all marine turtles it is long-lived. Having few predators once it reaches adult size, it may survive for well over 100 years.

Six huge species of sea-going turtles frequent Australia's tropical waters and coasts. All offer enough meat – more or less palatable – to feed dozens of people. But two at least have been responsible for mass poisonings overseas, with a death rate that is sometimes said to exceed 25 per cent. In southern Indian villages as many as 300 people at a time have been poisoned by eating a turtle, though the death rate there was not nearly so high.

Chemists know nothing of the nature of the toxin. Many symptoms of poisoning resemble those of ciguatera, but in the worst cases extensive damage is also done to internal organs such as the liver and kidneys. Victims who recover may suffer for months from ulcers and deep cracks in their mouths.

The toxin occurs in a turtle's blood as well as in its flesh and organs. It turns up intermittently and completely unpredictably. Without much doubt it comes from something in the diet. But these reptiles can eat almost anything: fish, aquatic plants, molluscs, prawns – even jellyfish. And they rove so widely, even into colder seas, that the cause is anybody's guess.

Hawksbill heads the danger list

Turtle shells – some bigger than bathtubs – are so useful in undeveloped communities that they are invariably kept after an animal is eaten. So in cases of poisoning overseas, the species to blame is usually obvious. The chief offender, especially in Sri Lanka and southern India – where incidentally turtles are legally protected – has been the hawksbill *Eretmochelys imbricata.*

Hawksbills are common in tropical Australian waters, particularly near coral reefs, though their nesting places seem relatively few and remote. They are mainly flesh-eaters but they also graze algae. Growing to about 1 metre in shell length, they were the species hunted for 'tortoiseshell' to make jewellery, ornaments and spectacle rims.

The reddish flesh of the hawksbill is not the most popular among traditional eaters of turtles. They are generally aware, too, of the possibility of poisoning. Tossing the internal organs and morsels of flesh to gulls, crows or dogs is not an act of generosity but a test. In Australian bush lore, acquired from Aborigines of northern coastal tribes, the hawksbill was to be treated with caution because it was understood to have a 'poison gland'.

More favoured for eating, but known to have caused deaths, is the green turtle *Chelonia mydas.* Also growing to 1 metre in shell length, it has nesting places on cays and coastal beaches all around tropical Australia. Adults seem mostly to be plant-eaters, although the young are carnivorous. Green turtles were those used for soup canning. It is interesting to note that when they became scarce in the hunting areas, hawksbill flesh was substituted. That may have been the industry's downfall.

Two other Australian turtles were blamed long ago for deaths overseas, but some authorities now believe their identification was a mistake. These are the loggerhead *Caretta caretta* (1.5 metres) and the rare leathery turtle *Dermochelys coriacea,* which has a tough skin embedded with bones rather than a true shell. It is the world's biggest turtle, reaching 3 metres. Papua New Guinea has the softshell *Pelochelys bibroni,* also blamed for poisonings.

All turtles are protected by law in Australia. But there is a dispensation for Torres Strait Islanders and Aborigines living in areas where turtles have been a traditional food – the big islands near Darwin, NT, for example, and the Queensland coast north of Cooktown.

Moving ponderously in shallow waters, turtles are all too easy to spear or to wrestle to the shore. The breeding females, dragging themselves up beaches on the spring night that they choose to lay their scores of eggs – the size of table-tennis balls – are obsessed by their task and entirely helpless. The young, creeping towards the sea when they hatch a few months later, make a feast for gulls. Australian measures to conserve such vulnerable animals include efforts to encourage turtle farming among the people who eat them.☐

The green turtle Chelonia mydas *is the best-known species because it was the one used to make turtle soup. Now very scarce in areas of the world where it was commercially exploited, it is still common in waters around northern Australia. It favours shallow seas with abundant marine grass growth on the seabed. The shell is usually greenish-brown with darker markings, but there is considerable variation within the species.*

TURTLE OR TORTOISE?

In Australia the word 'turtle' is applied to marine species and 'tortoise' to those found in fresh water. They are closely related and zoologically there is no such distinction. In most other countries, especially where similar animals live on land, 'tortoise' is reserved for the land-dwellers and all the others are called turtles. Confusingly, however, British sailors chose to call carvings from the shell of the hawksbill turtle 'tortoiseshell'.

Sharks and rays

Don't eat the livers of these restless prowlers. The toxin they could contain is a mystery – but not its effects.

With their greedy appetite for other fish and seals, sharks are often implicated in ciguatera and hypervitaminosis A poisonings. Occasionally they also contain a toxin of their own. In its effects it seems almost like a combination of the other two. But it is shared by the sharks' cousins, the rays. And rays are bottom-feeders with a totally different diet of molluscs and crustaceans. So the origin of the toxin is as baffling as its structure. It has not been isolated for analysis.

Elasmobranch poisoning, as scientists call it, causes severe illness when the liver of a toxic shark or ray is eaten. Other internal organs may also be highly toxic. The flesh is less poisonous, though eating it could cause vomiting and diarrhoea. The toxin is unaffected by cooking but is probably soluble in water, so repeated washings and rinsings of sharkmeat, changing the water every time, could reduce that risk. But eating the liver at any time is most unwise. Sharks marketed as 'flake' are legally limited in size to reduce the possibility of harm if they happen to be toxic.

In the worst cases, the first symptoms are like those of other food poisonings. They come on in 20-30 minutes. But then they progress fairly rapidly to disturbances of the central nervous system, causing loss of muscular co-ordination, visual difficulties, cramps and paralysis. The victim may become delirious before lapsing into a coma. Breathing failure is likely.

Hospital treatment is always required to maintain a victim's breathing and relieve some of the symptoms. But there is no specific drug to overcome the poison. A full recovery can take up to three weeks. □

NORFOLK ISLAND'S NIGHTMARE FISH

Fish caught off Norfolk Island, 1670 kilometres northeast of Sydney, occasionally have something in them that makes people hallucinate. Victims are not sure of what they see, or even think, for hours. They may feel they are threatened by others, or that they are about to die. Their sleep is invaded by nightmares. Mild symptoms of ordinary food poisoning may occur, but not always.

The illness seems to come from a wide range of fish, and visitors to the island are advised to take local advice on which species to eat and which to avoid. Residents call the likely culprits 'dream fish'.

A grey nurse shark Eugomphodus taurus *cruises past a bottom-feeding ray.*

The fiddler ray Trygonorhina fasciata *is common on sand flats in NSW coastal waters. It grows to 40 cm and lives on a diet of shellfish. Apart from the possibility of its liver being toxic, this ray is harmless to humans.*

ELASMOBRANCH POISONING

EARLY SIGNS Loss of appetite, nausea, vomiting, stomach pain, diarrhoea. Headache, joint pains, inability to stand. Tingling and numbness around mouth, burning tongue and throat. Fast pulse.

ACTION Induce vomiting with fingers down throat – but not if paralysis and weakness have already set in. If breathing is distressed give mouth-to-mouth resuscitation. Obtain urgent medical aid.

Port Jackson shark Heterodontus portusjacksoni.

Vitamin overkill

Animals feeding far to the south accumulate vitamin A in their livers.
Too much of it acts like arsenic.

Deficiency of vitamins – essential chemical compounds our bodies do not produce and which we can get only by eating – has long been known as a cause of ill health. Only recently, however, was it realised that there can be too much of a good thing.

Ample amounts of vitamin A are taken in normal diets. It comes in meat, butter, many vegetables and fish oils. Malnourished children since the 19th century have been dosed with cod-liver or halibut-liver oils. Some old-fashioned preparations tasted so revolting that many a child protested that it was being poisoned. Had the doses been massive, in bowlfuls rather than spoonfuls, the child would have been right.

'Seal liver poisoning'
Marine algae growing in the icy waters around Antarctica, and in the Arctic Ocean, produce unusual amounts of vitamin A. It ascends the ocean food chains, collecting in increasing quantities according to the size and age of predatory animals. Most is concentrated in their livers.

Sharks that roam in cold waters, or prey on fish migrating from the far south, are likely to have high vitamin A concentrations. So are big penguins, and marine mammals such as whales, dolphins and especially the pinnipeds – seals, sea elephants, walruses and so on. In fact the human poisoning syndrome referred to as hypervitaminosis A in medical books is commonly called seal liver poisoning.

The Australian sea-lion Neophoca cinerea *is the only pinniped native to Australia. It is found mostly on off-shore islands from WA to Bass Strait. Bulls – males – grow to 2.3 m and weigh up to 300 kg; the female (pictured) is considerably smaller and weighs up to 80 kg.*

261

In the Arctic, the livers of polar bears, wolves, foxes and dogs are particularly dangerous because these animals prey on fish and seals. In Australia the risk is remote, and likely to arise only if survival should depend on eating the seals that come to breed and nurse their young on our shores.

Sealing gangs in the early 1800s were often stranded and starved after mishaps to the sailing ships that supplied them. On rocky islets especially, their only reliable source of food was the carcasses of the seals that they were paid to club and skin. Their high death rate was probably put down at the time to scurvy – a vitamin deficiency. But the real killer would have been hypervitaminosis A.

A mere 80 grams – not nearly a meal for a hungry man – of seal liver contains enough of the vitamin to cause serious illness. It cannot be counteracted because the vitamin A has already been absorbed when symptoms appear, from 12 hours to a week after a single intake.

The syndrome resembles arsenical poisoning. It sets in with headache, drowsiness and weakness, soon followed by stomach pains and vomiting. Blurred or double vision may occur. The central nervous system is attacked. Victims may become confused or delirious, convulsed or paralysed, unable to speak or swallow. The skin darkens in places, starts cracking at the corners of the mouth, and lifts completely from the palms of the hands and soles of the feet. A victim of one heavy dose could recover after a few days – sometimes weeks – of such symptoms. But if toxic meals are eaten repeatedly, a likelier outcome is death.☐

THE AGONY OF XAVIER MERTZ

A mapping mission in Antarctica ended in death from hypervitaminosis A.

Antarctic explorer Douglas Mawson's lone, starving trek – three weeks and 150 kilometres on foot – is an epic of human survival. It tended to eclipse the tragic events that led to his ordeal, until medical curiosity revived them.

Mawson had two companions on his mapping mission across the glaciers fringing George V Land. First to die was 'Cherub' Ninnis, who plunged into a crevasse with a dog team and sledge on 14 December 1912. That sledge carried the party's tent, spare clothing and most of their food supplies.

Xavier Mertz, a fit 28-year-old – he was Switzerland's cross-country ski champion – was left with Mawson. They were 500 kilometres from their depot. In the best conditions it would be a 20-day journey. But they had only 10 days' food for themselves, and no seal meat for the six remaining huskies.

The dogs were fed on leather from worn-out boots, gloves and straps. Next day the weakest of them was killed and its meat shared between the men and the other dogs. The rest followed one by one until the last husky collapsed on Christmas Day. Nothing was wasted – even the dogs' paws were cooked.

On New Year's Eve Mertz rebelled against the dogmeat diet and was allowed tiny rations from the canned supplies. But by then he was weak and racked with stomach pains. His skin was peeling off. Several days later he could no longer walk and Mawson had to drag him on the sledge. On 7 January 1913, after hours of delirium, Mertz died.

Mawson struggled on with a scarcely believable determination, the skin lifting from his feet and much of his body, his hair falling out and his fingers and toes festering. Not until the late

Xavier Mertz (left of group above) with a sledging team of huskies. Obedient, yet savage fighters among themselves, huskies are derived from breeds of Arctic wolf. In Antarctica, where a few teams are still kept, they are fed on the flesh of Weddell seals (left).

1950s did Adelaide scientists Sir John Cleland and Dr Ron Southcott work out what had killed Mertz and put Mawson in such a pitiful condition: hypervitaminosis A, from the livers of the sacrificed huskies.

Pesticides in meat

Chemicals that help the farmer can find their way into food and linger in our bodies. National wealth, as well as health, is at stake.

It is grossly unjust that a 'dangerous' label can ever be put on placid animals raised for the table and tainted through human error. But a survey of animals that are poisonous to eat would be incomplete if it ignored the risks of agricultural chemicals in meat.

Manufactured compounds that protect crops from insect pests, kill vermin or discourage weeds have lifted farming efficiency and global food production. Only in the 1960s was it shown that they might not be altogether good. Residues of the most successful insecticide, dichlorodiphenyltrichloroethane – DDT – were found in people everywhere. From soils and plants and through animals, DDT had worked its way into virtually all of mankind.

No evidence was demonstrated that low doses of DDT had hurt anyone. But no scientist could prove that the dosing could safely continue. In principle, unnatural substances ought not to enter our systems – let alone stay there. DDT was banned in country after country.

Other and newer pesticides remain under suspicion. Many on the market have residues that persist, like DDT, through the food chain. The troublesome ones are chlorine-based. Until agricultural needs can be met in better ways, governments rely on regulations that control, but do not forbid, their use.

Australian meat producers are bombarded with safety advice and bound by rules which can be enforced by law. If they slip up badly they face the ultimate penalty: their livestock are quarantined and they can be out of business.

Feeding tests on the most sensitive laboratory animals indicate a level of pesticide consumption at which no apparent harm is done. That level is reduced – usually by 100 times – to set a human safety standard. Meat cannot be sold on domestic or export markets if maximum residue limits are exceeded. Fortunately it is relatively simple to check. Samples of fat from slaughtered animals are taken for laboratory analysis.

Even if a pesticide residue shows up at half the official safe level, the stock owner is counselled on what he may be doing wrong. If the legal limit is exceeded, government field officers are empowered to quarantine and test any of his stock. A 'freeze' on sale or movement off the farm continues until residues fall below the safety mark – and that can take years if the source of contamination is not discovered.

Where farmers may go wrong

All pesticides have to be government-approved. Containers must show the active components and any restriction on use. Chlorine-based preparations such as dieldrin, aldrin, heptachlor, HCB and BHC are prohibited in grazing areas and any other place to which livestock or poultry have access. Crops, stubbles and trash from soils treated with them must not be given as feed. Structures that have been treated – against termites, for example – must not be used to shelter animals or to store feed.

The most responsible farmer may strike trouble if he changes his procedures. Residues persist for years in soil, and stock are often contaminated when former cropping land is converted to pasture. Vapours from dieldrin-treated timber can taint feed 18 months later.

HARD LINE ON HORMONE BOOSTERS

Chemicals can be administered to stimulate hormonal activity in food animals. It makes them more efficient in converting plant matter to flesh, and speeds up their rate of gaining weight. Some of these chemicals, called anabolic agents, are similar to the steroids that athletes are tempted to take – illegally – to build up their muscle.

Tiny residues of anabolic agents, finding their way into human diets through meat, may seem harmless. We take much heavier doses naturally from meat, and from a wide range of the plants that we eat. And our own bodies produce them, in amounts often thousands of times greater.

But most of the commercial compounds are merely similar to natural hormones – they are not identical. While they have the same growth-promoting effect, they may have toxic properties in the long run. One that was formerly used in meat production overseas, diethyl stilbestrol, was also prescribed as a human medication for more than 30 years. Now it is known to have caused cancers.

The urge to employ anabolic agents on livestock is greatest where their husbandry is costliest – where they have to be specially fed for all or most of the year. In Australia the practice is forbidden in meat production. When natural grazing is abundant, the ban is no hardship to farmers.

Steroids were used in the production of Australian table chickens until the 1960s. They caponised – neutered – male birds so that hormonal activity went into weight gain instead of the development of sexual characteristics. Anabolic agents are also prohibited now in the poultry industry, where in any case they are no longer needed. Chickens are killed at the age of 7 weeks, before sexual development can affect their growth, so hormone treatments would be wasted. Overseas, treated poultry has caused precocious sexual development in little girls.

A flare-up of high residue levels can call for clever detective work. The source may be a tiny area – a spot where pesticides were mixed, or where a rusted drum once leaked. And sometimes the danger is imported. In one startling case, hundreds of hens in a Scottish breeding flock were killed by dieldrin and their eggs were highly contaminated – though dieldrin had never been used on the farm. Investigators traced it to a nesting box litter of wood chips, brought from a distant sawmill. The whole flock of 10 000 fowls had to be destroyed.☐

WHEN IT DOESN'T PAY TO SAVE

Keeping old pesticides, whether through thrift or forgetfulness, is storing up trouble. If they are more than 10 years old the usages and mixing recommendations shown on their labels are probably out of date. Some uses could be illegal.

Containers gradually break down. If the contents should leak out of a shed, or livestock get in, the owner does not merely risk the quarantine of his animals because of excessive pesticide levels in their meat. Stock – and young children, too – could be killed by direct poisoning.

Mercury in fish

Natural accumulation of a slow, subtle poison is worrying enough. To add to it by industrial carelessness invites disaster.

Health inspectors policing Sydney's fish markets swoop on a consignment of swordfish steaks. Samples are analysed and the sale of the steaks is forbidden. In the same week Victoria's health minister issues a warning to river anglers in a popular high-country fishing district. However many trout they catch, they should never eat more than two a week.

Both of those recent, headline-grabbing occurrences stemmed from the same concern: excessive amounts of mercury were detected in the fish. The danger is not uncommon. Nor is it new. But scientists and public health watchdogs have been alert to it for only 30 years.

Illness from breathing the fumes of evaporated mercury was well-known as a hazard for people working with the metal – in dentistry, for example, and in many manufacturing processes. Industries guarded against the occupational risks, but were less careful about disposing of wastes. The possibility that mercury could enter food chains and build up in lethal concentrations was never foreseen. One pitifully afflicted community brought it to light.

The tragedy of Minamata Bay

In 1953, fishing families near the southern Japanese town of Minamata began to contract a crippling disease. They were physically wrecked, impaired in sight and hearing, and psychologically disordered. By 1960, 43 out of 111 sufferers were dead. The victims included brain-damaged newborn babies – poisoned in their mothers' wombs.

Extraordinary amounts of mercury were found in the blood of the Minamata Bay villagers, and later in the fish or shellfish that they ate at virtually every meal. The source of contamination was soon obvious: industrial wastes discharging into the bay. But officials were tragically slow to stop the pollution.

Another Japanese outbreak, 1000 kilometres away at Niigata, killed six out of 41 victims in

---

OTHER METALS FOUND IN FOOD

While mercury prompts the most widespread worries, many other metals can contaminate animals that are relied on as food. They too can form organic compounds, accumulating in marine food chains as fish are eaten by bigger fish. Chronic poisonings from this cause are rare, however. Most of the metals present a more active danger when inhaled, or when eaten in plants grown in contaminated soils.

More than 20 metals are known to cause serious health problems. Kidneys and other internal organs are commonly affected. Damage to the brain and nervous system may also result from metal poisonings. In industrial processes and the disposal of factory wastes, particular care has to be taken with lead, cadmium, nickel, manganese, chromium and tin – all of them in heavy use worldwide.

Australia's worst inland trouble spot seems to be the upper Goulburn River, feeding into Lake Eildon. Trout were found here with 1½ times the legal limit of mercury.

- MANSFIELD
- *LAKE EILDON*
- Goughs Bay
- ALEXANDRA
- *Goulburn river*
- Eildon
- Jamieson
- MELBOURNE 110kms
- *Goulburn river*
- Kevington
- Knockwood
- Gaffneys Creek

1965. 'Minamata disease' erupted in Sweden, too. There the mercury was carried by fresh-water fish. The lakes where they matured were contaminated by the effluent of woodpulp and paper processing industries.

Alarm spread worldwide, especially where fisheries were economically important. Urgent efforts were started to monitor mercury concentrations in water, in fish and in people. Reliable techniques have now been perfected. But what the figures mean, in terms of human safety, remains open to debate.

Fish and the animals that eat fish – including humans throughout their evolution – must always have tolerated some mercury in their systems. Liquid 'quicksilver' when it is purified, the metal occurs naturally in solid compounds among many sorts of rock structures. Eroded into soils and then washed into rivers and oceans, the heavy particles settle on the bottom and gradually decompose.

How a metal becomes a food

Bacteria convert the chemicals in aquatic sediments, producing the basic substances of plant and animal life. One important group of bacteria make methyl – a primitive compound of carbon and hydrogen. And they can attach methyl to molecules of mercury. The new compound is organic and edible. Other microscopic creatures use it at the start of marine and freshwater food chains.

As animal eats animal the methyl mercury works its way up. In a process just like the amassing of ciguatera toxin, the biggest, longest-living and most voracious predatory fishes contain the greatest concentrations. High levels are typically found in sharks, tuna, swordfish, marlin and barracuda. Among mammals, toothed whales and seals are likely to contain large accumulations of the toxin.

Swallowed in a meal of fish, small amounts of methyl mercury are readily digested into the human bloodstream. Most of it collects in the kidneys, and within a few months it decomposes. The body absorbs the methyl components and the mercury, returned to its inorganic form, is excreted. Some of it comes out in our hair, in varying deposits that give analysts a way of measuring recent intakes.

Consuming tiny quantities of methyl mercury – doses that may have been natural since our remote ancestors first ate fish – probably does no harm. But now, throughout the world, levels of contamination in water and in fish are abnormally high.

A problem of man's own making

Centuries of coal burning and industrial processes have released enormous quantities of mercury. The mercury has drifted into the atmosphere, to return in rain and snow, or it has been dumped in discharges from a multitude of manufacturing activities. In Australia, rivers in some old mining areas are still mercury-polluted because the metal was employed in extracting gold from crushed ores.

Deliberate use of mercury has declined since the publicity given to the Minamata disaster. Where it remains essential, control of effluents is usually enforced by law. But accidents can still happen. And there are ominous indications that most of the damage was in any case done long ago: tests on big fish, caught last century and preserved in museums, show mercury contamination nearly as high as today's.

Commercial fishery grounds around Australia, and in the southern hemisphere generally, are less polluted than those of the north. And few Australians depend heavily on fish for their day-to-day nutrition. So our intake of mercury should be relatively low. A statistical analysis by CSIRO research scientist Dr Doris Airey, sifting through worldwide hair-testing figures, tends to confirm this. Americans, Chinese and Pakistanis, for example, appear to absorb twice as much mercury as Australians do. Britons and Japanese take three times as much.

WHAT MADE THE HATTERS MAD

Hat makers as far back as the Middle Ages were notorious for their moodiness and erratic behaviour. The saying 'as mad as a hatter' has been part of the English language for centuries, and the Mad Hatter of Lewis Carroll's *Alice in Wonderland* a favourite of generations of children.

Among the world's first factory workers, the hatters were perhaps also the first victims of the chemical side-effects of industrial processing. Day after day they breathed the fumes of mercury, used in the treating of wool to make felt. Emotional disturbance was just one early sign of chronic poisoning which would cripple them and condemn them to short, sickly lives.

Lewis Carroll's Mad Hatter, drawn by Tenniel.

WHERE A LUXURY DIET IS A LIABILITY

Lake Murray, in the Fly River region of Papua New Guinea, is the answer to a fish gourmet's prayers. It teems with true barramundi, the giant perch *Lates calcarifer*. People living by the lake eat all the barramundi they want – some of them three times a day.

But there is a catch. By some freak of geology the lake waters contain extraordinary levels of mercury. It must be natural because mercury is not used industrially or agriculturally for hundreds of kilometres around. University researchers from Port Moresby found that the bigger barramundi were so loaded with methyl mercury that only 600 grams of the fish need be eaten each day to reach the point where symptoms of serious poisoning are expected. Some of the villagers did eat that much, and tests on their hair confirmed that the mercury in their systems was above or close to the danger level. Intriguingly, none of the expected symptoms of advanced mercury poisoning were observed. The population remains under study, because scientists are convinced that the mercury must be having some hidden effects. Detection of such effects – or the discovery that the lake people have somehow developed an immunity – would have a tremendous impact on public health debate elsewhere in the world.

Chronic poisoning of the Minamata kind is caused by a prolonged intake of freakish amounts of methyl mercury. It can hardly happen in Australia, even among people who favour fish as the main part of their diet. Nor are fish likely to be the cause of the acute form of mercury poisoning, which leads quickly to kidney failure. This usually results from accidents in which industrial mercury is swallowed. In Iraq in 1972, for instance, 6000 people were poisoned and 450 died after eating grain that was intended for planting and treated with a mercuric fungicide.

How much is safe to eat?

Doctors are becoming more worried about the subtle 'subclinical' effects of mercury – the harm it may do in moderate doses while no ailment can be diagnosed. General vitality, fertility and resistance to other diseases may be reduced, some authorities believe. Psychological patterns may be altered. Intelligence, or at least the power to concentrate, could be affected. Among a group of US students from similar backgrounds, higher hair-mercury counts coincided with lower marks in class.

Unborn babies are protected from most of the ordinary diseases – the placenta acts as a barrier. But the Minamata tragedy showed that methyl mercury attacks foetuses, and particularly their developing brain cells. Now scientists wonder how many unexplained brain defects have been caused by mercury, carried in the blood of mothers who themselves showed no symptoms.

Amid so much medical conjecture, and with some experts in wide disagreement over what quantity of fish in a regular diet could cause trouble, international safety regulations call for extremely low mercury levels in fish offered for public consumption. The World Health Organisation's standard is one part in 20 million, measured in weight.

Australian health authorities, sympathetic to the arguments of the fishing industry, settle for a safe limit of one part in 2 million in local catches. They are guided by an estimate that 200 grams of fish every day – all of it contaminated to that extent, which is highly unlikely – would push the diner's blood-mercury content to no more than one-fifth of the level that is known to have poisonous effects.☐

The better the catch, the more a river angler may be at risk. Downstream of 19th-century goldmining areas where mercury was used, big fish could be heavily contaminated.

Metallic wastes from mining at Captains Flat, NSW, contaminated Canberra's Molonglo River for a century. Since 1974 the river has been restored from a polluted state (above) to purity (right) and fishing is safe.

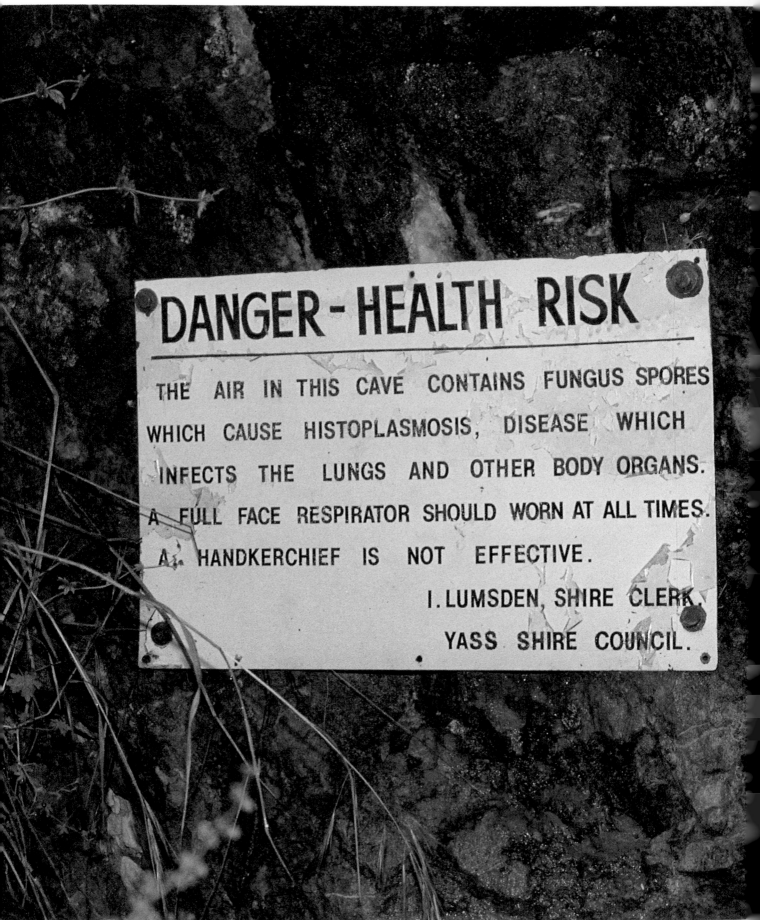

DANGER - HEALTH RISK

THE AIR IN THIS CAVE CONTAINS FUNGUS SPORES WHICH CAUSE HISTOPLASMOSIS, DISEASE WHICH INFECTS THE LUNGS AND OTHER BODY ORGANS. A FULL FACE RESPIRATOR SHOULD WORN AT ALL TIMES. A HANDKERCHIEF IS NOT EFFECTIVE.

I. LUMSDEN, SHIRE CLERK.

YASS SHIRE COUNCIL.

Warning notice outside bat-breeding cave at Wee Jasper, NSW.

Part 5

DISEASES CAUSED BY ANIMALS

*Dangerous animal-borne organisms are older than
the human race. What harm they do depends
on how we live and the care we take — for the spread of many
diseases is our own fault.*

Standing guard

In a climate that fosters epidemics, the price of public health comes high. Can we do more to help ourselves?

Human dominance of the animal world is least secure in the field of disease. Of all the pestilences that have scourged our race and interfered with its progress, only one – smallpox – can fairly be said to have been eradicated. Smallpox required no other animal agent, either as a host or as a carrier. The virus went straight from human to human, like influenza.

Every other disease that has wiped out vast populations – plague, malaria, yellow fever, typhus – is animal-related. Each is still active, along with hundreds of other animal-borne diseases that can destroy or blight human lives on a scale that is less disastrous, but no less a calamity for the people afflicted.

The latest to emerge, the virus responsible for the acquired immune deficiency syndrome, AIDS, may have been around for centuries. Monkeys are thought to be its natural reservoir. Other animals – biting insects, perhaps, or ticks – may have assisted in the original transmission of AIDS among Central African communities.

Dozens of other viruses are known to exist in animal hosts or carriers, without so far affecting humans. They have been described as 'diseases looking for somewhere to happen'. Shifts in population or changes in agricultural methods could be the trigger they need.

Sydney's Darling Harbour was for generations the main focus of efforts to keep out alien diseases. Emphasis has shifted to airports, but ships and their cargoes are still subject to inspection by port health authorities.

Lake Argyle, created in the tropical northwest by the Ord River Dam, has become a permanent reservoir of viruses.

A lucky country – but for how long?

Australia has been remarkably fortunate. Climatic conditions, especially on the tropical margins of the continent, favour the breeding of many of the most feared disease carriers. We have the yellow fever mosquito, for example – but evidently not the virus.

Throughout almost all of the human era, the worst ravages of animal-transmitted diseases were avoided in Australia simply because populations were small and widely dispersed. Compared with the density of settlement in tropical Asia, Africa and America, tropical Australia was virtually deserted. Now the north has many more people than ever before. With the expansion of agriculture, mineral exploitation and tourism, the invasion is accelerating. This new tropical population is not as scattered as the old. Many people live at close quarters in substantial cities. They are safer in one sense: medical services are easier to provide. But the rapid spread of epidemics is also made easier, if ever Australia's luck should change.

The other natural control on the spread of many animal-borne diseases was the dryness of the hinterland. As quickly as bore-sinking, damming and crop irrigation have made more parts of Australia liveable, pests and diseases have appeared that make them at times unliveable. That has been demonstrated over and over again in agricultural developments and horticultural experiments – and not merely in the tropics. Anywhere that standing water, summer warmth and people are associated, trouble brews.

In a quiet but costly battle to keep diseases from entering the country, Torres Strait is a frontline. More than 70 islands strung from the Gulf of Papua to Cape York, Qld, are the stepping stones of a traditional crossing route.

They serve not only people and whatever they carry – seen or unseen – but also animals and plants. Commonwealth medical officers based on Thursday Island tour the other islands by helicopter and boat, conducting health clinics and watching for new arrivals.

They are determined to break a chain of disease transmission that can reach all the way from Southeast Asia. Malaria is the most immediate worry. Northern Australia not long ago was a natural home for this affliction, and it is endemic throughout New Guinea. Irian Jaya has had epidemics of anthrax. Indonesia's eastern islands have rabies, and foot-and-mouth disease – which, if imported, would bring ruin to Australia's meat export industry.

Keeping track of the traveller

Major seaports have been under official health scrutiny for over a century, ever since a smallpox epidemic shocked the Australian colonies into joint action. But diseases can span the globe in hours now, thanks to jet airliners. A traveller can become infected in any foreign country and be in Australia before symptoms develop.

Non-Australian passengers can be refused admittance without an up-to-date certificate of vaccination against yellow fever. Port health officers have the power to make any arriving traveller undergo a medical examination. Quarantine stations are maintained at every international airport. Even Australians coming home can be held in quarantine if they show signs of a dangerous disease.

Entry-card details demanded of all air travellers include their intended address. If someone on their flight turns out to have a disease that could have been communicated, medical checks on the whole passenger list can be made quickly. Doctors throughout the country are helped to know all they can about exotic diseases. To a practitioner unaccustomed to tropical problems, for instance, malaria in its first stages can pass as flu.

Ships believed to be carrying dangerous vermin such as rats are prevented from berthing until they are fumigated. Airliner cabins and cargo holds are sprayed with insecticide at the first port of entry. (Qantas 747B cargoes are sprayed automatically from inside the plane as it leaves its last foreign port.) International terminals incorporate every possible device for screening out insects.

Animals imported intentionally are quarantined for long periods to allow for possible diseases to reveal themselves. Some animals are totally banned – pet rodents such as hamsters and gerbils, for example. Escaping and breeding in the wild, these could become reservoirs of plague and rabies if the diseases broke out here.

All of us are involved – through our pockets

Inside the country, compulsion is less easy to apply. Some animal risks – especially with pets – are accepted as a matter of personal choice. What portion of the total cost of medical services is incurred coping with animal-related diseases is impossible to estimate.

But the scale of the problem may be grasped from the example of just one disease, limited in its range and not a killer. For an 18-month public education campaign to cut the incidence of dengue fever in Queensland, federal authorities happily footed most of a $5 million bill.

Many diseases arise from occupational contact with livestock and slaughtered animals. The nation pays, both in lost export earnings and in primary industry departments spending time, money and manpower on preventive measures, education and inspection. Taxpayer-funded research organisations have to channel large parts of their budgets into trying to counter the diseases that farming families and meatworkers can catch.

In urban areas, where the risk of epidemics is greatest, councils have wide public powers delegated to them. Inspectors can interfere wherever a health hazard is reported – a rat infestation, for example – and order its removal. People can be fined if they do not comply, and council staff can enter their homes to do whatever is necessary. But the preferred method is friendly advice and help.

To open the following survey of diseases with a recollection of bubonic plague in Australia may seem unduly alarmist. But the story still holds lessons. What the outbreak at the turn of the century revealed was neglect by people in authority, unconcern among the well-off, and helpless apathy on the part of slum dwellers.

Nearly 90 years on, plague and other epidemic diseases have not been abolished. Nor have poverty and slums. No amount of official spending guarantees our safety if people are lax in their own standards of hygiene or do not care about those of their fellow citizens.□

Incoming mail and container loads are checked for foodstuffs that could contain disease organisms. As well as public health, the agricultural economy is at stake — part-cooked meats from Europe are a likely source of foot-and-mouth disease.

Poverty-stricken slum dwellers at the turn of the century could do nothing about their desperate overcrowding. But they had themselves to blame for the backyard squalor that encouraged the breeding of rats.

Bubonic plague

On the eve of proud nationhood, the slumland spread of a murderous outbreak put Australia to shame.

Leading lights of Australian society and politics had much on their minds at the dawn of the 20th century. The colonies were agog with preparations for their federation and the ceremonial pomp that it promised. Agriculture was pulling out of a 20-year depression and optimism was high among the well-to-do.

If they knew of the wretched poverty of inner-city dwellers they were removed from it, on pastoral properties or in suburban villas made accessible by new tramways. After all, the poor did have roofs over their heads. But they were crammed in one-up-and-one-down terraces, let and sublet and left to fall into ruins. Rats infested these slums.

Adelaide had the first case of plague. The *Sydney Morning Herald* reported it as 'not the slightest occasion for alarm'. But officials in Sydney were secretly panic-stricken. They had already found dead, plague-infected rats at Darling Harbour, near where two ships from New Caledonia had berthed. A quiet clean-up of the waterfront was started, too late.

Though Sydney's first victim was isolated on 19 January 1900, the disease was not proclaimed and public sanitation measures were not enforced until March. In the meantime many people had died. Vigilante groups launched their own attacks on the slums — venting their spleen mostly on a hapless Chinese community.

Sydney's population had doubled in just 20 years. Revolting conditions in the slums apparently surprised government inspectors. The city council was blamed for neglect of its public health responsibilities. The chief medical officer told his premier of 'maladministration now for the first time revealed to the general public'.

Once it was clear that plague had spread beyond the dock areas and the Rocks, the government was forced to show its hand. Sanitation teams swarmed into other slum districts, first with disinfectants and often, later, with sledgehammers and firestarters. But by 19 August there had been 303 cases of plague and 103 deaths.

Every other mainland city had its victims. Reappearing spasmodically until 1909, the epidemic killed 530 out of 1212 people afflicted in Australia. Parts of Sydney were cleared and rebuilt, and councils approached their public health duties with a new zeal.

But in 1921 it was as if none of the lessons had been learned. Plague-infected rats came ashore from the freighter *Wyreema* at Sydney, Brisbane and ports to the north. They were in animal fodder. Sydney had 35 cases of plague and 10 deaths, Brisbane 59 cases of plague and 28 deaths.☐

Sydney's worst slum dwellings were set on fire and demolished.

Plague through the ages

The worst outbreak of this ancient scourge of mankind was the Black Death.

Rats by the thousand were collected in the dockside warehouses of Sydney — most of them already killed by plague. Their discovery stirred public health authorities to launch a clean-up campaign that came far too late.

Plague cannot start here. It has to be imported. Once it is in, we have the animals to carry it on. Chains of infection can be long and complicated, but the links are nearly always fleas.

Australasia is the world's only major inhabited region where wild animals do not harbour the plague bacterium *Yersinia pestis*. The natural host animals are more than 200 rodent species, ranging from ground squirrels in the US to marmots in Asia and Europe, gerbils in Africa and cavies – guinea pigs – in South America. Some hunting animals preying on the rodents also become infected. This 'wild plague' seldom concerns humans.

Three kinds of human plague

Human plague occurs mostly when the infection reaches the rats that like to live closest to us, infesting buildings and garbage or travelling in ships' cargoes. Epidemics break out when people are crowded together and unsanitary conditions allow too many of these rats.

Typically the troublemaker is the black rat *Rattus rattus*, often called the ship rat. This is probably mankind's oldest and most constant companion. But other rats can spread plague if humans are bitten by their fleas. And house pets can pass on rat fleas even though they are unlikely to be infected themselves.

Many rodent fleas can carry plague. But Australia's outbreaks, and most of the catastrophic epidemics overseas, are blamed on the Oriental rat flea *Xenopsylla cheopis*, which can carry plague germs in its saliva. It is relatively uncommon here. In South America there is evidence that the human flea *Pulex irritans* transmits plague from person to person. Some people suggest that this also happened in the Black Death of the Middle Ages.

Plague's main form is bubonic. Lumps appearing in the armpits or groins of victims used to be called 'buboes'. They are swollen lymph glands. Septicemic plague, which infects the bloodstream more than the lymph system, is rare. Neither form is directly contagious, but both can lead to pneumonia. Then the germs can be passed on by coughing. When the disease is widely spread this way it is known as pneumonic plague and may be almost 100 per cent fatal if not treated.

Bubonic plague has a much lower mortality rate, even without medical aid.

Six years that transformed Europe

Accounts of plague stud the histories of ancient Middle Eastern civilisations. Early in the Christian era the disease hastened the decay of the divided Roman empire and brought on the Dark Ages in Europe. From records that began to reappear in the 11th century it is clear that plague outbreaks were a familiar occurrence in Western Europe. The Black Death was not the first emergence but the worst.

The Black Death pandemic – an epidemic that spreads through country after country – probably originated in east Asia. It was certainly associated with the conquests of the Mongols under Genghis Khan, and it killed millions of Chinese in the 1330s. In 1346 it broke out among Tartar tribes besieging a Black Sea port. European traders fled in rat-infested ships.

In two years plague was carried to most continental ports and as far as Britain and Ireland. It moved overland rapidly. By the time the infection burned itself out in 1352 – presumably because few rats were left to harbour it – an estimated 25 million people had been killed. Continental Europe lost at least a third of its population, and Britain perhaps two thirds. Social order broke down as superstition and religious mania gripped the survivors.

Flea was last to get the blame

People noticed that rats died in large numbers during many outbreaks of plague, but thought the animals were merely fellow-sufferers – not part of the cause. As late as the 1660s, when Daniel Defoe was documenting the Great Plague that killed 68 000 Londoners, it was believed that the disease originated in rotting organic matter and rose from the ground in a gaseous 'miasma'.

Once the role of rats was established, in the late 1870s, scientists turned their attention to fleas. They were thought at first merely to pass plague germs from rat to rat. Their direct responsibility for infecting people by biting was not proved until after the turn of the century, in Sydney, by J. Ashburton Thompson, NSW chief medical officer during the plague years.

The last great epidemics of plague were in Manchuria and Siberia in the winters of 1910-11 and 1920-21. In those regions direct contact with the wild hosts was involved – marmots are hunted for their pelts – and pneumonic plague developed. Both epidemics stopped as soon as spring came and people moved out of their crowded quarters, but about 70 000 were killed.

Plague today

Not all the countries known to have plague report their cases to the World Health Organisation. Sketchy figures show a global incidence of hundreds of cases a year, with a death rate usually lower than 10 per cent. South America in some years accounts for more than half of the world total. The only country where plague is known to have reached epidemic proportions in recent times is Vietnam, in the chaos that accompanied the collapse of the southern regime in the early 1970s.

Most bubonic plague victims, if treated soon enough, respond to antibiotic drugs. A vaccine is available for medical workers who place themselves at risk in dealing with pneumonic plague. Prevention is simple in well-organised communities: get rid of rats.□

TRADE PATTERNS LAID THE PATH

Australia had ample warning that plague was on its way. Patterns of international trading made its arrival by ship inevitable. But plague was seen as an Asiatic disease. Its spread through Asian ports, after killing perhaps 100 000 people in Hong Kong in 1894, was considered no cause for concern in the south.

Alarm was not raised among officials until the end of 1899, when Sydney authorities had to refuse a plea from New Caledonia for inoculation serum. They found they had far too little for their own purposes. No move towards rat eradication in housing areas was made until the epidemic was in full swing.

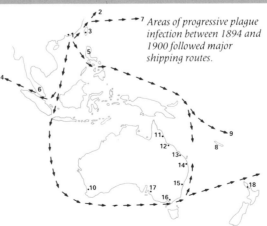

Areas of progressive plague infection between 1894 and 1900 followed major shipping routes.

1 Hong Kong 1894
2 Japan 1896
3 Taiwan 1896
4 India 1896-8
5 Philippines 1897
6 Malaya 1898
7 San Francisco 1899
8 New Caledonia 1899
9 South America 1899
10 Fremantle 1900
11 Cairns 1900
12 Townsville 1900
13 Rockhampton 1900
14 Brisbane 1900
15 Sydney 1900
16 Melbourne 1900
17 Adelaide 1900
18 Auckland 1900

Australian encephalitis

Birds spread the virus, mosquitoes inject it. Unlucky victims – perhaps only one of thousands bitten – suffer brain disease.

Far inland in the heat of early summer, waterfowl finding flooded ground make a joyous sight. Waders and swimmers appear from nowhere to splash down among a flush of new plant life. They call excitedly to their fellows and quickly set about the business of nesting. Most are nomads, ranging far in their quest for food and breeding only when a supply is assured.

The same conjunction of standing water and warmth favours the breeding of mosquitoes. They are stay-at-homes, for they do not need to fly far. Meals of blood from biting the birds give them the nourishment they need to continue breeding. And their larvae provide some of the birds' food.

Sometimes, from somewhere, a virus enters the relationship. It inhabits a group of mosquitoes and disappears when they die – unless by their biting they help it to find another host. Then the virus multiplies in the bloodstream of the bitten creature, most often a bird. But the new host can just as easily be a native mammal or a farm animal – or a human being if people are nearby.

Birds are not affected by the virus. In a few days their systems conquer it and they become immune. In the meantime, however, if they are bitten by new, uninfected mosquitoes the multiplied virus is passed on. And if during that period the birds fly to other feeding grounds, the virus makes a geographical leap that its mosquito carriers could not have achieved.

In various parts of the world more than 150 viruses are naturally transmitted by arthropods – mostly insects, but also arachnids such as ticks. Scientists call them 'arboviruses', short for arthropod-borne. Overseas, the yellow fever virus is the one most feared.

Australia's encephalitis arbovirus, which has counterparts in Japan, China, the US and the USSR, can lead to a dangerous inflammation of human brain tissues.

Encephalitis can also develop in rabies, or through invasion of the brain by amoebas. It may be a complication of other viral diseases that are not associated with animals. Herpes, mumps, measles and poliomyelitis can take this course in rare cases.

Pre-school children are most at risk

Humans are almost as resistant as birds are to the mosquito-borne virus. When there is an outbreak of encephalitis, blood tests on healthy people in the same district show that huge numbers of them possess antibodies to the virus. They too have caught it, but their immune systems have overcome it.

Researchers believe that for every case in which encephalitis develops, as many as 3000 adults may be infected without harm. Presumably some people are infected repeatedly at intervals throughout their lives, renewing their immunity. Among children under five, however, the chance of severe illness comes down to one in hundreds. The young are more prone to the disease not merely because they are smaller but also because the first exposure is the most dangerous. Newborn babies, however, may retain some protection from their mothers' antibodies.

Though the disease affects relatively few people the death rate is high. In early epidemics it was close to 50 per cent. Under modern medical treatment fewer than 25 per cent of victims die, but the outlook for survivors of severe attacks is poor. Some form of brain damage, leading to permanent mental and physical impairment, is likely.

Australian encephalitis presumably took its toll of the families of early settlers. Its effects could not have been distinguished from those of some more common afflictions. The nation did not become aware of any special menace until epidemics in 1917 and 1918 killed nearly 100 people in NSW, Victoria and Queensland. Most victims lived near the floodplains of the Darling and Murray Rivers.

Investigating these epidemics, John Cleland succeeded in 1920 in isolating a virus and

---

AUSTRALIAN ENCEPHALITIS

EARLY SIGNS Severe headache, neck stiffness and pain. Aversion to bright light. Fever, weakness, loss of appetite.

ACTION Obtain urgent medical aid.

AREAS AT RISK

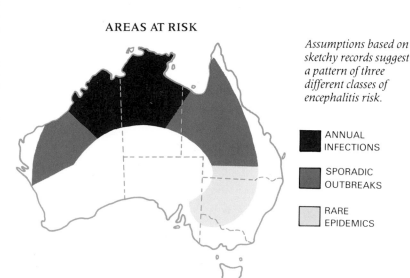

Assumptions based on sketchy records suggest a pattern of three different classes of encephalitis risk.

ANNUAL INFECTIONS

SPORADIC OUTBREAKS

RARE EPIDEMICS

The encephalitis virus seems to infect the rufous night heron Nycticorax caledonicus far more than any other bird. This heron, formerly called the nankeen night heron, roosts by day and comes out at dusk to feed busily on insects, fish, frogs and crustaceans. It is found anywhere there is permanent water.

Alice Springs

The distribution of Culex annulirostris *is probably wider than indicated here, especially in Central Australia, but comprehensive records do not yet exist.*

Surprising new evidence from the north

A theory was formulated that the virus was of tropical origin, though the disease was not known in the north. Frank Macfarlane Burnet suggested that the virus had its home in the permanent swamps of western Papua and that migratory birds brought it south in springtime, interacting with mosquitoes wherever they rested to keep alive a chain of infection and re-infection.

The key to epidemics in the southeast seemed to be prodigious spring rains over central Queensland and western NSW, filling the vast catchments of the Darling-Murray system. By summer, as the floods worked downstream, normally parched lowlands of the Murray Valley were covered by sheets of standing water. In these rare years, with substantial numbers of human beings close by, bird and mosquito populations exploded.

Erratic occurrences since the mid-1950s suggest that the spread of the virus cannot be so simply explained. In one dry summer many cases of encephalitis were scattered through the southeast. Yet the following year, when conditions seemed certain to cause an epidemic, nothing happened.

An outbreak of 60 cases in 1974 took six lives in the southeast. But long after it passed its summer peak there, isolated cases appeared in Queensland, the Northern Territory and Western Australia. Newly alert in the northwest, medical authorities discovered that encephalitis was an annual occurrence with the coming of the monsoonal 'Wet' to the Kimberleys, WA, each February.

Encephalitis was also found to be widespread in most years farther west in the Pilbara, and it has since shown up as far down the west coast as Carnarvon. Recognised at last as a disease of national consequence, the disease has been renamed Australian encephalitis.

Many earlier cases in the north must have gone unrecognised. They would have been scattered, bearing no apparent relationship to southern epidemics. And symptoms in the north are generally less severe. They indicate meningitis – an attack on the brain covering – rather than inflammation of the brain itself.

demonstrating its effects on monkeys and sheep. But no one knew where the virus had come from or how it had entered the victims' systems. Scientists were none the wiser when outbreaks recurred in 1922 and 1925. Then what they called 'Australian X-disease' seemed to disappear completely.

It returned to the upper Murray Valley at the end of 1950. In the first two months of 1951, 19 out of 45 victims were killed, and intensive studies began on what now became known as Murray Valley encephalitis. Mosquitoes were found to be responsible. Then antibodies were discovered in the bloodstreams of waterfowl.

CHICKENS STAND SENTRY DUTY

White leghorn chicks, newly hatched in sterile conditions, are trucked every November to farms dotted along the Murray Valley and in the irrigation areas of NSW. They are sentinels, stationed there to give the first warning that the encephalitis virus has arrived.

Samples of the chickens' blood are taken every week throughout summer and sent for laboratory analysis. If antibodies against the virus are found in a flock, authorities put emergency mosquito control plans into action in that area and the public are warned to protect their children.

Blood samples are also taken from farm livestock by agricultural field officers, and from native birds and mammals by wildlife rangers. In the far north of Western Australia, sentinel chicken flocks protect mining and agricultural communities.

Frequent exposure to mild infections may lead to greater immunity among Aborigines and among Europeans with long experience of the tropics. But the presence of the disease raises concern for the health of the communities that have mushroomed out of recent developments in mining and agriculture.

Water management comes under suspicion

'A virus reservoir of world magnitude' was created by the Ord River dam and irrigation project, according to the late Professor Neville Stanley, head of microbiology at the University of Western Australia. Mosquitoes taken in 1976 from the Ord, just east of the Kimberleys, carried more viruses than any others in his experience. Blood samples from residents showed that 90 per cent had been infected by the encephalitis virus, though no cases were reported.

Storage of water on a large scale, anywhere in inland Australia, is an invitation to waterfowl and mosquitoes. Where engineering works regulate the flow of rivers, breaking the characteristic cycle of drought and flood, conditions may be created that allow a permanent stock of viruses to build up.

The 1974 episode in the Murray Valley was the first in which Australian encephalitis was studied while an outbreak was still in progress. Researchers were able to establish that the mosquito species that carries the virus in the epidemic-prone areas of the south is nearly always *Culex annulirostris*. But whether any particular bird deserves to be singled out for blame is debatable.

Of dozens of waterfowl species that can harbour the virus, the rufous night heron seems by far the most likely to be infected. Blood sampling revealed antibodies to the virus in more than 90 per cent of this species. Less than 10 per cent of cranes and waterhens had been infected, and only about 5 per cent of ducks. By their sheer numbers and their nomadic habits, however, ducks could be as important as the herons in the geographical spread of the virus.□

RABBIT SCARE BROUGHT WRATH ON SCIENTISTS

Mystery made the 1951 encephalitis epidemic all the more frightening. The role of mosquitoes was fairly well established, but no one knew how the insects came to be infected with the virus. Suspicion fell on the Commonwealth Scientific and Industrial Research Organisation.

The epidemic coincided with the Murray Valley's first explosion of myxomatosis disease in rabbits. Released in 1950 in a CSIRO-devised eradication campaign, the myxoma virus attacked the rabbits' brains. And it was spread by *Culex annulirostris* mosquitoes. People in Mildura, Vic, became convinced that it was myxomatosis killing their children.

Even a hospital superintendent joined the onslaught on the scientists. 'The CSIRO was playing with fire when it let this virus loose,' he charged. Prominent among the defenders was CSIRO adviser Dr Frank Macfarlane Burnet, later knighted and awarded the Nobel Prize for Medicine. His assurances that myxoma could not harm humans fell on deaf ears. An argument with a Mildura councillor ended with a dare: If Burnet was so certain, he should take a dose himself.

Burnet took up the challenge along with Dr (later Sir) Ian Clunies Ross, head of the CSIRO, and Dr Frank Fenner, who was in charge of myxoma research. In a demonstration that would have made compelling television today, all three men were injected with enough virus-infected serum to kill hundreds of rabbits. Their survival in good health was announced in the federal parliament and the public outcry waned.

Macfarlane Burnet accepted a frightened man's dare.

Headlines from the Sydney Daily Mirror *of 1 March 1951 reflected the frustration of medical researchers.*

The Sleeping Girl

One tiny victim of encephalitis touched a nation.

Hearts were touched throughout Australia in 1951 by the plight of Heather Williams, 'the Sleeping Girl'. Three years old when encephalitis broke out in the Murray River town of Wentworth, NSW, Heather became infected and spent nine months in a coma.

Worse was to follow. When she regained consciousness she was found to be brain-damaged and partially paralysed. Heather was confined to a wheelchair, and died at 16.

Viral encephalitis is sometimes called 'sleepy sickness' because extreme drowsiness is a common symptom. But it should not be confused with 'sleeping sickness', an even more dangerous African disease caused by the bite of the tsetse fly.□

'She was found to be brain-damaged and partially paralysed.'

MOSQUITOES: PUBLIC ENEMY NUMBER ONE

People can hardly be expected to see any good in mosquitoes. With the diseases they carry they have been the worst enemy of mankind – killing, according to some estimates, up to half of all the humans who ever lived. And apart from the danger, they are insufferable pests.

But the benefit of mosquitoes in nature is clear enough.

They are food – and in a country like Australia they represent tonnes of it a year. At all stages of their lives they are virtually defenceless. Eggs and immature, water-dependent forms are gobbled up by fish, frogs, waterfowl and rival insects. Adults on the wing are prey to birds, bats, frogs and spiders.

SURVIVORS AGAINST ALL THE ODDS

As well as being the prey of many creatures, mosquitoes are at the mercy of the elements. High winds and extremes of temperature and humidity take a heavy toll. Egg-laying sites are often fragile. Rare is the mosquito that reaches adulthood, let alone one that survives for a month after. What enables these delicate insects to perpetuate their line is a high degree of adaptation to different surroundings – Australia has about 300 species – and a compulsive dedication to breeding.

Only the females bite and suck blood. Their sole purpose is to obtain proteins and vitamins for egg production. The males, if they eat at all, draw nectar or sap from plants. They rarely fly except to mate.

The whining sound of a female's fast-beating wings works as a mating call. Different species in a given area have different whining tones. When a sound at the right frequency passes the feathery antennae of a male, fine hairs vibrate in unison. This movement excites sensory cells at the antenna bases. Nerve impulses are sent to the brain and flying mechanisms are triggered. Involuntarily, the male takes off and heads for the source of the sound.

Tracking down a meal of blood

Mosquitoes see well in the poorest light. But to locate their favourite blood-feast target from a distance, the females use other highly developed senses. They backtrack warmer currents of air and detect carbon dioxide from breathing, along with faint chemical traces from mouth and body odours. It may be variations in the chemistry of skin secretions and breath odours

Mosquito eggs and larvae crowd the surface of a quiet pool as a newly emerged adult Culex annulirostris *prepares to fly off. The population of just a square centimetre of water can make a hefty meal for a small fish — or a squadron of potential disease carriers.*

that cause one person to be bitten less than another. Recent research suggests that anopheles mosquitoes actively seek malaria carriers because the disease organisms make blood easier to draw up. Humans are not a first choice as blood providers, except by the few species that have adapted to lurking in and around houses. Many others bite people if they happen to be in range, though their normal preference is wild mammals, livestock or birds. And mosquitoes are by no means restricted to warm-blooded hosts. Frogs, snakes and goannas – even mudskipper fish – are bitten.

Choosing a laying site

Mating only once, usually on the wing, well-fed females lay batches of 100 or more eggs every week or so. After each feed of blood they rest for some days while the eggs develop. Then they find a laying site. The choice varies widely according to the species but one condition is constant – there has to be water or at least a damp surface. The water need not be stagnant – quiet reaches of streams can be used. Coastal mosquitoes may choose evaporating rock pools, despite their high salt content.

Some species enjoy permanent pools overgrown by plants. Others seek highly specialised container habitats – holes and forks in trees, leaf

A rainforest frog of the east coast, Litoria lesueurii, *provides a meal of blood. Various mosquito species have their own feeding preferences, with humans seldom the first choice.*

axils and other parts of plants where water collects. One species even has the audacity to lay its eggs in the trapping pools of insect-eating pitcher plants, and gets away with it. Increasingly, though, people have helped the container breeders. Some of the worst disease carriers take advantage of artificial receptacles around houses, gardens and waste land.

The hatching time is variable, depending on temperature and moisture. In the last stage of development, embryos grow a cutting instrument to pierce the tough outer layers of their shells. Sucking in water, they bloat their bodies and the shells burst open. Larvae emerge, looking much like maggots.

Growing in a world of water

In larval form the mosquito is an expert swimmer, feeding on bacteria and decaying plant matter. Brush-like appendages beat the water, sweeping food into its mouth. Nothing about its anatomy suggests the flying insect that is to come. It does, however, breathe fresh air, hanging with its rear end at the surface. Many species have a siphon to aid breathing.

Larvae go through four moults of their outer skin as their size increases. After the fourth moult, perhaps 2-4 weeks from hatching, they

---

BITING PARTS ARE HIDDEN

A mosquito's long proboscis, pointing down towards a bite victim's skin, never pierces it. This is merely a protective half-sheath, concealing a set of fine, sharp-pointed probes and a sucking tube. The sheath peels back as the probes – called stylets – penetrate.

The end of the sucking tube is forced into the wounded skin and muscles in the mosquito's throat start a pumping action to draw up the oozing blood. At the same time a tiny amount of saliva goes into the wound. It contains agents to delay clotting and make the blood flow more easily. Sometimes it also contains disease organisms.

Itching and swelling of a mosquito bite are not a sign of any disease but a response to substances in the mosquito's saliva. Some people have an excessive sensitivity and suffer prolonged discomfort. They can be treated with antihistamine drugs.

The top picture on the right, below, shows a mosquito drawing back its proboscis to expose the sucking tube as the stylets penetrate human flesh. The bottom picture shows the sheath drawn back fully and the sucking tube deep in flesh.

Mosquito starting to feed

Mosquito engorged with blood

---

cease feeding. In this pupal stage, which takes about two days, they breathe through organs that jut from their mid-sections like trumpets.

At last the hard pupal casing splits down the back and the adult emerges, soft-bodied and helpless. It rests on the water surface long enough to dry out, then flies off to renew the cycle. Females seek blood immediately but males cannot mate for a day or so. The antenna hairs through which they are summoned do not work until their sex organs have taken up a functioning position.

The lifespan of the female is medically important because it determines whether or not the insect can be dangerous. Mosquitoes are not born with viruses or parasites in them. They must first take a meal from an infected host animal, then live long enough for the organisms to become infective to another host, such as a human. That generally takes 10-14 days.

EGGS THAT CAN WAIT FOR RAIN

Parched for months or years, the near-deserts of Central Australia are still not free of mosquitoes. Just as ephemeral plants have adapted to delay their seed germination until moisture comes along, specialised mosquitoes have evolved to beat drought.

Inland species of the anopheles and culex groups simply hide and wait. Unusually long-lived, they lurk in the cool depths of rock crevices, goanna burrows and rabbit warrens. Females with fully developed eggs delay their laying, and go on taking blood, until rainwater collects. Travellers can get bitten at the height of a dry season.

Aedes mosquitoes have taken their desert adaptation much farther. The females have a slender, pointed tip to their abdomen. They can extend it into an egg-depositor, enabling them to bury their eggs in mud or drying soil at the very end of a wet season.

The eggs need humidity for only a day or two, in which time the larvae are fully formed inside their shells. Then the shells harden and the larvae can lie dormant for years. When eventually an egg is covered by water, the reduction in oxygen triggers the hatching of the larva. But not all the eggs in a batch will hatch at the first wetting, which may be a false alarm. To ensure that the species survives, some eggs hold out for a second, a third or even a fourth flooding.

CHARACTERISTICS OF THE MAIN MOSQUITO FAMILIES

ANOPHELES GROUP	AEDES GROUP	CULEX GROUP

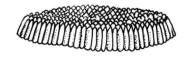

Eggs loosely clustered

Eggs scattered

Eggs in raft

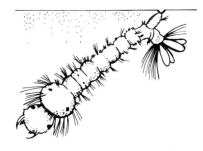

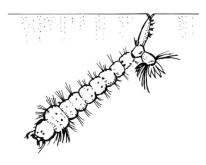

Larva lies flat below surface; no siphon

Larva hangs down in water; short siphon

Larva hangs down in water; long siphon

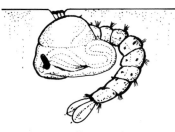

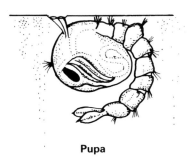

Pupa

Pupa

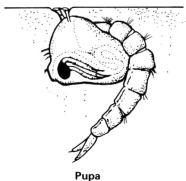

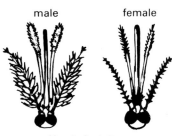

male   female

Pupa

Head of adult

male   female

male   female

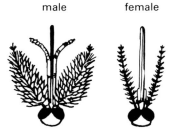

Head of adult

Head of adult

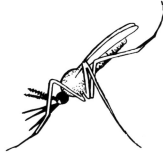

Posture raised, almost vertical

Posture flat

Posture raised, horizontal

Anopheles farauti will be the principal culprit if mosquitoes spread malaria in Australia again. A breeder in swamps and well vegetated lagoons of the far north, it is fortunately restricted in range. The females seldom fly more than 500 m from their breeding site in the quest for blood.

Aedes aegypti, spreader of the dengue virus, breeds close to dwellings in any artificial containers that catch water. It bites by day, in shade, or in the early part of the night. A low flyer, it generally bites around the ankles and rarely goes into houses.

White bands on the legs and a white ring around its proboscis distinguish *Culex annulirostris,* the mosquito most often responsible for encephalitis epidemics in the southeast. It favours quiet water with sunlight and some vegetation, including temporary pools when rain runs off roads. Its most active biting period is sunset.

Germ warfare

Microbes discovered in a Middle East desert make the most promising weapon so far against dangerous mosquitoes.

Total eradication of mosquitoes is out of the question. Even if it were technically and economically feasible, such a drastic interference with nature could prove environmentally damaging. All that can be hoped for is control of mosquito populations, wherever they are in dangerous contact with human communities.

Only physical engineering can be said to have much permanence. Watercourses are diverted, swamps are drained and reclaimed, and other ground depressions where seasonal rainwater collects are filled in, often at heavy cost. Mosquitoes are denied their habitats, but so are birds and fish. And other earthmoving projects, for water storage and irrigation, create new breeding opportunities for the mosquitoes.

Oil products such as kerosene have been widely used – especially when they were cheaper – for spreading on the surface of stored water. They form a film that prevents mosquito larvae already in the water from breathing. But the treatment has to be repeated every season.

Chemical insecticides have never been sufficiently selective. They kill many other harmless and desirable insects, and sometimes frogs, birds and mammals. They may have long-lasting residues that are potentially harmful to humans, so their use near livestock and crops is undesirable. Complicated machinery may be needed to apply them, and mishaps can be hazardous. Most importantly, perhaps, mosquitoes can develop a resistance to whole groups of insecticides.

How science tries to put nature to work

Most research now goes into possible methods of biological control, taking advantage of natural processes. Fast-breeding fish that specialise in eating larvae – particularly members of the guppy family – have been introduced with success in many areas of confined water. They eat other insects, however, and may destroy native fish species. Parasites are a potentially useful field of experimentation. Certain fungi, worms and single-celled animals invade insect larvae. Some of them can be mass-produced, but proof of their practical effectiveness and safety may be many years away.

Mosquito breeding can be sabotaged – at least under laboratory conditions – by subjecting adult males to radiation. They can be sterilised, in which case when they mate with normal females the eggs are infertile. By another technique the chromosome structure of

Water in any container — flower vases, plant-pot saucers and disused lavatories indoors; buckets, paddling pools and discarded household goods outdoors — can provide a breeding site for mosquitoes and should be emptied whenever possible, or at least changed once a week. A clean garden can become mosquito-infested in three warm, wet weeks — perhaps while you are on holiday — if water is allowed to collect anywhere. Nearby waste ground makes the situation much worse, as overgrowth shelters semi-permanent breeding sites in many sorts of rubbish.

Electronic insect killers are a safe, effective way of keeping your garden free of flying pests such as flies and mosquitoes. A powerful black light bulb attracts flying insects which are then electrocuted by grids surrounding the bulb inside the enclosure.

the males' genes can be altered. Then their progeny·will be sterile. But neither method is likely to be of much practical use in the wild.

A discovery in Israel in 1977 has led to the best form of biological control available so far. Investigating the disappearance of mosquitoes from seasonal pools in the Negev Desert, scientists found a modified strain of the bacterium *Bacillus thuringiensis*. It attacked the larvae of mosquitoes, blackfly and bloodworms – but nothing else. Another microbe, *Bacillus sphaericus*, was found to have a similarly selective capability. Fermented cultures of the two organisms were rigorously tested in the early 1980s, under World Health Organisation auspices. They proved fully effective and environmentally safe. Limited quantities of a British preparation, based on toxic crystals extracted from the Negev germs, were used successfully in Australia. Production by a Queensland company began in 1985, with the enthusiastic support of public health authorities.

Crystals of protein in the new larvicide kill their targets within 24 hours and remain effective for two days or more. Dams and other big storage areas can be sprayed from the air or from the ground, and domestic rainwater tanks can be treated by hand.

A researcher collects larvae from a sluggish creek close to Jabiru, NT. Identifying their species helps predict likelihood of disease.

A HUMAN SACRIFICE MAKES THE BEST LURE

Various devices have been invented to catch mosquitoes. But when biological researchers or public health officers need to know what species bite where, and in what numbers, nothing works better than their own skin. When a mosquito settles to bite, it is sucked up through a tube into a trap of metal gauze. Simpler tubes are held in the mouth and operate with a sudden intake of breath. In others a small, push-button motor provides the suction.

Tethered animals are used as collectors in some surveys. Window traps can be employed to monitor mosquito movements in and out of houses. Vehicle-mounted traps, though they collect all flying insects, can help by showing the preferred flight times of mosquitoes.

Traps that employ light as a lure collect many mosquitoes, but even more of other insects. Their efficiency is reduced further in urban areas that are lit at night. The most complicated and expensive trapping machines use a combination of light, carbon dioxide emission and motorised suction. They are essential when large numbers of live mosquitoes are needed quickly for virus studies in an epidemic.

Though male mosquitoes do not bite, their detection in traps can be important in disease prevention. It may reveal an unsuspected breeding site.□

Field workers for the Queensland Institute of Medical Research use battery-powered aspirators — much like vacuum cleaners — to suck mosquitoes from their resting places.

Mosquitoes lured by an incandescent bulb are drawn live into a gauze trap by the suction of a battery-driven motor. Live specimens, needed in virus research, are even more strongly attracted if a device emitting carbon dioxide is added to the trap.

Ross River fever

Adult women are the main sufferers from a mosquito virus that seems to be spreading all over Australia.

Disease has given the Ross River, which provides the city water supply for Townsville, Qld, an undeserved notoriety. Though Ross River fever was first observed in the Townsville district last century, it is well known along the Murrumbidgee-Darling-Murray river system of the southeast. Lately it has turned up in Western Australia, the Northern Territory and even northern Tasmania.

The fever virus is carried both by the inland *Culex annulirostris* mosquito – also responsible for Australian encephalitis – and by *Aedes vigilax*, a common and far-flying coastal species that breeds in salty rock pools and mangrove swamps, biting day or night. Given moisture and warmth, there is virtually nowhere that the virus cannot be spread.

In natural environments the virus has alternative hosts in birds and mammals that are bitten by the mosquitoes. Domestic animals are also infected. In urban areas it seems that mosquitoes can transmit the virus from person to person without the need of any other animal host. In fact the most rapid and intensive spread of the disease occurs in towns, especially if nearby land is subject to summer flooding. Community epidemics in the irrigated crop-growing areas of the southeast can take hold more quickly and affect more people than influenza would. Sometimes Ross River fever has put more than half of a town's population out of action at the one time. Few victims, fortunately, suffer severe complications.

Crippling pains in every joint

Children are least affected by the virus, and adult men far less than women. The virus is much like the one that causes rubella – German measles – and early symptoms may be identical. Like rubella the Ross River virus can infect unborn children, but no evidence has been found of their being harmed. There is no vaccine against the virus and no specific cure.

Effects can range from a short-lived skin rash to a severe brain inflammation resembling meningitis. But the usual form of the disease is a prolonged bout of joint pains and swellings that doctors call epidemic polyarthritis. The onset is usually within a week or two of infection.

A rash often appears a few days before the arthritic pains are felt, though sometimes it is delayed until well after they have started. Small spots and lumps occur commonly on the trunk and limbs, and less frequently on the face and hands. The rash is seldom itchy but the blemishes can stay tender for weeks. In rare cases there is a severe headache with a high fever and chills.

Joint stiffness, swelling and pain set in fairly suddenly. Victims often wake to find the weight of their bedclothes intolerable – and then they find that they cannot move without greater pain. All the joints are affected but the pain is most intense in the fingers, wrists and ankles. Extreme fatigue is a by-product of the disease.

Recovery usually takes 2-4 weeks, though illness dragging on for more than three months is not uncommon. Intermittent symptoms resembling rheumatism have persisted for more than two years in a few cases.

ROSS RIVER FEVER

EARLY SIGNS Rash (not always), mild fever. Joint stiffness and pain.
ACTION Give aspirin or other mild painkiller and encourage rest. Obtain medical advice.

THE TOURIST'S REVENGE
Australians travelling to the Pacific Islands often complain of picking up diseases – especially dysentery and other stomach upsets. Now it seems that they have turned the tables. Ross River fever – not previously known outside Australia – broke out in Fiji in 1979. It afflicted more than 30 000 people – one in 20 of the population. The disease has since spread throughout the South Pacific.

The malaria threat

All the ingredients remain in the north for the spread of ruinous epidemics. And infections are imported day by day.

Australia's eradication of malaria was certified by the World Health Organisation in 1981 – even though the previous year's tally of cases had set a record. Those infections did not count because all 628 cases were traced to foreign origins. What could have happened if any of them had been picked up by Australian mosquitoes is a matter of conjecture. But the possibility is real, and cause for concern.

Just before the WHO clearance was announced, Queensland's health minister warned his parliament that malaria in its most deadly form could sweep the north at any moment. He was launching a campaign to boost public

MALARIA
EARLY SIGNS Severe throbbing headache, violent shivering, high temperature.
ACTION Administer anti-malarial tablets if available. Obtain medical aid.

health surveillance measures and to make doctors more alert to the early signs of the disease. A few weeks later a senior Northern Territory health official specialising in insect dangers forecast trouble in new bauxite and uranium mining communities, because of their closeness to mosquito breeding grounds and because many workers came from malaria-infected areas overseas.

Three factors govern the chances of malaria re-establishing itself in northern Australia: how vulnerable the region is, how receptive it is, and how readily the disease can be combated. On all counts the risk is high, and is increasing.

Papua New Guinea and Irian Jaya, virtually on our doorstep, are hotbeds of the worst form of malaria. In PNG, where spending on mosquito control was cut during the 1970s, malaria accounts for tens of thousands of deaths and 90 per cent of child mortality. The disease is rife on Timor and the home islands of Indonesia, in the Solomon Islands and on outlying islands of Vanuatu. Of travellers arriving in Australia each year, it is estimated that more than 250 000 have recently been in countries with malaria.

Our receptive zone forms a band north of latitude 19°. Climate and landscape favour prolific breeding by the mosquito most likely to transmit malaria. The disease was endemic across this band last century. Little has changed – except that human populations have increased by scores of thousands. Some mosquito control measures are in force at major centres. People going to work in high-risk districts of the Northern Territory, if they have just been in a country with malaria, are put on drug courses. But the drugs are becoming ineffective.

A microbe that breeds in our blood
Malarial diseases are caused by tiny, single-celled animals called plasmodia. They are nursed in the bodies of anopheles mosquitoes and they breed in the bodies of humans. The

continued existence of these parasites relies on a cycle of transmission from one to the other.

About 10 days after a mosquito sucks blood from an infected person, juvenile plasmodia enter its saliva glands from its stomach. For the rest of that mosquito's life, anyone bitten by it receives plasmodia in an injection of saliva. The parasites travel through the bloodstream to the liver – taking only about 20 minutes – and remain dormant there for 5-10 days.

Suddenly there is a discharge of plasmodia back into the bloodstream. Taking on a sexual form, they invade red blood cells and multiply rapidly. As each cell is destroyed they go on to others. This is when the victim is overwhelmed by fever. A pounding headache usually sets in, with violent shivering and vomiting. Sometimes the victim becomes delirious. Symptoms are varied and they often mimic other illnesses.

An attack may last for an hour or a whole day, then disappear. But the fever comes back at more or less regular intervals as new batches of plasmodia are discharged from the liver. The victim becomes steadily more anaemic and weak from the loss of red cells. Untreated people can easily succumb to other diseases. But those who normally enjoy good health usually recover after a fairly quick succession of fever bouts. The illness may recur much later, however. No immunity is gained, though re-infections may produce less severe symptoms.

Four species of the parasite are known. Only two cause concern in the Australian region. *Plasmodium vivax* causes relapsing malaria. Its victims usually suffer bouts of fever every second day in the initial phase, with occasional recurrences months or even years later.

Plasmodium falciparum is much more feared. Malarial attacks are frequent but irregular, or the fever may be continuous. This form of the parasite can invade the blood vessels of the brain, causing cerebral malaria. The death rate is very high, especially among children. New Guinea is one of the world's main reservoirs of the *falciparum* organism.

Australia has several species of anopheles mosquitoes but the one chiefly implicated in malarial infection here is *Anopheles farauti*. Knowledge is limited because there has not been a significant epidemic allowing scientific study since 1942, among soldiers based at Cairns, Qld. A more widespread species, *Anopheles annulipes*, may have been responsible for outbreaks in more southerly areas.

Drug defeated by a mutant strain
Many drugs, starting with quinine, have been developed to kill plasmodia. Courses of tablets are prescribed for people to take before they go to a dangerous area, while they are there and for a period after they leave. In the past these

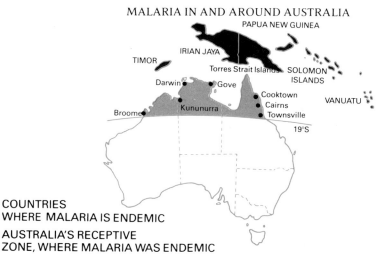

MALARIA IN AND AROUND AUSTRALIA

PAPUA NEW GUINEA
IRIAN JAYA
TIMOR
Torres Strait Islands
SOLOMON ISLANDS
Darwin
Gove
Cooktown
VANUATU
Cairns
Broome
Kununurra
Townsville
19°S

■ COUNTRIES WHERE MALARIA IS ENDEMIC

▒ AUSTRALIA'S RECEPTIVE ZONE, WHERE MALARIA WAS ENDEMIC

were highly effective, both in preventing infection and in curing the disease. *Plasmodium falciparum* was best countered with a drug called chloroquine. But the parasite has produced a new strain that resists it. Thought to have originated in Thailand, the mutation has been found in New Guinea. It can be fought with other drugs, but not with the same safety for pregnant women. More resistance is expected. In 1984 the director of the Queensland Institute of Medical Research, Dr Chev Kidson, predicted that all known drug treatments would be useless against new strains of plasmodia within five or 10 years.

Mosquitoes and malaria cannot be abolished with current knowledge – only controlled with enormous effort and expense. And the battle can never be allowed to cease. Sri Lanka had 3.5 million malaria sufferers in 1940, more than half of its population. With international help an all-out campaign was mounted against insects, using DDT. By 1950 there were only 600 000 sufferers – and in 1962 just five. Then DDT went out of favour and Sri Lanka, letting its guard down, stopped spending money. Twenty years later the annual toll of malaria had shot back up to 2 million.

Australia can regard itself as safe only if other countries are – especially our neighbours to the north. They lack the resources to mount continuing control campaigns of guaranteed effectiveness. What financial or technical aid Australia could offer would be a few drops in an ocean of infection. But help of a more practical kind may be emerging from our laboratories.

Genetic trickery offers a ray of hope
Smallpox was beaten with a vaccine. Immunisation of entire populations is probably the only way that malaria will ever be overcome. The cycle of man-mosquito-man infection would be broken once and for all. The search for a malaria vaccine has gone on for a century without success. Now genetic engineering may offer the answer. In a combined project of the Walter and Eliza Hall Institute and the Commonwealth Serum Laboratories in Melbourne, the Queensland Institute of Medical Research and a Sydney company, Biotechnology Australia, biochemists have tampered with the molecular structure of the protein coating of plasmodia.

They take a tiny particle and replace it with a subtly altered genetic combination. The parasite is tricked into going on multiplying. But the surface protein produced by all its descendants will be hostile to the real thing. The altered protein can be used to attack the natural one.

Sabotaged plasmodia are injected into bacteria to multiply. The most common of 'stomach bugs' are used. Unlimited quantities can be shipped in flasks to research laboratories. Serums made from the cloned protein are tested on monkeys – the only animals apart from humans that can get malaria. It may be many years before the human usefulness of a serum is proved.

The thrust of Australian research is against plasmodia in their blood stage, after they leave the liver but before they adopt sexual forms and invade the red cells. Work in the US, well advanced, is directed at the juvenile stage while plasmodia are travelling to the liver. One way and another, scientists are optimistic that vaccines will succeed against at least some strains of the parasite.□

WHERE
ANOPHELES FARAUTI
MOSQUITOES
ARE FOUND IN
AUSTRALIA

The discrepancy between Australia's malaria receptive zone and the known distribution of Anopheles farauti *in this country is due to the fact that other* Anopheles *mosquitoes capable of transmitting malaria occur farther south.*

Dengue fever

Old-timers called it 'breakbone'. It vanished for over 25 years, but now it is back with a vengeance.

A generation of North Queenslanders forgot about dengue, and ignored the house-pest mosquito that carries it. After an epidemic between 1953 and 1955 the disease seemed to disappear. Isolated cases occurred in the late 1970s, most of them among people who had just been overseas. But in 1981, soon after the opening of international air services to Townsville, dengue came back to stay.

More than 3000 people, from Townsville north, were infected with the virus. Some came down with a fever that racked them with joint and muscle pains for days or weeks. No one died of it – this form of dengue is not a killer. But out of Asia and into the Pacific has come a different strain of the virus. It causes dengue haemorrhagic fever. The usual excruciating symptoms are followed by shock and internal bleeding, with a high death rate from kidney failure.

Blood tests on Queensland schoolchildren have shown that 40 per cent were infected with the dengue virus at some time – long after the mid-50s epidemic. Obviously the fever developed in very few of them. But the high rate of infection is disturbing. In Southeast Asia there is a strong supposition that the haemorrhagic form takes hold when a child has been infected before. Some researchers call it double dengue.

In its non-fatal form dengue was common last century, occurring not only in the tropical zone but also well to the south. Gold prospec-

DENGUE FEVER

EARLY SIGNS Muscle and joint pains, high temperature, headache. Occasionally a skin rash.

ACTION Take aspirin or other mild painkiller. Seek medical attention.

WHERE
AEDES AEGYPTI
MOSQUITOES
ARE FOUND
IN AUSTRALIA

Aedes aegypti is now believed to occur only in Queensland, although it was present throughout the mainland of Australia at the end of the 19th century. The whole of Queensland is shaded on this map, but in reality the species probably only occurs in isolated areas.

tors and bushmen nicknamed it 'breakbone fever' because of the agonies they suffered. Whole mining communities had epidemics. One man died of a haemorrhagic fever in the 1890s but it need not have been dengue – many other foreign viruses cause similar symptoms. One could have come with the immigrants who flocked to the goldfields.

A dangerous household pest

Queensland public health authorities are in no doubt that the state's increased flow of international travellers makes the introduction of dengue in its haemorrhagic form probable, rather than just possible. Dr Brian Kay, head of mosquito epidemiology at the Queensland Institute of Medical Research and also director of the state's disease vector control unit, is outspoken about the risk. He points to recent epidemics in Cuba and Mexico, where tens of thousands of people were stricken when the haemorrhagic strain made unexpected appearances. Both countries' economies were disrupted, their hospital systems were put into chaos, and more than a hundred people died.

'It would be unrealistic and downright foolish,' he says, 'to think the virus won't show up here. What we have to do is try and make sure that it doesn't spread. And the only way we can do that is to get to the people themselves –

individual householders.

Aedes aegypti, the mosquito that carries the dengue virus, breeds exclusively around houses. It has evolved to exploit humans for their blood – and for their untidiness and carelessness. We provide all the water containers in which its eggs are laid. In a typical yard and garden there could be a dozen such breeding sites, and on a block of waste ground, hundreds.

The species was once widespread throughout NSW and was also recorded in Western Australia and the Northern Territory. Now it seems to be confined to Queensland, as far south as Brisbane. Dr Kay believes that a closely related species with similar habits but a more southerly distribution, *Aedes notoscriptus*, may also be a carrier of dengue.

In 1984-85 Dr Kay was jointly responsible for spending $5 million, largely contributed by federal health authorities, on a public education campaign to control household mosquitoes. Nearly 500 officers toured Queensland, giving school and community lectures, pointing out dangerous sites and disposing of rubbish. Their efforts were backed up by a blitz of advertising.

'We can't hope to eradicate the mosquitoes completely,' he says. 'The best we can do is keep on making people aware. But if someone in a neighbourhood doesn't care, everybody could be in trouble.' ☐

Goodbye to filariasis

A curse of the northeast coast was beaten after 80 years. But a similar parasite increasingly menaces dogs.

A filarial worm Wuchereria bancrofti *emerging from the sucking tube of a mosquito.*

Elephantiasis, a gross, crippling swelling of the legs or genitals, is filariasis at its worst. It results from years of repeated infections by parasitic worms that are carried by mosquitoes. In Australia that rarely happened. Filariasis here usually meant recurrent bouts of fever – not unlike relapsing malaria – and soreness of the lymph glands.

The disease showed up in the 1850s, presumably brought in with the flood of fortune seekers who were lured by gold discoveries. It is widespread in the world's tropical regions, taking a particularly heavy toll in Africa. Of three kinds of tiny, hairlike filarial worms that can live in people only one, *Wuchereria bancrofti*, seems to have reached Australia.

Worms multiply in the lymph systems of infected people and the juveniles migrate to the outer bloodstream. Mosquitoes biting a filariasis sufferer pick up the immature worms and they develop in the insects' stomachs. After a few days, a mosquito bite can pass the parasites on to someone else.

Urban populations on or near the east coast, for the whole length of Queensland and well into NSW, were exposed to filariasis. The carrier was almost certainly *Culex fatigans*, the most common of pest mosquitoes inside houses. Up to the 1920s the infection rate in some places was higher than 10 per cent.

The cycle of infection between mosquitoes and people was broken partly by the increased use of screening on doors and windows, and partly by the prompt administration of a worm-killing drug that was developed. It was not difficult – victims could be treated as hospital outpatients. By the end of the 1930s the disease was scarcely to be found in Australia.

Because the carrier mosquito remains abundant, there is a theoretical risk that travellers could reintroduce filariasis. Southeast Asia has one of the other types of filarial worm – especially hard to eradicate there because it can live in monkeys as well as people. But there is little chance of a serious spread in modern communities.

Dog heartworm threat moves south

Owners of dogs in the eastern states, and particularly professional breeders, have every reason to be alarmed by another mosquito-borne parasite, the dog heartworm *Dirofilaria immitis*. Immature worms live in huge numbers in an infected dog's circulatory system and the adults lodge in its heart. Fully grown, they can be 30 cm long – and 200 or more of them can be in the heart at the same time. In such forces they simply block the flow of blood, causing liver damage if not heart failure.

Regular dosing with a drug can prevent the disease. Untreated dogs in tropical Queensland are about 90 per cent infected. Brisbane has an infection rate of about 50 per cent. To the south the incidence is lower but steadily rising, with infection having spread through NSW and into Victoria. Once the disease is well advanced, the drug treatment itself can kill a dog.

Many mosquitoes may carry dog heartworm. In Queensland the prime culprits are thought to be *Aedes vigilax*, the far-flying seashore breeder, and *Culex annulirostris*. In NSW the prolific domestic species *Culex fatigans* and *Aedes notoscriptus* come in for most of the blame.

Humans can get dog heartworms from mosquito bites, but they cannot develop or reproduce in our blood. At worst, dog heartworms in people may form cysts in the walls of arteries, requiring surgery to remove. Lung cysts, mostly harmless, are more common.□

When the itch is worse than the bite

Allergic reactions, and infections from scratching, trouble more people than any disease that blood-sucking insects carry.

Excessive sensitivity to insect bites is common, particularly among newcomers to Australia. March or horse flies and Australia's many kinds of biting midges – often called 'sandflies' – can cause more of a problem than mosquitoes. That is because they bite in swarms, and some people react badly to just one bite.

Severe allergies to the saliva of mosquitoes and other biting flies can progress like the allergy to bee or wasp stings. They become worse with each attack. People at high risk from mosquitoes are given courses of immunotherapy with extracts from the insects that affect them. Desensitisation to midges is much more difficult. Milder sufferers are treated with antihistamine drugs, and their sensitivity usually subsides as a few years pass.

Sometimes the first swarming attack can be overwhelming. A 51-year-old man fishing at Fraser Island, Qld, in 1982 collapsed almost immediately after he was set upon by March flies. Soon he could scarcely breathe. Had his companions not succeeded in summoning a helicopter, he would probably have died.

Infections from the scratching of itching bites can have severe effects, especially on children. Antibiotics and prolonged confinement to bed may be needed before they clear up. If children are unusually troubled by bites, parents are advised to make sure that their fingernails are kept short, and to provide an antiseptic soap for washing hands. It is far better, however, to use repellents and take every other possible measure to prevent bites. Nearly all children grow out of their sensitivity, but parents should not take it for granted.

Compared with other parts of the world, Australia is fortunate in its biting flies. Many are dangerous to livestock but none, apart from the mosquitoes, is known to carry organisms harmful to humans. March flies or horse flies in Africa, for example, cause a parasitic disease much like filariasis. Similar flies in Australia are apparently clean.

The true sandfly *Phlebotomus* rarely bites people at all in this country and it seems disease-free. Yet the same insect overseas carries viruses, as well as a microbe that from North Africa to China and in much of South America causes *kala-azar* or Dumdum fever. It attacks bone marrow

The bite of a March or horse fly Tabanus *is unlucky for some.*

and if untreated leads to a slow and painful death.

Some of the biting midges, which many Australians refer to as sandflies, can carry viruses that cause disease in farm animals. None of these species is known to bite humans – or if they do, to infect them. But midge-borne viruses came under intense scrutiny in 1985, after an unusually high incidence of defects among children born two years earlier at Coffs Harbour, NSW, was revealed.

One of the biting midges sometimes carries a virus that causes brain and joint abnormalities in lambs and calves. Sheep farmers know the effect as 'floppy lamb'. The summer of 1983-84 was a bad season for insect-borne viruses in NSW – Ross River fever was rife. To make sure that the midge virus could not, after all, be dangerous to humans, University of NSW researchers began a years-long programme of blood testing to check for antibodies. They emphasised that this work was but one of many avenues of investigation into the Coffs Harbour defects, and that there was no particular reason to fear midges.

Tick paralysis

Hospitals are equipped to rescue children from a fearsome toxin. But simple alertness works better.

Ticks are subtle in their parasitic raids on our blood. Their bites are seldom felt at first, and the animals are rarely seen without a deliberate search. Undisturbed, they feed for days and inject a poison that can put their host — especially if it is a small child — in mortal danger.

Paralysis from tickbites, though little publicised, ranks as a major killer on the east coast of Australia. It caused more than 20 known deaths in the first half of this century. Many other tickbite deaths in that period and before must have been put down to 'infantile paralysis', now known as poliomyelitis. Eighty per cent of the victims recorded in NSW between 1904 and 1945 were under four years of age.

Pioneering settlers knew of the threat that ticks posed to their dogs, and vaguely of some danger to humans. But the paralysis tick *Ixodes holocyclus* was not scientifically identified until 1899, and only towards the end of the 1930s was its life cycle fully explained.

Ignorance remained, even among doctors. In one of NSW's last fatal cases, diagnosed as poliomyelitis, the tick was found only when the child's body was prepared for a post-mortem examination. The parasite was covered by an adhesive dressing — placed on it by the doctor who first attended the child. Much more recently a pathologist was sent a tick, cut from a patient's flesh by a surgeon who mistook it for a skin growth.

Thanks to scientific advances, human deaths from tickbites seem to be a thing of the past. Most doctors in infested areas are alert to ticks as a possible cause of paralysis. Hospital intensive care procedures are usually sufficient to keep victims alive until the effects of tick toxin wear off, and in the gravest of cases an antitoxin can be injected. But there is no room for complacency.

As recently as 1972, Brisbane doctors were baffled by the paralysis of a two-year-old girl. More than a day after treatment was first sought for her, a tick was still feeding. It was a male nurse, sponging the little girl's hair, who came across a lump on her head and discovered the parasite. Only then was antitoxin called for — and not enough could be found in Brisbane. A supply had to be flown from Sydney.

Suffering that could be averted

Other young victims undergo terrifying ordeals before they are saved. They lose the use of their limbs and the power to speak, swallow or breathe. Connected to breathing machines, fed intravenously, wired up to monitoring devices, they spend a week or two in intensive care. Their full recovery takes weeks more.

They could be spared all that — and the burden on hospital services could be lightened — if older family members were more aware of the danger of ticks.

Paralysis ticks are sometimes called scrub ticks, bush ticks, dog ticks or bottle ticks. They live in bushland along the humid eastern belt between Cairns, Qld, and the Gippsland Lakes in eastern Victoria. Their range is not continuous, and in some areas it reaches no more than 15 kilometres from the coast. But where mountain rainforests extend well inland, so do the ticks. Gardens in woodland suburbs of the coastal cities may be infested, especially if they are visited at night by bandicoots. Sydney's North Shore has the worst reputation of any heavily populated area.

Adult ticks may be active at any time of year if temperatures are high and plants hold plenty of moisture, although in the southeast heavy infestations are most often experienced in October and November, after ample spring rains. Drying northwesterly winds, droughts and bushfires impose a natural control on tick populations, and a district can be trouble-free

---

TICK PARALYSIS

EARLY SIGNS
Unsteadiness in walking. Lethargy, loss of appetite. Sleepiness, general weakness.

ACTION Seek urgent medical attention. Begin search for tick and remove quickly if found.

BETTER WEAPONS ARE ON THE WAY

The war against tick paralysis is being waged in Brisbane laboratories.

Antitoxin to overcome tick paralysis has been made since the 1940s from serum taken from beagle dogs. They are conditioned to withstand heavier and heavier infestations of ticks — up to 32 at a time. It is a long process that produces good results, at a high price, on afflicted pets. But the antitoxin is used reluctantly on humans, in only the gravest of cases. It is fairly weak so substantial quantities have to be injected, and too many people have a severe allergic reaction to protein originating in dogs.

Brisbane staff of the Commonwealth Scientific and Industrial Research Organisation have succeeded in producing a more potent serum more quickly in rabbits, and in modifying it so that it promotes immunity without any toxic effects. They are on the way to making not only an improved antitoxin for paralysis victims, but also a vaccine that could be given to prized breeding animals and to humans at special risk. A better way of desensitising allergy sufferers, instead of using the now-abandoned process using extract from tick saliva glands, is also in prospect.

The last major step is to develop a method of mass-production. This calls for the breaking down and separate identification of all the components of the toxin protein. Then techniques of genetic engineering should be able to do the rest. Researchers are confident of achieving their aims in the 1990s.

for years on end. But in a favourable summer the ticks can suddenly return. Then they may be more dangerous than ever, because parents raising young children may never have been confronted with the problem before.

Ticks do not consciously hide, but they seek shelter and remain longest in places where they are least likely to be brushed off. The scalp under a covering of long hair and the narrow clefts behind the ears make ideal sites for them. They may also lodge in folds of skin and in body orifices such as the ears, nose and vagina.

If children in eastern coastal regions play in shrubbery or long grass that remains moist in early summer, it is a wise precaution to check them for ticks once a day. A newly lodged adult tick is oval, flat and yellowish, and about the size of a match head. As it feeds it swells to the size of a pea, darkening in colour.

A child should be examined closely, all over the scalp and neck and behind the ears. A lump anywhere should be carefully investigated – sometimes an engorging tick is concealed by puffiness around the bite. Bushwalkers are advised to examine themselves thoroughly and have companions check their scalps and the skin creases at the back of the neck, especially if bumps can be felt there.

In a season that seems to favour ticks, medical attention should be sought at the first sign of unexplained weakness in a child's legs. Any possibility of contact with ticks in the previous week should be mentioned to the doctor. This is particularly important if the child has just been brought from a tick-infested area to a district where the parasites are not normally encountered.

What happens when a tick bites

Immature ticks of both sexes may bite humans but they rarely cause trouble. Only the adult female is responsible for serious paralysis. Grasping the skin with claws under its head, it inserts a barbed tube called a hypostome. Feeding alternates with spells of salivation and rest – and it is the salivation that does the harm.

Nutritious blood components are concentrated in the tick's body and unwanted fluid is injected back into the host, along with secretions of the tick's saliva glands. These include chemicals to aid the flow and digestion of blood. But there is also a powerful neurotoxin. Its action in causing paralysis seems to be accidental, for it is no use to the tick. Native animals in tick-infested regions are usually immune to it.

The toxin is not produced in significant amounts during the first two days. After that, as a tick's body mass increases, so do the size and output of its saliva glands. Small children are likely to be affected from the third day. Their walking becomes unsteady and they lose energy and appetite. Often their voices change and they may have difficulty in focusing. A child may sleep for unusually long periods and be difficult to rouse. A general feebleness becomes noticeable. There is normally no fever, unless the tick also carries germs and the bite is infected. But there could be inflammation around the bite site, helping to locate the tick. Lymph nodes are often enlarged near the bite, causing swelling and tenderness.

An adult female paralysis tick Ixodes holocyclus, *half-engorged, feeding on a human arm.*

On about the fifth day of the tick's meal – usually the second or third day of obvious illness in a child – muscle weakness spreads to the arms. Removal of the tick now, unless medical treatment is also given, is unlikely to arrest the progress of paralysis. Throat and chest muscles are soon stricken. Unaided, a young victim may die of respiratory failure.

Lesser doses of the toxin sometimes cause a temporary paralysis concentrated in the region of the bite. A bite behind the ear, for example, can produce a palsy on that side of the face, lasting for some days. In adults the usual effect is dizziness, headache and a general weakness and lethargy. Victims' sight may also be interfered with – they may experience double vision, aversion to light, or an inability to control darting movements of the eyes.

Other ticks with toxic effects

Southerners received bad news in 1986 when the *Australian Medical Journal* reported a proven case of profound paralysis caused by *Ixodes cornuatus*. This pale-coloured species is found in high country in Tasmania and Victoria and along the NSW south coast. A three-year-old Melbourne boy, bitten during a camping holiday, was gravely ill for nearly a week. And the usual antivenom had no apparent effect. The bite of an adult male *Ixodes cornuatus* – rare in the case of the main paralysing species – brought severe pain to a doctor in the Dandenong Ranges, near Melbourne. A four-year-old girl, bitten in northern Tasmania, was acutely ill with stomach cramps, vomiting, headache and neck pains, though she showed no sign of paralysis.

Illnesses blamed on Western Australia's ornate kangaroo tick *Amblyomma triguttatum* are generally mild. But it attaches itself with unusual tenacity. Sometimes the skin has to be cut to remove an engorged, buried tick.

The kangaroo tick of the east, the soft-bodied *Ornithodoris gurneyi*, has a bad reputation among bushmen although it is seldom encountered. It ranges from western NSW through western Queensland to the Gulf of Carpentaria. It feeds for only a few minutes. But it has been responsible for severe irritation, headaches, attacks of vomiting, temporary blindness and loss of consciousness.

Kangaroo soft ticks were blamed for the death of a police constable in 1963, although it was dehydration that killed him. Kenneth Ryan, 24, was driving from Richmond in northwestern Queensland to Brisbane to be married. In a district that he knew well, he left his car to walk to a station homestead, seeking help with a flat battery. The temperature was about 45°C.

Constable Ryan's body was found close to a waterhole, 8 kilometres from his car. The station manager described the area as 'thick with ticks'. He recalled bites that he had received himself, leaving him temporarily paralysed and helpless. If the policeman had been bitten, the manager said, he would have collapsed and lain at the mercy of the sun.

An allergic reaction that can kill

Some people who are bitten on more than one occasion experience an increasing sensitivity. This develops into an allergy to the saliva of any tick – even those immature forms that do not usually cause illness. The reaction may consist merely of a widespread swelling and itching of the skin, and perhaps blistering. It can be controlled to some extent with antihistamine drugs. But in the worst cases there is a risk of choking and shock, which may lead to collapse and death.

Severe allergy sufferers, warned by previous attacks, learn to avoid situations that put them in contact with ticks. If they cannot eliminate all risk, they carry antihistamines and sometimes adrenalin. Some success in reducing the sensitivity of serious allergy victims was achieved with immunotherapy, using an extract from tick saliva glands in courses of injections taking many months, but this is no longer used.

During damp summers in the southeast, any disturbance of taller plants – while clearing lantana, for example – can produce a shower of tick larvae. Their bites may set off a maddening rash that lasts for hours. Called 'scrub itch' in Brisbane, it is a form of allergic dermatitis.□

Where the paralysis tick Ixodes holocyclus *may be found. The range is not continuous and the actual occurrence of the parasites varies with local weather.*

Second time lucky

Recognition of his plight saved the life of an allergy sufferer.

'As the doctor attended him he lapsed into a deep coma.'

Bitten on the neck by a tick, a 35-year-old allergy sufferer remembered how severely he had reacted to a bite 10 years before. He had his father drive him immediately to a doctor in the nearest town. Within 20 minutes he was dizzy and struggling to breathe.

As the doctor attended him he lapsed into a deep coma. He became black in the face, with saliva and other secretions frothing from his mouth. The doctor called an ambulance station for oxygen and a suction pump, and injected adrenalin. For minutes more the patient seemed to be on the point of death, but gradually his breathing gained strength and his pulse returned. He regained consciousness 45 minutes later and recovered fully in a few hours.

Decisive action by the doctor undoubtedly saved a life in this case, reported in the *Australian Medical Journal* in 1966. But it was the victim's own appreciation of his plight, anticipating an extreme allergic reaction well before it happened, that made the doctor's success possible. The man later went onto a course of immunotherapy with tick extract.□

Complaint was in vain

A 'lump' on her head ended in a child's death.

'All her limbs hung limply and her breathing was strained.'

Parents of a Sydney five-year-old, killed by tick paralysis in 1927, had warning enough. During a picnic near the shores of Middle Harbour, a tick was removed from another child in their party. Later their daughter complained of a lump on her head. But the tick went unrecognised.

Seven days after the picnic, the little girl had difficulty in walking. A day later, on admission to hospital, all her limbs hung limply and her breathing came in strained sighs. She was drifting into coma, her lips blue and dripping with the saliva and mucus that she could no longer swallow. A tick 'about the size of an almond' was found in the child's scalp, according to an *Australian Medical Journal* review of the case. She died the same day.□

Paralysis in pets and livestock

Dogs, cats, cattle and sheep are all at mortal risk from the paralysis tick.

Native animals that are frequently bitten by paralysis ticks are usually immune. Pets, farm stock and other introduced animals can acquire immunity – even a super-immunity that makes them overproduce antibodies to the toxin – under controlled conditions in laboratories. But it seldom happens naturally, unless their exposure to the toxin is remarkably gradual. More often the first heavy dose is fatal.

Dogs are especially susceptible. If they have not had contact with paralysis ticks before, or not for a long time, just one of the parasites feeding for its full period is likely to kill them. Cats are perhaps more resistant, and are certainly better at getting rid of ticks when they wash themselves.

Early signs of paralysis in a dog usually show about a week after a tick has attached itself. The dog loses its appetite and its ability to bark, and starts to lose control over its hind legs. After a few hours the legs give way when it tries to walk. It vomits anything it is given to eat or drink. Full paralysis and death may follow within a day, though some dogs recover after showing the most severe symptoms.

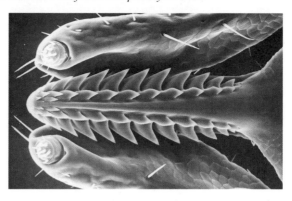

Barbs on a tick's feeding tube — called a hypostome — make it difficult to dislodge once the animal has started its meal of blood. At each side are the modified claws that the tick uses to grip the skin of its host before it starts to feed.

A cat may give an earlier signal of its plight with a noticeably altered 'miaow', before it loses its voice altogether. The progress and symptoms of paralysis are much the same as in a dog, but cats are far less likely to vomit.

Veterinary treatment with antitoxin and supportive drugs has had a high rate of success with cats and dogs. It is an expensive procedure, however. Owners taking dogs into east coast bushland are advised to check them afterwards for ticks. If dogs are kept in tick-infested districts, regular inspections should be combined with weekly baths using a recommended chemical solution.

Cattle and sheep, because of their greater size, can withstand heavier doses of paralysis tick toxin. But huge losses – sometimes more than half of a herd or flock – have been sustained when stock not previously exposed were put to graze on infested paddocks where dozens of ticks at a time could feed on each animal. Newly born calves succumb easily to the toxin of just a few ticks. Apart from taking extreme care in their choice of calving paddocks, cattle farmers can avoid much of the danger by bringing forward mating so that calving occurs at the end of winter instead of later in spring.□

TICK TYPHUS MOVES SOUTH

Bites of the paralysis tick can cause a mild form of typhus. The tick sometimes carries *Rickettsia australis*, which are among the smallest bacteria known. Illness sets in some days after a tickbite. It is marked by a rash of scattered spots up to 1 cm in diameter, usually accompanied by a headache. Fever, flu-like symptoms and tenderness of the lymph nodes follow. The disease runs its course in two weeks or so, but can be cured more quickly with antibiotics.

Tick typhus emerged as a problem during World War II, among troops training for jungle warfare. It was first called North Queensland tick typhus, but the main focus of the disease was later found to be in the southeast of the state.

No cases were detected south of Lismore, in northern NSW, until the late 1970s. Then in the space of seven months a little girl and a man were infected after tickbites in northern suburbs of Sydney. It now appears that tick typhus must be considered a possibility throughout the east coast range of the paralysis tick *Ixodes holocyclus*.

Ticks of several species can also carry the rickettsial bacteria that cause Q fever. They transmit them to livestock, so in Australia that disease is usually contracted in slaughtering work rather than from a tickbite.

HOW TO REMOVE A TICK

Never try to pick off a tick with your fingers. Squeezing it will increase the amount of toxin it injects. Use tweezers or small scissors to remove a tick without touching its body. Slip them gently under the tick's head, pressing into the bite victim's skin if necessary, and close them on the feeding tube. A firm, steady pull will draw up the tube and the tick will come away intact.

Killing the tick first, by dousing it in kerosene or turpentine, used to be recommended. There is no advantage in this if you have a suitable tool to remove the tick.

TICKS ARE BORN TO BE PARASITES

The world's 700-odd species of ticks are arachnids, in the same animal class as spiders. But the ticks are wholly adapted to a life of parasitism. Their mouths are specialised to pierce the flesh of host animals and suck up blood. And their life cycle has a variable timetable, depending on the weather and the availability of hosts.

Australia has more than 70 species, mostly natives but a few introduced with livestock. Thirteen of them are argasid or soft-shelled ticks. The rest are ixodid ticks, carrying a hard plate on their backs. This shield covers most of a male tick's back, but only a small area behind the head of a female.

SECURING A FUTURE ON THREE MEALS A YEAR
The life cycle of the paralysis tick

Engorged female — now up to 13.2 mm long and 10.2 mm wide — drops to the ground and spends 3 weeks in moist vegetation before laying 2500-3000 eggs. Adult female dies after laying eggs.

Adult female sucks blood from host no. 3 — which may be any animal, including humans — for 6-20 days. This is the danger period, when the toxin from the paralysis tick's saliva enters the host.

Eggs incubate in moist leaf litter or under bark for 40-60 days, depending on temperature and humidity, then hatch into larvae.

Adult ticks climb vegetation and await host no. 3. Males wander about on hosts looking for females to mate with but do not suck blood from the host. Adult males die after mating.

Six-legged larvae or seed ticks, just visible to the naked eye, climb vegetation and congregate on the tips of leaves.

Larva attaches to host no. 1.

development 2-20 weeks

Nymph moults into 8-legged adult tick. Unfed adult female is oval-shaped, approximately 3 mm long and 2 mm wide.

female   male

EGG

ADULT

LARVA

Host no. 1 is usually a native animal such as a bandicoot, possum or kangaroo. Larva inserts mouthparts into host and sucks blood for 4-6 days.

Engorged nymph — now about 1.15 mm long — drops to ground, into moist vegetation where second moult occurs.

NYMPH

moult and development 4-16 weeks

Engorged larva drops to ground, into moist vegetation where first moult occurs.

moult and development 4-36 weeks

Host no. 2 may be another native animal, or a dog, cat, goat, human etc. Nymph sucks blood for 4-7 days.

Nymph crawls up vegetation and awaits host no. 2.

Larva moults into 8-legged nymph, approximately 0.5 mm long when it emerges.

In southeastern Australia the development of the paralysis tick from egg to adult takes about a year. Variations in temperature and moisture can hasten or delay development at any stage. Under laboratory conditions, female ticks have been kept alive for more than two years.

All ticks go through three stages of development. First the eggs hatch into a six-legged larval form, sometimes called a seed tick. The larvae moult to become eight-legged nymphs. Emerging from a further moult, adult ticks are ready to breed. In the southeast the paralysis tick *Ixodes holocyclus* takes about a year to complete the process — but as little as three months in damp tropical conditions.

For each transformation to take place, from larva to nymph and from nymph to adult, a long meal of blood has to be taken from some vertebrate animal — a warm-blooded one in the case of most ticks. Adult females need an even bigger feast of blood to produce their eggs. The adult males wander about on host animals but do not feed from them. They may take some blood by piercing the bloated bodies of females, before or during mating.

Choice of hosts is wide
Some ticks return to the same host for each of their three meals. Others use two or, like the paralysis tick, three hosts. Their choice is wide. It is not at all unusual for three different animals to provide blood for one tick. One host

could be a marsupial, another a bird and the third a dog – or of course a human.

A tick's senses are primitive – it has no eyes or antennae. At each stage of its development a simple instinct makes it climb plants and wait. The moment it is brushed by fur or feathers – or by human hair, skin or clothing – it clings, using pincers on tiny claws under its head. Then it crawls to a sheltered feeding site.

Larval ticks, hardly visible without a magnifying glass, seek their first host animal between two and 20 weeks after hatching. They spend about a week feeding, usually around March in the southeast. Then they drop off into moist plant litter and await their moult.

Anything from one to nine months later, but most often around July, nymph ticks about the size of pin heads spend a further week feeding from the second host. Again they drop to the ground to digest their meal and develop their final form.

Adult ticks are active in a further one to four months – usually around October-November in the southeast. A male may spend months on a host's body, waiting to chance on a mate. The females feed for a week if they can, and sometimes for up to three weeks. Then they fall to the ground, shelter under leaves and rest for about a fortnight while eggs develop.

Between 2000 and 3000 eggs are laid during the following month, then the female dies. By this time its mate is also dead. In about two months the eggs hatch to renew the cycle.

A fully-fed adult female Ixodes cornuatus, *a high-country tick species of the southeast that has only recently been shown to be capable of causing life-threatening paralysis.*

BANDICOOTS TAKE THE BLAME

Nosing through damp leaf litter in search of grubs and worms, bandicoots present themselves as prime candidates to be the feeders and carriers of paralysis ticks. Unfortunately these rat-like marsupials, otherwise inoffensive, are the animals chiefly responsible for bringing the ticks from the bush to the backyard.

Bandicoots, themselves immune to the tick toxin, have taken readily to manmade environments, sheltering wherever there is good ground cover. They nest on urban wasteland, in rubbish dumps and in stormwater channels. At night they may invade home gardens in their quest for food.

Dog owners who have lost their pets to ticks sometimes call for the eradication of bandicoots. But bandicoots are protected native animals, and they play an important role in insect control. Householders who are anxious to discourage them should keep their gardens free of plant litter, snails and slugs. If bandicoots are a serious menace, state wildlife services will issue permits for their trapping and removal.

Conservationists suggest that if there were no bandicoots to act as the principal hosts, the infestation of other wild animals, pets and livestock would increase. Paralysis ticks enjoy a liberal choice of hosts. Among animals that are commonly exploited are kangaroos and wallabies, possums, koalas, dingoes, spiny anteaters, native and introduced rats and mice, ducks and magpies, along with all the domesticated species, including chickens.

Paralysis ticks are restricted to warm-blooded vertebrates. But even the reptiles do not escape – other tick species infest lizards and snakes.☐

A northern brown bandicoot infested with paralysis ticks.

Lyme disease

Symptoms come and go in a condition so bizarre that the linking organism –
carried by ticks – escaped notice.

Two concerned mothers prodded scientists into the belated discovery of the Lyme syndrome. At first investigators thought they had a rare new disease on their hands. Now they know that it is widespread, occurring in North America, Europe and Australia. But the symptoms are so diverse and erratic that until the 1980s specialists treated different aspects without appreciating the connection, let alone looking for a common cause.

Parents in the small American rural township of Lyme, Connecticut, were alarmed in 1975 by what seemed to be a freakish epidemic of rheumatoid arthritis among their children. A campaign by two mothers drew the attention of medical researchers at Yale University. They found 39 children and 12 adults suffering from a strange arthritic condition that affected different joints at different times but was mostly felt in the knees. Apparently unique, it was given the name Lyme arthritis. Its onset seemed to be seasonal, in summer or early autumn.

Further investigations produced a surprising fact: more than 90 per cent of the victims had earlier had a skin sore that took the form of an expanding red ring. Many of them recalled tickbites beforehand. The skin condition was known as erythema chronicum migrans, or ECM. It had been reported only once before in the US but it was common in Europe. In Sweden it was associated with mosquito bites. ECM had not been considered a serious complaint. Even untreated, the lesions disappeared in a month or two.

Publicity of the Lyme findings brought to light scores of other cases, in 24 states of the US and in Canada. Along with ECM and arthritis, a correlation of other symptoms began to show up. The agonising headaches and neck stiffness of meningitis or encephalitis were commonly reported, with flu-like symptoms and swellings of the liver, spleen, lymph nodes and testicles.

Nerve attacks that change behaviour
Weeks or months after the early symptoms were noticed, many victims suffered neurological disturbances. Common effects were the loss of control of movements, interference with vision and hearing, nausea and vomiting, loss of memory and concentration, sleepiness, emotional instability, depression and other behavioural changes.

In a small minority of cases the heart was attacked. Muscle inflammation led to varied weaknesses and abnormalities, and always to some degree of distortion of the heart rhythm. Many symptoms resembled the effects on the heart of rheumatic fever. Because the Lyme syndrome poses the risk of a drastically altered heart rate, its consequences can be serious.

Along with the ramifications of the disease, some victims suffered a recurrence of ECM skin lesions. In the baffling array of symptoms that were reported there was a common thread. They cleared and relapsed repeatedly. That suggested something circulating in the bloodstream. But the heart problems did not often recur.

Researchers had no clue to the organism responsible, though there was little doubt that ticks were the carriers in the US. A virus was suspected. Then many of the symptoms were found to respond to antibiotics such as penicillin and tetracycline, indicating a bacterial cause.

Australian interest was no more than academic at first. The ECM skin complaint was virtually unknown here, and the tick species originally blamed in the US was absent. But in 1980, just one day after a leading immunologist, Professor Robert Clancy of the University of Newcastle, NSW, had given an address on the subject to local specialists, Australia's first known case of Lyme disease was detected in the Hunter Valley.

A 21-year-old coal company employee was taking samples in bush near Branxton, halfway between Maitland and Singleton. He was bitten on the leg by an unidentified creature. Later he recalled having brushed away what may have been a tick. What followed proved to be the world's first classical example of the full-blown Lyme syndrome, monitored by doctors in all its phases of remission and relapsing.

Within a few days of the bite the victim had a circular rash that spread from the lower leg to the middle of his thigh. It disappeared and reappeared, sometimes with secondary sores breaking out on his shoulders and face. After five months the symptoms of a relapsing arthritis set in. Two months later he was stricken by headaches, memory loss and difficulties with his movements. His general behaviour and emotional state changed markedly. A drastic acceleration of his heart rate was also recorded.

For the next two years the disease followed a typical course, with episodes of illness gradually becoming milder and less frequent. A complete recovery seemed likely. But six years after the bite the victim's fitness to work remained in dispute, with a court action pending.

LYME DISEASE

EARLY SIGNS Red spot enlarging to form ring with clear centre; ring continues to enlarge. Fatigue, headache.

ACTION Seek medical attention.

The hunt is on for Australia's carriers

Many more cases of ECM skin rash have emerged in the Hunter Valley. In some, other symptoms of the Lyme syndrome have followed. Treatment with antibiotics keeps all of the symptoms under control. But anxious efforts are being made to establish what carries the disease organism in the district, and how it can be avoided.

American work in 1982-83 identified the organism as a previously unknown spirochaete – a spiral bacterium – that was given the name *Borrelia burgdorferi*. It was isolated from the skin and blood of ECM patients and from the fluid surrounding their brains and spinal cords. The bacterium was also found in the gut tissues of *Ixodes* ticks. In each of the three US regions where Lyme disease is most prevalent, a different tick species was involved. The organism has since been found in another genus of ticks, *Amblyomma*. And it is known to have inter-mediate hosts, including deer and mice.

The same spirochaete bacterium has been isolated from Australian patients. One of the tick species blamed in the US, *Ixodes pacificus*, is also found here. But it is not necessarily the Hunter Valley carrier. The possibility that mosquitoes or even fleas are involved cannot be ruled out. Nothing is known of intermediate hosts. Preparations began in 1986 for a fullscale testing of ticks and other animals in which the organism could be found.

Australian research into the Lyme syndrome has confirmed that it results from an unusual action of the body's immune system. Antibodies succeed in destroying the spirochaete itself, but then they absorb its soluble parts and set up complexes of altered cells that circulate in the body, provoking the various relapsing symptoms. A genetic factor seems to determine whether this type of reaction will be mild or severe in individual victims.□

The heartbreaker of the north

An introduced tick has blighted the lives of countless Australian cattle farmers.

Cattle ticks are not directly dangerous to human health. In their weakening and destruction of beef and dairy herds, however, they have ruined the lives of countless farming families and brought many an investor to bankruptcy. In tropical Australia especially, they dictated the terms of development for generations.

Introduced into the Northern Territory from Indonesia in the 1870s, the hot-climate cattle tick *Boophilus microplus* has spread to infest nearly 1.5 million square kilometres of grasslands from west of the Kimberleys in WA to northern NSW. It lives through all of its growth stages on the one host animal. Cattle are the usual hosts, but horses and sheep also suffer.

After good rains the ticks are so prolific that more than 10 000 in various stages of development may be found on a single beast. Infested in such numbers, cattle lose condition and can die from blood loss alone. But the ticks also carry parasites of their own — single-celled organisms that develop in their gut tissues and mature in their saliva glands.

Of three fairly similar organisms the cattle ticks can carry in northern Australia, the most common and harmful are *Babesia argentina*. They invade red blood cells, then burst them when they reproduce by dividing in two. Vast numbers multiplying in the bloodstream of cattle cause what is known as tick fever. The urine of infected stock runs red with the fluid from ruptured cells, and many animals die.

Calves overcome the disease more readily than adult cattle. Then they seem to have a

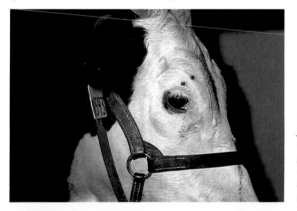

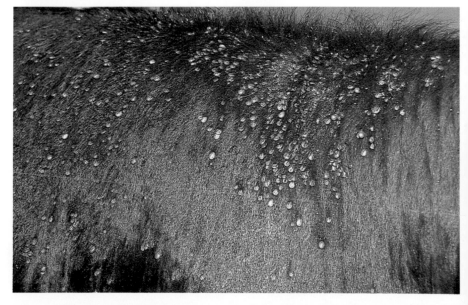

The cattle tick Boophilus microplus *feeding on the head of a cow (left). The beast pictured below is infested with so many of the parasites that it is unlikely to survive.*

Brahman cattle, introduced from Indonesia in the 1870s, brought the cattle fever tick to Australia. Brahmans and other crosses with the Asiatic zebu, such as the Santa Gertrudis, have a resistance to the disease that is not possessed by European breeds.

The cattle tick Boophilus microplus now infests nearly 1.5 million square kilometres of grassland from WA to northern NSW.

lifelong immunity. So the greatest loss of life occurs when the ticks and their parasites enter areas where mature herds have not been previously exposed. When that happens, the cost in stock losses and declining condition can run into tens of millions of dollars in a season.

Strict controls are imposed in northern NSW on the southward movement of stock. A vaccine is used on the most valuable breeding animals in areas that may become vulnerable. In the north, the cattle industry now relies, almost exclusively on breeds descended from crosses of European cattle with Indian zebus. The zebu, like the water buffalo, has a natural resistance to ticks and tick fever.

Another tick, the three-host *Haemaphysalis longicornis*, occasionally causes problems in the cooler cattle-raising districts of the southeast. Various people call it a bush tick, grass tick, bottle tick or New Zealand cattle tick. Not known to carry disease, it can kill stock through blood loss but more commonly it simply prevents them from thriving as they should. Infestations occur mainly along the coast but in ideal seasons they are experienced well inland – in the upper Murray Valley, for example. They are dealt with by repeatedly dipping or spraying cattle with chemical solutions.□

Q fever

Goat slaughtering opened fresh avenues for a virulent germ that can pollute the air in abattoirs.

Infection starts in the bush, among ticks and their bandicoot hosts. Ticks can pass Q fever to livestock, apparently without harming them. But when the stock are slaughtered, meatworkers can catch the disease – just by breathing. On their first exposure to as few as 10 airborne bacteria they can contract Q fever.

Old hands in east coast abattoirs know Q fever all too well. The majority have had it, recovered with treatment and are now immune. For most sufferers in Australia the infection is like a severe influenza, with more than the usual aches and pains. If diagnosed promptly it is easily overcome with antibiotic drugs. But victims may be off work for weeks, and complications such as pneumonia and hepatitis develop in some cases.

Q fever is caused by minute rickettsial bacteria – similar to typhus germs – called *Coxiella burneti*. They are passed in tick saliva to a wide variety of wild and domestic animals. Multiplying bacteria concentrate in the udders, placentas and birth fluids of pregnant mammals. They may infect milk, urine or dung.

When infected beasts are butchered, the bacteria disperse easily through contact or in the air. The hygienic use of high-pressure hoses to wash carcasses actually helps the germs because they can be carried in fine droplets of spray. In dry air they float with dust.

Q fever is well recognised as an occupational hazard for newcomers to the meat industry. Not only slaughtermen but also maintenance workers, cleaners, managerial staff and even visitors to meatworks are sometimes at risk. All practical precautions are taken when pregnant animals are slaughtered, and a vaccine was undergoing field trials in 1986.

A change forced by drought

Because of the immunity gained after one bout of Q fever, and alertness to the risk in regions where livestock are commonly tick-infested, the disease gave rise to few worries until the end of the 1970s. Then a widening drought cut the supply of prime stock to meatworks. To bolster their output, many companies turned to slaughtering feral goats for export.

Elimination of roaming herds of goats – officially declared as noxious animals, competing with stock and native wildlife for ever-scarcer food – was applauded. Rounded up mainly in northwestern NSW, they were trucked to urban works in Victoria and South Australia, as well as elsewhere in NSW. Many of the nannies were pregnant.

Infection from cattle or sheep had been rare in many of the works that went over to butchering goats. Some employees had been dealing solely with pigs, which do not harbour the bacteria. When the goats arrived, Q fever flared up in epidemic proportions.

Alarm at the outbreaks prompted a revealing survey of Q fever patients at Melbourne's Fairfield Infectious Diseases Hospital. Dr Denis Spelman reported in the *Australian Medical Journal* that out of 111 sufferers admitted from 1962 to 1981 – most in 1979 – 102 had recently worked in abattoirs. Nearly all were slaughtermen. Two other victims were farm workers. Four more people, including the only woman, had contracted the disease abroad.

In many countries, Q fever is prevalent not only in the meat industry but also in rural communities. There it seems to stem from living at close quarters with livestock. It is probably associated most with hand-milking and the drinking of unpasteurised milk, and sometimes with the use of dung as fuel.

In foreign cases up to 50 per cent of victims contract pneumonia. But among Fairfield Hospital's 111 patients, that occurred in only eight cases. Three of those were the travellers and two were the farm workers. So the abattoir employees contracted pneumonia in only three out of 102 cases. Earlier figures from a Queensland survey were similarly low.

The disparity is so great that scientists suspect that different strains of the bacteria are at work here and abroad. That theory is given more weight by a marked difference in incubation periods. In Australia illness may set in a month or more after the last known contact with a source of infection. In Europe and America, incubation time is less than three weeks.☐

Herds of feral goats Capra hircus *were rounded up and captured in the late 1970s.*

THE Q STOOD FOR QUERY

Solving the puzzle of Q fever stands as the most significant triumph of infectious disease research carried out solely in Australia. What seemed at first to be just a local problem emerged as a previously unrecognised disease of worldwide importance.

Brisbane doctors, comparing notes in 1935, found that many of them were treating workers from the same abattoirs for the same odd disease. It was something like a prolonged attack of flu and something like typhoid. Because its occurrence had the hint of an epidemic, the doctors handed the problem over to the Queensland health department. A government bacteriologist, Dr E. H. Derrick, spent a year testing samples of the patients' blood on guinea pigs. He established that all the meatworkers had been infected by the same organism. But he had no idea what it was.

Guinea pig serum was sent to Melbourne's Walter and Eliza Hall Institute, where Derrick had worked briefly. Frank Macfarlane Burnet launched a study of the spleen tissue of inoculated mice. He was searching for cell changes indicating a virus. Instead he came across clusters of tiny rickettsial bacteria. Their species, *Coxiella burneti*, took his name. The honour was not without its price – he was the world's first laboratory victim of the disease.

In their private correspondence Derrick and Burnet had called the disease 'abattoirs fever', but they realised that that name could have explosive industrial repercussions. Burnet's suggestion of 'Queensland rickettsial fever' was rejected as an unfair slight on the state, which later knowledge proved it to be. Stumped for any other idea, they settled on Derrick's Q (for query) fever.

Soon after the results of Australian research on the bacterium and its antibodies were published, the organism began to show up widely overseas. Only just in time, the work of Derrick and Burnet and their associates served to explain what would otherwise have been mystifying epidemics in Europe during World War II. Thousands of troops on both sides were struck down with Q fever, especially in Greece and Italy.

Farm campaign is beating brucellosis

A nationwide programme of eradication seems to be succeeding.

Brucellosis is a bacterial infection with many similarities to Q fever, both in the ways that it is caught and in the way that human victims are affected. The big difference is that brucellosis is harmful to the livestock that carry it.

Cattle are the animals mainly threatened, though sheep and goats are also prone to the disease. Infection is usually transmitted from cows to heifers during the calving process. It can also be passed on in milk, or in the birth discharges that drop onto calving pastures. The bacterium involved, *Brucella abortus*, was so named because it often induces miscarriage in cows. Nearly half the calf production of an infected herd can be lost through miscarriage, stillbirth or failure to thrive after being born. Infected male calves sometimes prove infertile.

People can be infected through direct contact with cattle, especially at calving time, or by drinking unpasteurised milk. But the bacteria are more commonly picked up by breathing dust from animal hides or droplets from birth fluids and washed carcasses.

Though farmers and veterinarians who assist calvings are at risk, infections have been greatest among abattoir employees. About 25 per cent of a sample of Melbourne meatworkers, surveyed in 1978, had antibodies to *Brucella* in their blood. The incidence was highest among works drovers, rather than slaughtermen and processing hands.

Human brucellosis — sometimes called undulant fever — causes intense headaches and recurrent attacks of fever with heavy sweating. Its accompaniment of joint pains, fatigue and weakness can continue for weeks or months if the disease goes untreated. In extremely rare cases there may be complications involving the lungs or heart. Antibiotic drugs are highly effective if they are used promptly.

To safeguard the agricultural economy, rather than as a public health measure, a nationwide programme of brucellosis eradication was launched in the mid-1970s. It has been an immensely costly exercise, largely funded by the cattle industry through slaughtering levies. All herds have been blood-tested repeatedly, infected cattle have been killed and their owners compensated, and huge numbers of healthy stock have been vaccinated.

Eradication measures were completed in most regions by 1986. Years of monitoring will follow, concentrating on milk and carcass sampling and spot-checks on herds. Official declaration of Australia as brucellosis-free appears likely before the target date of 1992.□

Scrub typhus

Chiggers are so tiny that they feed unseen. As a calling card,
they leave their hosts a shattering fever.

Size has no bearing on the harm that disease-carrying animals can do. Some minute forms of life are all the more dangerous because they cannot be seen and their attacks are not felt. In World War II microscopic chiggers — the larvae of tropical rat mites — laid low more Allied troops, training in northern Australia or fighting in New Guinea, than all the efforts of the Japanese. But for antibiotic drugs, newly available at that time, the loss of manpower through illness would have been disastrous.

Mites are essentially the same as ticks. Only their much smaller size sets them apart. Mites are more diverse in character, however. Australia has more than 2000 species, few of which have been the subject of much study. By no means all of them are parasites, or parasitic in all the stages of their life cycles. Many eat plant matter. Some have adapted to rely on a particular manmade food — cheese, for example, or

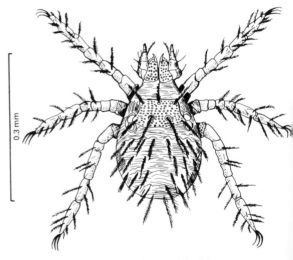

The larva of a trombiculid mite of the genus Leptotrombidium.

wine corks. We even have a group of predatory mites that kill other species.

Mites of the trombiculid family are responsible for scrub typhus. They are widespread from southern Asia to northern Australia. The species most to blame here is *Leptotrombidium deliense*. Adults are free-living in moist soils, but the chigger larvae feed on the blood of mammals. Like ticks, they wait in low vegetation, ready to cling to whatever brushes past. Their usual hosts are native rodents.

Throughout their lives the mites may carry rickettsial bacteria that are passed from generation to generation through the eggs. If mites happen to be wiped out in an area — by a bushfire, for example, or through the use of insecticides — rats and mice not only provide the transport that brings in a replacement population but also act as reservoirs for re-infection with the bacteria.

People 'sleeping rough' in tropical bush are likeliest to suffer an infestation of chiggers. Soldiers cutting vegetation to make trails, or for fuel or camouflage, would have brought down showers of larvae on their hair, skin and clothing. Chiggers are too small to be seen with the naked eye. And unless the host is susceptible to the form of allergic dermatitis Queenslanders call scrub itch, the bites are not felt.

Chiggers feed for a few hours, then drop off. If they are infected, the rickettsial bacteria are pumped into the host's bloodstream with their saliva. After normal attacks by a small number of infected chiggers nothing happens for 7–10 days. Then black-scabbed sores appear at the bite sites. The victim suffers an agonising headache, and eventually fever and a more general rash develop.

Scrub typhus — known as tsutsugamushi fever in Asia — is not as dangerous as the classical louse-borne typhus but it can be a most serious illness, prostrating its victims for a fortnight or more. Antibiotics such as tetracycline are effective in reducing the severity of the fever, especially if they are used early in the course of the disease.☐

SCRUB TYPHUS

Scrub typhus transmitted by Leptotrombidium deliense *occurs along the eastern seaboard of northern Queensland and in southeastern Papua New Guinea.*

When mites get under our skin

Soft moist areas of the human body make ideal egg-laying sites for a parasitic mite.

Scabies is a universal complaint, so common in old times that the British simply called it 'the itch'. It is a distressing rather than dangerous skin reaction to the activities of a parasitic mite, *Sarcoptes scabiei*. At worst scabies provokes such a frenzied scratching that the skin is torn and open to bacterial infection.

If unhygienic conditions allow it, the scabies mite goes through its whole two-week life cycle in and on the human body. The adult female seeks a soft, moist place. Groins, armpits, the folds in front of the elbows and behind the knees, and the webs between the fingers and toes are favoured sites. The mite bores through the outermost layers of skin and then burrows along underneath them, sometimes as far as 5 cm. Eggs are laid in the burrows and immature mites shelter there or in hair follicles.

None of this is felt at the time. The reaction occurs about a month after the first infestation, in a fiercely itching rash far from the mites' breeding area. Tiny blisters may come up closer to the site of infestation. Scabies is simply overcome by applying a prescribed lotion, all over the body, after bathing.

Every living person, except for the newborn, harbours another parasitic mite, *Demodex folliculorum*, in large numbers. They live at the bases of our eyelashes and inside the oily sebaceous glands of our skin. Their remains can clog our eyes, or provide tissue that bacteria can invade if a blackhead is carelessly squeezed, but they seldom cause any other trouble.☐

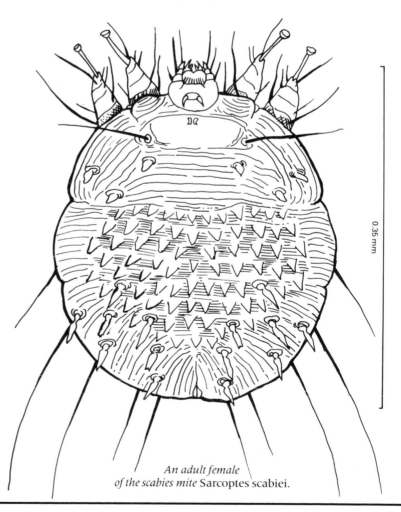

0.35 mm

An adult female of the scabies mite Sarcoptes scabiei.

LITTLE TO FEAR FROM LEECHES

For most people, a feeling of disgust is the only problem brought by bloodsucking leeches.

Revulsion at the discovery of leeches ruins many a rainforest walk. People are distressed or affronted to see the animals' bloated bodies, or the runnels of blood that they leave behind. But normally these opportunist feeders, in grabbing what may be their only meal in months, do no harm at all.

Leeches are worms, adapted with a sucker at each end of their body. Not all are parasites — many aquatic species eat insect larvae, snails and other worms. But the terrestrial 'bush' leeches rely on the blood that they extract in brief encounters with mammals, birds, frogs and reptiles. Under the head sucker they have twin jaws that break the flesh, and a proboscis that is inserted for feeding. Bites are not usually felt.

Abundant in the rainforests of the east coast and Tasmania, and also in monsoon vineforests on the Northern Territory coast, adult leeches of various species reach 2-8 cm in length. They are generally dark with streaks of a brighter colour along their backs. Unfed they are slim and wiry, and in their manner of moving they look just like many garden caterpillars, looping their way along by arching their backs and drawing their ends together. They move much more quickly than caterpillars, however.

Once a walker collects them by brushing against ground ferns or other low-growing foliage, they retreat from the light, disappearing most often into socks and footwear. They feed for about 10 minutes, then release their grip.

While feeding, a leech injects a substance that stops the blood from clotting. Blood may trickle from the tiny wound for a minute or two after the leech has gone, but this is no cause for alarm.

Pulling off a leech after it has started to feed — its body quickly swells to a slug-like form — runs the risk of tearing the flesh. The wound could ulcerate or become infected. It is better to leave the leech until it has finished. If the

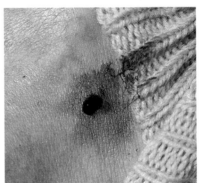

unwilling host cannot bear that, some heat applied close to the leech, with the smouldering tip of a cigarette or a twig, should cause it to drop off. The only realistic health risk from leeches is to people who develop an allergy to the saliva that is injected. That problem is uncommon and it usually arises gradually, so that sufferers learn to increase the precautions they take against leeches or else avoid them altogether. Cases of life-threatening allergy can be treated with courses of immunotherapy.

Unfed leeches are slim and wiry (far left). They cling to low-growing foliage until a host arrives. They suck blood (top, above) from the host for about ten minutes, swelling rapidly to a slug-like size, before dropping off. The bite usually oozes blood (bottom, above) after the leech has fed.

Louse-borne typhus

Lucky is the country that never suffers this scourge. It picks on the victims of poverty, warfare and natural disasters.

Typhus of the classical kind, transmitted only by human body lice, is a disease of deprivation and squalor. Rampant in Europe from the Middle Ages, it came early to the colonies of Australia in the jammed living quarters of sailing ships. Occasionally it took hold in urban slums and in mining camps, fostered by the same unhygienic conditions that allowed the spread of bubonic plague.

Epidemics of typhus break out when people are unwashed and wretchedly overcrowded — especially if they have to share bedding or clothing. Often the victims are also weakened by undernourishment. That may be part of the reason why louse-borne typhus, though clinically little different from tick typhus and mite-borne scrub typhus, takes a much more dangerous course. It leads frequently to pneu-

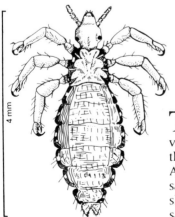

4 mm

Pediculus humanus

monia, then heart failure. Untreated, 20–25 per cent of victims may die.

The small bacterium that causes typhus, *Rickettsia prowazekii*, may well have disappeared from Australia. But its carrier, the louse *Pediculus humanus*, is still with us. Travellers careless of their cleanliness could reintroduce infection and create a new reservoir of the disease among local lice. But in normal Australian conditions, its spread would be most unlikely.

With public health standards enforced and antibiotic drugs available, typhus itself is less to be feared than the circumstances that could promote an epidemic. In the 20th century the disease has come to be associated with war, or with natural catastrophes such as famines and earthquakes. When populations are driven from their homes and social organisation breaks down, typhus gets its chance.

In the trench warfare of World War I, more Frenchmen are said to have died of typhus than were killed by Germans. With eastern Europe in turmoil at the end of that war, tens of millions of Russians, Poles and Romanians contracted typhus. Estimates of the death toll range up to 3 million. Western Europe shook off the disease — until the persecuted minorities of Nazi Germany were herded into concentration camps. Death rates among refugees after World War II were lower, thanks to improved medical techniques. The modern hotbed of typhus, largely because of wars and famines, is Africa.

In emergencies that force the crowding together of many people for long periods, the provision of washing facilities takes a high priority in the prevention of typhus. Living areas can be sprayed with insecticides and bedding and clothing fumigated.

If the disease has already broken out, treatment with antibiotics such as tetracycline is highly effective. Medical teams and relief workers can be vaccinated.

Body lice in many countries may also carry spiral bacteria of the *Borrelia* family, related to those that cause lyme disease. Victims suffer a prolonged, intermittent illness known as relapsing fever or famine fever. Extremely debilitating but seldom fatal, it is not known to have occurred in Australia.☐

Oriental rat flea sucking blood   *Female rat flea*

TYPHUS FROM A FLEABITE

Fleas can transmit a relatively mild form of typhus, similar to scrub typhus, that they pick up from rats or mice. Called murine typhus, it seems to be permanently established in some Australian country towns, though cases of the disease are uncommon. The main host may be the introduced brown rat *Rattus norvegicus*.

Murine typhus is prevalent in Asia, Central and South America and around the Mediterranean. The bacterium responsible is *Rickettsia typhi*. Many scientists believe that this was the original typhus organism, from which the other rickettsias developed to take advantage of different hosts and carriers.

Three compelling reasons to wash

Body, hair and pubic lice make life a misery for the unwashed.

Lice of three different kinds will live on us if we let them. All are completely specialised to human hosts — they go to no other animal. And they die within hours if we do not accommodate them. So keeping free of these parasites is a matter of routine hygiene. People who fail to keep themselves clean invite extreme discomfort, loss of sleep, and the risk that scratched bites will become infected.

All lice feed on birds or mammals. Though wingless they are insects, so they differ in many ways from the spider-like ticks and mites. But they are similarly small and flattened in form, with mouth parts adapted to chewing fragments of skin and feathers or to sucking the blood of their hosts.

Hundreds of species of lice have so far been identified in Australia. In reality there are probably thousands, each with its own parasitic niche in the animal kingdom. All kinds of birds have lice. And the only land mammals that escape seem to be bats and the monotremes — the platypus and spiny anteater. One louse species even spends months at sea, embedded in the skin of elephant seals.

The three lice that parasitise people are drawn by human skin odours and temperature. Once aboard a host they instinctively hide from light and fasten at a feeding site. Each has its own area: *Pediculus humanus capitis* makes for the hair on the head, *Pediculus humanus corporis* for the general body surface and *Phthirus pubis* for the pubic hair or the armpits.

All three feed exclusively on blood. The

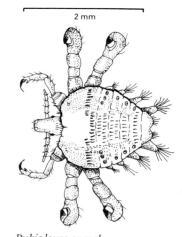

2 mm

Pubic louse or crab Phthirus pubis.

forward part of their heads is formed into a rounded beak containing three piercing stylets and a small, soft proboscis with internal teeth that are pushed out to grip the host while the louse feeds. Meals are taken frequently — as often as 12 times a day.

Female body lice may be as big as match heads. Head lice are somewhat smaller, as are pubic lice — often called crabs because of their broader shape and prominent claws. They live for up to a month, with a cycle from one generation of eggs to the next of a little over two weeks. Each female lays as many as 300 eggs — cemented to hairs by head and pubic lice, or loosely scattered in clothing by body lice.

The main consequence of infestation is an intense itching. Pubic lice cause the most severe irritation. They partly bury themselves in hair follicles. These become inflamed and pimples filled with pus may develop. The urge to scratch is irresistible, often leading to a rash of second-

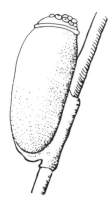

An egg (nit) of the head louse Pediculus humanus capitis — *here magnified 90 times — firmly cemented to a human hair. Eggs hatch within 10 days.*

ary infection resembling impetigo, with blisters and yellow-crusted sores.

Lice are spread by close physical contact and sometimes by the swapping of clothing or by the multiple use of bedding. Pubic lice are transmitted almost exclusively during sexual intercourse, but it has been established that they can be picked up from lavatory seats. It is a sensible precaution to wipe or cover the seat of a public toilet before using it.

In overcrowded living conditions, lice infestations can be prevented by scrupulous personal hygiene. In most Australian households the only problem — common even in the cleanest of families — is head lice brought home by schoolchildren. Frequent washing of the hair in hot, soapy water, followed by fine-combing, usually solves it. Chemists sell preparations to defeat more persistent infestations, but medical advice may be needed in the case of secondary bacterial infections.☐

Leptospirosis

Many animals can be carriers. Rats are traditionally blamed, but dairy cattle cause most of the trouble in Australia.

Intensified methods of mechanised milking, bringing cows together in large numbers, heighten the risk of leptospirosis. Shed workers must be protected with special clothing.

Leptospirosis has been poorly understood in Australia. Misled by medical books giving a British or American view, many doctors think of it as a rat-borne infection that damages internal organs — sometimes fatally — and is always marked by jaundice. In that form, known as Weil's disease, it used to be an occupational risk run by urban sewer workers, rat catchers and miners.

Weil's disease is virtually unknown in mod-

ern communities. But leptospirosis takes other forms — often unrecognised. In Australia it is a disease of rainy rural areas. Rats can still be the carriers, especially in tropical Queensland. But so can many other wild mammals, harbouring different strains of the germ. A far more important source of infection is livestock. Australians contract the disease mostly from cattle.

Doctors are supposed to notify state health authorities of all cases of leptospirosis. Fewer than 100 are reported in most years. But pathology laboratories find hundreds more from blood tests, often after the patients have recovered. Still other cases probably go undiagnosed. Mild infections are easily mistaken for influenza.

The effects of livestock-borne leptospirosis — transmitted by sheep, pigs and goats as well as cattle — are seldom as drastic as those of the classic Weil's disease. But they can prostrate many victims for a week or more. Prompt treatment with antibiotic drugs is effective, and in the mildest cases recovery takes place even without them. But heavy infections can advance to a stage that requires hospital care.

A germ with 150 varieties

The organism responsible for leptospirosis is a slender spiral bacterium, *Leptospira interrogans*. It occurs in at least 150 varieties, relatively few of which come into contact with people. Until recently each variety was classed as a separate species. A related bacterium, *Leptospira biflexa*, also has about 150 varieties — all harmless. Some are used in testing the blood of suspected sufferers because they provoke a reaction from antibodies that were stimulated by the disease.

Bacteria grow and multiply in the kidneys of carrier animals. They are excreted in urine, passing easily to other animals that graze fouled pastures or drink surface water. If a young animal survives its first general infection, but its immune system for some reason fails to kill all the bacteria, it becomes a new carrier.

Human infection comes not from swallowing the bacteria but usually because the bacteria penetrate cuts or abrasions in the skin. Infection occurs in water that is polluted by animal urine, or through contact with the infected tissues of slaughtered stock. There is no evidence of transmission from person to person — in swimming pools, for example.

Seasonal floodwaters on tropical croplands are an occasional source of infection. Within a fortnight of freakish flooding around Innisfail, Qld, in 1967, more than 20 people were in hospital with leptospirosis. Canefield rats were automatically blamed for that outbreak. But other wild hosts of the bacteria could have been involved. Because the carriers are diverse, the number of different varieties of bacteria is greatest in tropical infections.

Only one or two varieties are commonly implicated in the livestock infections of the south. There the spread of bacteria from animal to animal is fostered by intensive farming in high-rainfall districts. Sample studies suggest that 40–80 per cent of sheep and cattle in such areas contract the disease, though not nearly as many remain as carriers.

The risk to people is highest in dairy farming — especially in milking sheds where cows are regularly massed together — and at meatworks. After examining the occupational pattern among known victims, Professor Solomon Faine of Monash University medical school, Melbourne, estimated that Victorian dairy farmers have one chance in 10 of catching leptospirosis during their working lives, and meat inspectors one chance in four.

What the disease can do

Leptospirosis shows itself between three and 12 days after infection, with a sudden onslaught of fever, headache and vomiting. A rash of flat spots usually appears for a couple of days. Without treatment painful spasms of the neck and back set in, suggesting meningitis. Sharp abdominal pains may also be felt. Bright light is unbearable. Delirium and temporary mental disorders have been reported in some cases.

Urination is reduced throughout the illness and waste products build up in the kidneys. At an advanced stage tissue damage sets in, but it

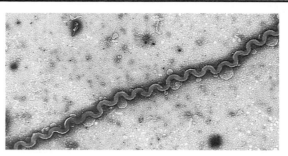

The organism responsible for leptospirosis is the slender, spiral bacterium Leptospira interrogans. *It occurs in 150 different varieties, only a few of which come into contact with people.*

can be reversed under hospital treatment. Life-threatening failures of the kidneys or liver are unlikely with Australian forms of leptospirosis.

People whose jobs put them at high risk usually wear protective boots and clothing. They can be further safeguarded by flushing and drainage systems that prevent livestock urine from accumulating. A vaccine has been developed but it is not licensed for use in Australia.

Milking sheds and cattle pens must be hosed clean quickly to prevent the accumulation of urine.

People who contract leptospirosis are immune afterwards, at least to the particular bacterial variety that infected them.

An animal vaccine is widely used in dairying districts, where leptospirosis can cause heavy losses in milk production. There seems to be little chance of complete eradication, however. The cost would be astronomical because sheep, too, would have to be treated. And the existence of many varieties of the bacteria in wild animals makes re-infection all too likely.☐

LEPTOSPIROSIS

EARLY SIGNS Fever, headache, vomiting. Rash of flat spots.
ACTION Obtain medical attention.

GIVE A DOG A BAD NAME...

In Britain and the US, dogs have a reputation for transmitting dangerous forms of leptospirosis not only in their urine but also in their saliva. They carry their own bacterial variety, *canicola*, and sometimes the worst rat-borne variety, *icterohaemorrhagiae*.

Dogs in Australia are completely free of *canicola*, and rarely carry any other form of the disease. It is wise, for several reasons, not to allow a farm dog to lick you. But urban pets are extremely unlikely to be carriers of leptospirosis.

Anthrax bides its time

Germs form spores if an infected carcass is opened. Drying like dust,
they can wait 50 years for a chance to kill again.

A bull died of anthrax soon after drinking from this water trough in Victoria. The trough is being disinfected with formaldehyde.

ANTHRAX

EARLY SIGNS Pimple developing into blister, becoming bloodstained. Rising temperature.

ACTION Obtain urgent medical attention.

AUSTRALIA'S 'ANTHRAX BELT'

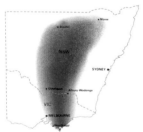

Anthrax is now confined to an 'anthrax belt' running through the centre of NSW and Victoria. Cases mainly affect sheep, cattle and pigs, although occasional infection of humans, goats, dogs and horses is reported.

Anthrax rivals rabies as the world's most widely feared animal disease. A mass killer of livestock, it can also be deadly to people in contact with infected carcasses. And it can flare up where it is least expected, having lain in wait for as long as half a century.

Bacillus anthrax was the first organism to be proved responsible for an infectious disease. Its rod-like forms were seen under a French scientist's microscope in 1849. Cumberland disease, as anthrax was known until then, was already wiping out sheep, cattle and horses in NSW. It seems to have arrived in Sydney in 1847, probably in a shipment of hay. Infections spread in Victoria, Queensland, Tasmania and South Australia. Only the Northern Territory and Western Australia were spared.

Because spores of the germ can lie dormant in soil for decades, total eradication of anthrax may be impossible. But the careful disposal of dead animals and the vaccination of survivors have turned the tide against it. Since the 1930s infections have been confined to a belt running from the middle of northern NSW to western Gippsland in Victoria. There are still many outbreaks each year within that belt, usually involving only small numbers of animals.

Sheep are most often the victims in the northern part of the belt, generally in summers that are drier than average. Cattle, more commonly infected in the south, may succumb at any time of year but most outbreaks occur with above-average rainfall after a drought. Pigs are also readily infected but anthrax in horses, goats and dogs is fairly rare. Wild animals are prone to the disease, but there is no recent evidence of their contracting it.

How the germs are spread

If the carcass of any animal killed by anthrax is opened, exposure to oxygen causes the germs to form tiny spores. These can be spread by carrion-eating birds and scavengers such as pigs and dogs, or blown in the wind when they dry out. Virtually indestructible in natural conditions, the spores can wait indefinitely for an opportunity to resume their breeding form. Theoretically they could germinate spontaneously, given a perfect combination of soil type, temperature and moisture. But usually they are eaten or inhaled by grazing livestock.

In cattle, sheep and goats, the bacteria multiply so rapidly and invade the bloodstream in such huge numbers that farmers rarely have any warning of trouble. Death follows less than 30 minutes after the onset of fever. Laboured breathing, staggering and sometimes blood in milk or urine are the only signs.

Anthrax acts less swiftly in pigs, dogs and horses. They can be saved, if prompt veterinary help is available, by a combination of antibiotic drugs. Dogs and pigs with anthrax show signs of drowsiness and have swollen throats. Horses may also have swellings in the chest, abdomen or legs, and suffer diarrhoea.

Animals killed by anthrax may show a dark discharge from the mouth, nose or anus. But a definite diagnosis can be made only by blood testing, using smears taken by veterinarians. Carcasses must never be opened or moved.

Warning signs that dot the shoreline of Gruinard spell out the danger to the public — but they are frequently ignored by foolhardy trespassers.

Once the disease is confirmed the owner is legally bound to destroy all carcasses by burning them, or by deep burial if there is a fire ban.

Farmers must report anthrax outbreaks to agricultural inspectors. All surviving stock are vaccinated and quarantined for some weeks. Owners and their neighbours are urged to start programmes of annual preventive vaccination. Immunity develops in less than two weeks.

The human effects of anthrax

Human anthrax in its worst form was often referred to as woolsorter's disease. Victims are overwhelmed by a sudden pneumonia — presumably because they inhaled the bacteria spores. This form is likely to be fatal without treatment. But it was always uncommon.

Anthrax in its more usual form used to be called 'malignant pustule'. It is contracted by touch and shows two or three days later as a small, itchy pimple, generally on a hand or arm. A blister quickly develops, containing a clear fluid at first but becoming bloodstained. Fever and a severe general illness, dangerous if untreated, soon follow. Meanwhile the skin sore grows to become a deep ulcer, perhaps the size of a 50 cent coin, with more blisters forming around it.

Responsible animal husbandry and legal controls have made human infection by anthrax extremely rare. Australia's few sufferers in recent times contracted it when animals were slaughtered while the bacteria were incubating, or during post-mortem examinations of animals that died for no apparent reason. Such occurrences need not be feared provided that treatment is sought promptly. Antibiotic drugs are highly successful against anthrax.□

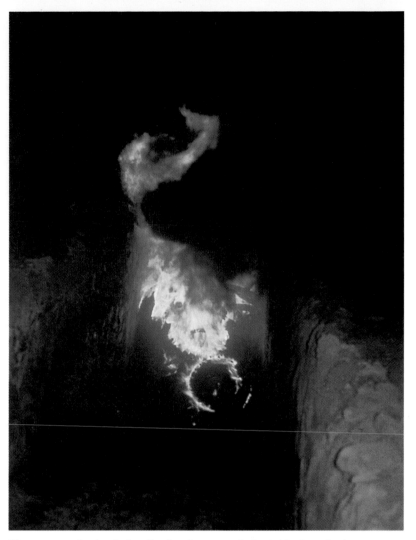

The carcasses of animals that die of anthrax must be buried in deep pits, burnt, or both. Chloride of lime is then spread over the site.

Gruinard Island, off the rugged northwest coast of Scotland, is still contaminated with anthrax almost 50 years after spores were deliberately released there.

BEAUTIFUL, REMOTE — AND DEADLY

'Anthrax Island' was the scene of a lethal wartime experiment.

During World War II, scientists from Britain's Porton Down bacteriological research unit exploded a total of six 'bombs' full of anthrax spores over sheep tethered on an island off the northwest coast of Scotland. The experiments were part of Britain's first foray into the field of biological warfare, prompted by fears of similar research being carried out in Nazi Germany. Gruinard, a small deserted island, was chosen because of its remoteness — although at its narrowest point the strait separating the island from the populated mainland is less than 300 metres wide.

It was the first time that scientists had had an opportunity to observe the effects of a large quantity of lethal micro-organisms unleashed on a natural environment. The results were horrifying — not only because of the immediate death of the sheep, but because Gruinard remained heavily contaminated after the war, and is still contaminated today. The anthrax spores' ability to survive more or less indefinitely has led one mainland resident to declare bitterly that Gruinard is now 'a contaminated monster. The spores are sitting there quietly, perhaps increasing in intensity, and they're never going to go away.'

Cows' milk caused the King's Evil

Scrofula, a painful and unsightly disease, once afflicted people from all walks of life.

Bovine tuberculosis is hardly heard of in Australia, thanks to childhood vaccination, milk pasteurisation and the slaughter of tens of thousands of infected cattle in the 1950s. Water buffalo herds roaming the Northern Territory wetlands may be the last significant reservoirs of the disease. Rarely will any child be seen with scrofula — the King's Evil.

Cattle can carry their own strain of the germ that causes human TB, *Mycobacterium tuberculosis*. It is transmitted in their milk. Because germs are swallowed rather than breathed, the human effects are different from those of the normal form of the disease. Instead of attacking the lungs, the bacteria cause growths in the lymph nodes of the abdomen, or more commonly of the neck.

Infected glands used to be removed surgically. Now combinations of antibiotic drugs overcome the disease. It is painful and debilitating rather than dangerous. It is also most unsightly if the neck is affected, for the swollen glands discharge through the skin. The weeping sores were called scrofula.

The disease and its repulsive manifestation used to be common at all levels of European society, especially among children. Before the nature of bacterial infection was understood and the disease traced to cows' milk, scrofula was regarded as a curse. Among the English there was a superstition that only the monarch, by touching an afflicted person, could lift this curse. Hence its other name, the King's Evil. The last monarch actually to oblige favoured subjects by touching their children was Queen Anne, who reigned from 1702 to 1714.

Other strains of the same TB germ can thrive in birds, fish and rodents. The bird strain is adapted to higher than human body temperatures and the fish strain to much lower temperatures, so neither causes trouble. The rodent strain, which in Europe is mainly associated with voles, can live for a while in humans but it does not cause disease. In fact it can be used as a vaccine to spur the production of antibodies against human TB. □

Roaming herds of water buffalo in the NT wetlands are probably the last significant reservoir of bovine tuberculosis in Australia.

Hydatid cysts

A tiny worm breeds innocently in country dogs. Passed to people, its eggs form growths that mount a slow, secret threat.

Dogs should be kept well away from areas used for slaughtering.

Australia's most dangerous worm is one of the smallest, the dog tapeworm *Echinococcus granulosus*. Typical of successful parasites, it does nothing to harm or annoy its natural host. Alternative carriers are less fortunate. In livestock, wild animals and sometimes humans, the consequences of swallowing the eggs of the tapeworm can be grave.

Hydatid cysts are older than history. They have probably been killing people since sheep were domesticated and dogs were first used to herd them. The use of the offal of slaughtered sheep to feed the dogs — understandable when farmers knew no better — has perpetuated the cycle of worm infestation for thousands of years. In Australia the disease has a further dimension because it has a wild cycle, between dingoes and the wallabies that they hunt.

An infected dog excretes the eggs of its tapeworms. Other dogs may pick them up directly, but large numbers of the eggs can be eaten by grazing animals. When those animals die or are slaughtered, later generations of dogs become worm-infested by feeding on their internal organs or other raw meat from their carcasses.

Human hydatid infection results most often from swallowing the tapeworm eggs in food or water that has been fouled by dog faeces. But eggs can also cling in a dog's coat and be picked up when it is handled, or they can be in its saliva after it licks itself. They can even be breathed in by someone who nuzzles a dog. Hydatid cysts in children are commonly in the lungs, through inhaling.

What happens if an egg is swallowed

In the human digestive system a tapeworm egg incubates quickly. The larva hooks itself to the intestinal wall and starts burrowing through. Entering the bloodstream, it circulates until a lodging place is reached. In most cases this is the liver. Less usual sites include the kidneys, the soft cores of bones, and the brain.

Denied a suitable environment for its normal development, the larva becomes a hydatid cyst — a fluid-filled bladder that grows steadily bigger. A cyst expanding in the abdominal

---

FOUND BY ACCIDENT

Car seat belts can have an unexpected effect. Occasionally a high-speed traveller, apparently unscathed after a minor collision or emergency braking, complains of sharp abdominal pains. Doctors discover a hydatid cyst — burst by the sudden pressure of the seat belt, or enlarged by the entry of blood from a slight internal injury.

---

cavity may easily reach the size of an orange, or even a melon. The body's reaction is to encase it with fibrous tissue. Instead of entering the adult worm phase and producing eggs, the cyst multiplies by forming inward buds. These are junior cysts with embryonic tapeworm heads, capable of independent life if the parent cyst is broken. Often there are hundreds of them.

What happens to the host depends on where cysts are located and to what degree important functions are impaired. Serious illness can develop fairly quickly, with delicate surgery perhaps the only hope of preserving the victim's life. Jaundice, because of pressure, is sometimes a clue to the existence of an abdominal cyst.

But deep-seated cysts often go on growing for years, unsuspected. Apart from a vague

feeling of illness or discomfort, there are no symptoms. Such cysts may be discovered by accident during unrelated investigations or surgery. They are always removed at the first opportunity, even if they are giving no trouble, because they are bound sooner or later to burst and send infection to other parts of the system.

Dealing with hydatid cysts

Though drugs are effective in controlling the worms in dogs, none has been found to work against the cysts in humans. The only answer is surgery. In the past it was often self-defeating

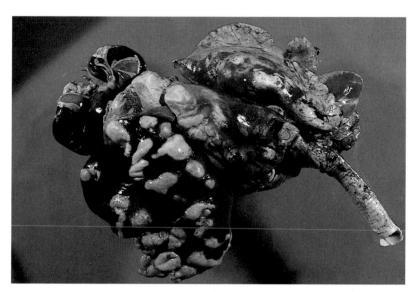

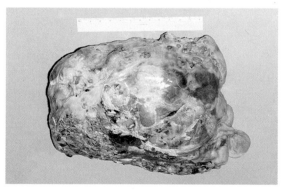

Multiple hydatid cysts of the liver, lungs and heart of a sheep (above) and of a cow's liver (left). The dog tapeworm Echinococcus granulosus (below), here magnified 24 times, is only 2–4 mm long. These specimens have been dyed for photographic purposes.

because offspring cysts escaped and re-infection set in. Modern techniques, using an Australian-invented suction device, have an excellent record of success.

Apart from sensible hygiene in dog-owning families, the most important preventive measures are the worm-treatment of dogs and control of what they eat. Legally enforced programmes have had striking success in New Zealand and in Tasmania, which in the early 1960s had the worst rate of human infection in the English-speaking world. Inspection and worming were made compulsory for all dogs and feeding them uncooked offal was prohibited. The incidence of worm infestation in Tasmanian dogs was cut by five-sixths in 10 years, and the demand for surgery on cyst victims fell by the same proportion.

Human infection remains a significant problem on the mainland, where dog tapeworms may never be eradicated because of their wild

> ### ADVICE FOR DOG OWNERS ...
> - Worm your dog regularly.
> - Never feed it raw sheepmeat or offal.
> - Don't let it lick cooking utensils or plates used by people.
> - Don't let children share food with the dog.
> - Keep it away from the droppings of other dogs.
>
> ### ... AND DOG LOVERS
> - Avoid close contact with farm dogs.
> - Wash your hands after patting any dog you do not know is worm-free.

existence among dingoes and wallabies. Nearly all hydatid cyst cases are reported from sheep stations. But city dwellers should not be complacent. All dogs should be wormed regularly. It is particularly important to treat a dog that is brought from a rural area, or a town dog that travels into farming districts or the outback with its owners. ☐

Tapeworms in food

Just waiting to be eaten, they find their natural home in our intestines.
The answer: cook or freeze all meat.

Nature, giving all its creatures a chance, intended us to harbour tapeworms. These parasites have evolved to take a dormant form in all of mankind's traditional sources of animal protein. But we are their real hosts. Only we can sustain them in their adult state — sometimes for years — and help them by distributing their myriad eggs.

In unsanitary conditions, tapeworm eggs are broadcast in the faeces of infested people. Other animals — especially livestock, insects and fish — can pick them up and become temporary hosts. In an exact reversal of the cycle that causes hydatid disease, the worm larvae do not develop fully but form cysts. If live cysts are eaten in food, more people acquire tapeworms.

Tapeworms, also called cestodes, form a big group in the 6000-species flatworm family. They are encountered all over the world. The one that causes most trouble, a fish tapeworm growing to 9 metres and living for 20 years, is unknown in Australia. Among the most widespread here is the dwarf tapeworm *Hymenolepis nana*, which uses insects as intermediate hosts. Infestations used to be common among Aborigines who relied on grubs and beetles for much of their diet.

These days only the beef tapeworm *Taenia saginata* causes much concern in Australia. Fastened to the intestinal wall and feeding on blood, it has an insidious effect that may not be noticed — even though the worm could be reaching a length of 5 metres or more. But some

An early 18th century illustration of a tapeworm by Nicolas Andry.

SIMPLE MEASURES TO AVOID DISEASE

Tapeworm larvae, nestling in the connective tissues of meat, are never easy to spot. Even if they are shaken loose when the meat is cut up, they look like fragments of muscle or nerve. They stay alive and can be passed to other food.

- After cutting up meat, wash the knife and the surface that you used before any other food touches them.
- When you cook meat, especially pork, make sure it gets a strong heat right through.
- If any meat is to be used raw without being finely ground, first deep-freeze it for at least 24 hours.

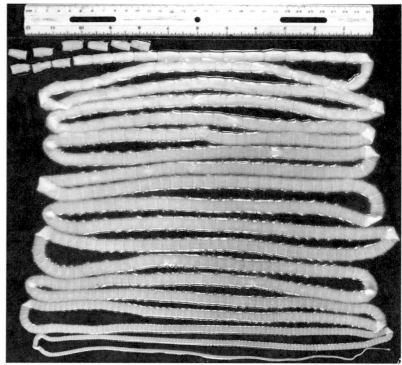

victims suffer a marked decline in general health and vigour, becoming anaemic and losing weight. Children, however well fed, can fall behind in growth. Diarrhoea and abdominal discomfort occur occasionally as warning signs. Doctors can prescribe a medicine that kills beef tapeworms. Then they are digested without further trouble.

From waterfleas to feral pigs

Far more to be feared is the pork tapeworm *Taenia solium*, because it can also take a larval cyst form in humans, lodging in vital organs or the brain. It is extremely rare in Australia. A much smaller worm of pigs, which causes a persistent disease called trichinosis overseas, is not found here. But another tapeworm, *Spirometra erinacei*, is frequently encountered in the meat of feral pigs.

Spirometra is not a natural parasite of man. It bores through the intestinal wall to form painful and occasionally dangerous cysts in fibrous tissue near the stomach muscles or under the abdominal skin. The condition, called sparganosis, requires surgery.

The main hosts of *Spirometra*, in which adults can grow to 1 metre, are dogs, dingoes, cats and foxes. Eggs from the hosts' faeces hatch into aquatic embryos that are eaten by waterfleas. Inside these, the embryos develop to the first of two larval stages. Infested waterfleas may be gobbled up by frogs and tadpoles or swallowed by all kinds of mammals and reptiles as they drink. In these next intermediate hosts the worm develops to its second larval stage, called a sparganum.

People can contract sparganosis through swallowing waterfleas in unboiled water from outback pools and creeks, or while swimming. But the disease comes much more commonly from eating undercooked or carelessly handled meat from feral pigs. Along with the natural

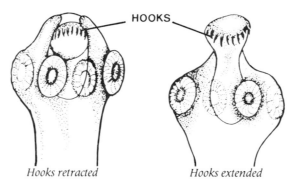

HOOKS

Hooks retracted   *Hooks extended*

hosts, pigs do not only swallow spargana themselves, but they also scavenge the bodies of other animals harbouring them.

In sparganum form — a thin white ribbon about 6 cm long — the tapeworm can live for at least two weeks in the main compartment of a home refrigerator. Abattoirs slaughtering feral pigs — mostly for export — are required to deep-freeze the carcasses for many days.□

The beef tapeworm Taenia saginata (above) grows to more than 5 metres. Its head has four suckers but no hooks. The head of the dwarf tapeworm Hymenolepis nana (left) has a circular row of 20–30 small hooks that can be pushed into the flesh of the host or retracted into the worm.

The dwarf tapeworm Hymenolepis nana grows to a maximum of 10 centimetres.

Larvae of the tapeworm Spirometra erinacei in a human breast.

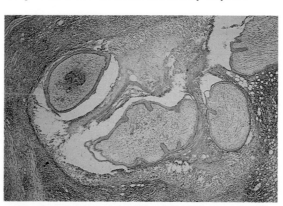

Hookworm disease

Penetrating the skins of barefoot children, hungry parasites can hamper physical and mental development.

Infestation with hookworms occurs when people defecate or spread faeces as fertiliser on ground where others can walk or lie. The disease that results — usually caught in childhood — may condemn its victims to lives of sickliness, stunted growth and low intelligence. Yet it is simply prevented and easily cured. Not before time, it has become rare in Australia.

The worms are still common in moist coastal regions of Queensland and the Northern Territory. Two kinds, *Ancylostoma duodenale* and *Necator americanus*, were introduced last century, probably by Chinese or Pacific Islands labourers. Tragically, the spread of the hookworms coincided with the herding of Aborigines into mission settlements.

From their roaming tradition the Aborigines brought a conservationist approach to hygiene. They took care not to pollute water sources, but left their excreta to fertilise the ground after they moved on. Now they were forced to stay in one place, poorly supplied with sanitation facilities and unaware that alien organisms had arrived that would make their old habits dangerous.

Hookworm disease became rife in Queensland, not only in the native people but also among agricultural workers of all races. A government campaign in the 1920s was fairly successful in ridding Europeans and Chinese of it, but less effort seems to have been made with Aborigines. In 1960 about two-thirds of all the people in settlements had hookworms. The incidence was lowered dramatically during the following decade, thanks to a revitalised Aboriginal health programme.

Tiny invaders that wait in soil

Hookworms are members of the roundworm or nematode family, which has more than 10 000 species. All are tubular and pointed at both ends, and about half of them are adapted to parasitise plants or animals, in water as well as on land. And they are remarkably prolific — well-fed adults can lay hundreds of thousands of eggs a day.

Hatching in warm, moist soil, minute hookworm larvae bore into any human skin that stays in contact with them, usually the soles of the feet. They travel through the bloodstream and lungs to the small intestine, where they attach themselves with four hook-like teeth or two cutting plates. Feeding on blood, adults grow to about 1 cm in length. Their eggs stream into the bowel and are excreted.

Rashes sometimes break out where the larvae have penetrated. Victims of heavy infestations suffer anaemia, often with abdominal pain and alternating cycles of diarrhoea and constipation. Bronchitis is common. When the infestation is prolonged in children, they are deprived of the vitamins needed for growth and brain development, and their sexual maturity is delayed.

Sanitary methods of sewage disposal eventually break the cycle of hookworm infestation.

THE LIFE CYCLE OF HOOKWORMS

Hookworms are parasites that suck blood from the lining of the small intestine.

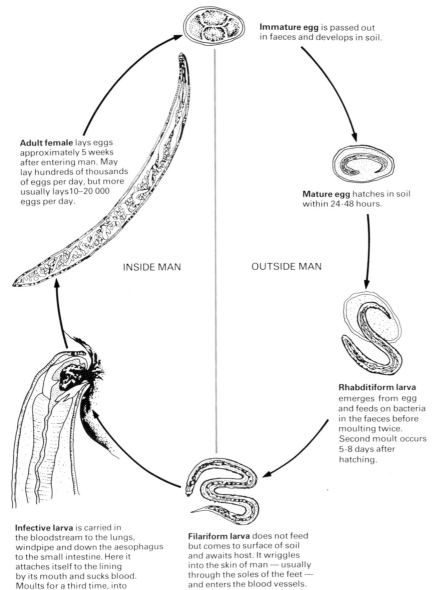

Immature egg is passed out in faeces and develops in soil.

Adult female lays eggs approximately 5 weeks after entering man. May lay hundreds of thousands of eggs per day, but more usually lays 10–20 000 eggs per day.

INSIDE MAN

OUTSIDE MAN

Mature egg hatches in soil within 24-48 hours.

Rhabditiform larva emerges from egg and feeds on bacteria in the faeces before moulting twice. Second moult occurs 5-8 days after hatching.

Infective larva is carried in the bloodstream to the lungs, windpipe and down the aesophagus to the small intestine. Here it attaches itself to the lining by its mouth and sucks blood. Moults for a third time, into an adult.

Filariform larva does not feed but comes to surface of soil and awaits host. It wriggles into the skin of man — usually through the soles of the feet — and enters the blood vessels.

Existing victims can be treated with a drug that quickly kills the parasites. Their larvae are long-living in the soil, however, so it is unlikely that the worms will ever be entirely eradicated. People who believe they are at risk have little to fear if they wear shoes.☐

SLUGS NURSE A WORM THAT ATTACKS THE BRAIN

Rats and slugs play their part in a worm disease that causes a dangerous swelling of the brain and spinal cord coverings. Called angiostronglyosis or eosinophilic meningitis, it is encountered widely in Southeast Asia but very rarely in Australia. A few cases have occurred in coastal Queensland.

A tiny roundworm from Asia, *Angiostronglyus cantonensis*, has rats as its principal hosts. In their adult form the worms live in the arteries between the rats' hearts and lungs. Excreted larvae are swallowed by slugs, among other animals, and develop to an infective stage inside them. The slugs are eaten by more rats.

People can acquire the worm larvae through accidentally eating slugs on food. It is also possible that some freshwater prawns carry them in an infective form. Finding no home in the human bloodstream, the larvae infect the watery fluid that surrounds the brain and spinal cord. Illness follows in about two weeks, with a mild fever, vomiting, agonising headaches, neck stiffness and aversion to bright light. Skin tingling and a highly exaggerated sensitivity to touch are often experienced.

The disease is potentially fatal if enough of the worm larvae are swallowed. But usually it runs its course in a few days or weeks and an otherwise healthy victim recovers under hospital care. A drug is available to relieve the symptoms.

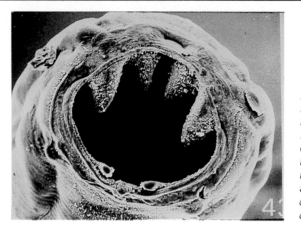

A scanning electron micrograph of the mouthparts of Ancylostoma duodenale, *showing the four hook-like teeth.* Necator americanus *has two cutting plates instead of teeth.*

Threadworms are no disgrace

Threadworm infestation is a common childhood complaint, even in the cleanest of homes.

Schoolchildren are almost inevitably infested with threadworms — also called pinworms or seatworms — at some time. They present no danger, and do not reflect on the cleanliness of a child or its home. Eggs of the threadworm *Enterobius vermicularis* are easily passed among children at play, on the hands or in clothing, until eventually they are carried to the mouth.

While a sufferer sleeps, female threadworms crawl from the intestines to the anus, where they lay thousands of eggs. They set up an itch so intense that no child can refrain from scratching. Some eggs lodge under the fingernails and may be passed to the mouth, increasing the infestation. The complaint can continue indefinitely and may cause abdominal pain, but it is easily treated by doctors.☐

Fever from lung invasion

The common human roundworm thrives in conditions of poor sanitation and hygiene.

Given conditions of long-neglected hygiene, the common human roundworm *Ascaris lumbricoides* is capable of damaging health anywhere in Australia. Compared with the hookworm it is a giant, growing to 30 cm or more and looking much like an earthworm. Its eggs are swallowed in food or water that has been contaminated by the fouling of soil.

Ascariasis, the disease caused by chronic infestations with this roundworm, may not show any obvious symptoms. Its effect in robbing children of nourishment and stunting their growth is slow and subtle. But adult worms can block the small intestine, causing abdominal pain and vomiting. Larvae of *Ascaris* occasionally start a lung infection marked by fever, coughing, wheezing and headaches. If there is reason to believe that a child has the parasites, its faeces should be examined — a worm may be visible.

Improved sanitation soon cuts the incidence of ascariasis. Doctors can prescribe a drug for victims that does not kill the worms but paralyses them, so that they can be excreted with the aid of a laxative. The eggs last for years in damp soil, so new cases can be expected. But the disease will not take hold seriously unless standards of hygiene slip again.☐

Toxoplasmosis

Cats foul soil with an organism that infects one person in every three.
The main danger is to unborn babies.

Blood studies indicate that more than a third of all the people in the world catch toxoplasmosis. The illness is generally so mild that most people are unaware of it. They may simply feel off-colour. But in rare exceptions the consequences are horrifying. If a pregnant woman is infected, her baby may be aborted, stillborn or born with brain and eye disease.

Toxoplasmosis has been recognised for most of this century as a parasitic disease harmful to almost any mammal and to many birds. It is notorious as a cause of lambing losses in the sheep-raising industry, and of congenital blindness in wild animals. But researchers were slow to grasp the extent of its incidence in humans — and even slower to discover its main carriers.

Only in the 1970s was it established that cats — pets, alley cats and feral tribes alike — are the natural hosts. In cats' intestines, a single-celled organism *Toxoplasma gondii* undergoes its full development and founds new generations. The process does the cats no harm. But in other animals the parasite multiplies in the wrong places, forming cysts that can damage the brain, eyes and internal organs.

Plants play a role in infection

Microscopic organisms are discharged in the faeces of a cat. They are not infective when the droppings are fresh, but after a day or two in soil or litter — lightly covered over by fastidious cats — they form spores. These can remain alive in moist soil for three months or more. They are spread by worms, beetles, snails, cockroaches, flies and so on, and easily transferred to shooting vegetation such as grass.

Grazing livestock swallow the parasites with their food. Meat, if not thoroughly cooked, often contains live cysts. Cases of severe toxoplasmosis from eating raw meat are relatively common in parts of Europe. In France, infections of the unborn are said to occur once in every 1000 pregnancies — about 35 times higher than the Australian rate.

Low-growing, uncooked vegetables such as lettuce can also carry the parasites. But most of the trouble here stems from handling soil, and later transferring the organisms to food or directly to the mouth. So it is a commonsense precaution — and an essential one for pregnant women — to use gloves when gardening or handling cat litter, and to wash the hands before eating or touching food. Vegetables and fruit to be eaten raw should be washed.

Middle three months are worst

Many toxoplasmosis infections do not cause any fever or other acute illness but are long-

A cat playing in a garden makes a pretty picture— but danger lurks unseen. Anyone touching soil or plants that could have been fouled by cats at any time in the previous three months should wear gloves. This precaution should always be taken by pregnant women.

lasting, without noticeable symptoms. It seems that mothers-to-be have nothing to fear from chronic infections. Danger arises only if the disease is contracted while the foetus is developing. The foetus is at its most vulnerable in the fourth to sixth months of growth.

Organisms multiply quickly in the liver and spleen of an infected foetus, causing tissue destruction and sometimes enlargement of those organs. Cysts commonly form in cells of the brain, eyes, liver, lungs and heart muscles. The brain can be enlarged by the accumulation of fluid, suffering damage through pressure inside the skull. Or the head may be abnormally small because the brain is not fully developed. Blindness occurs because of inflammation of blood vessels around the retinas.

Toxoplasmosis contracted later in life is rarely a serious disease. It may reveal itself as a flu-like fever, sometimes with such soreness and enlargement of the lymph nodes that it is wrongly diagnosed as glandular fever (mononucleosis). Cancer of the lymph system may even be suspected. Sometimes toxoplasmosis takes a latent form that causes trouble only if the body's immune system is broken down by another disorder — such as AIDS — or by the immunosuppressive drugs needed after organ transplant surgery.

Antibiotic drugs are successful against many infections, including some contracted in pregnancy. People can be infected more than once but in normal health they develop an increasing resistance. Promising work towards a vaccine was reported in the mid-1980s. It will call for techniques similar to those used in producing experimental vaccines against malaria.□

KEEPING A CAT FREE OF MICROBES

Cat owners can try to ensure that their pets never harbour the organisms of toxoplasmosis. But the necessary control of diet and activity verges on the inhumane:
- From birth, a cat must never be allowed out.
- Indoors, it must be prevented from hunting flies, moths, cockroaches and mice.
- It must never eat raw meat.

Keeping your own cat indoors does not make your garden any safer — other cats will come visiting.

Direct contact with a pet cat brings no risk of toxoplasmosis. The organisms do not become infective for at least a day after they have been discharged into soil or litter.

Concern about toxoplasmosis has given fresh force to calls for the extermination of unwanted cats. Many authorities recommend a system of licensing and compulsory immunisation, followed by the deliberate spreading of feline enteritis among stray urban populations. But even those measures would not solve the problem of feral cats in the bush.

SAD DILEMMA FOR TRANSPLANT TEAM

Heart transplant patient Karen Scoble (pictured) carried latent toxoplasmosis. It was the disease, rather than her cardiac problems, that killed her. Karen, 16, received her transplant at St Vincent's Hospital, Sydney, in July 1984. She was re-admitted three months later with symptoms of bladder inflammation. Doctors identified the infection as toxoplasmosis but could do little about it — Karen was on drugs that suppressed her immune system to prevent the rejection of her new heart. The disease progressed rapidly; she died within a week.

Toxocariasis

Crawling babies are at risk of becoming infested with worms from the egg-ridden droppings of dogs and cats. Some infections have led to blindness.

A desire to see pet dogs and cats vigorously healthy ought to be reason enough to dose them regularly against worm parasites. If owners need further convincing, they might ponder the consequences of spreading toxocariasis in their community.

The small roundworm *Toxocara canis* commonly infests untreated dogs. Puppies are born with the parasite if their mothers have it. The closely related *Toxocara cati* is found to a lesser extent in cats. Some other animals harbour their own varieties of the worm. But it is mainly pet dogs, allowed and sometimes encouraged to void their bowels in public parks, that pose a threat to people — especially children.

Worms reach maturity in the intestines of dogs, where they feed on blood. Their eggs, discharged in many thousands every day, are invisible in the faeces. They are not immediately infective, needing some days in soil to incubate. They can remain viable for years, blowing about in dust until they find suitable soil conditions.

There is seldom any danger in handling a dog at home, or in being licked or nuzzled by it, unless the dog has been sniffing ground that was fouled earlier. Infection is far more likely to take a direct route from the ground to the hands of a child at play, and then to its mouth. Infants at their crawling stage are particularly exposed.

Confusing symptoms — or none at all
Unable to find their natural environment in the human body, *Toxocara* fail to develop beyond

A bat of the Pteropus *species eating a soursop. Bats were discovered to be contaminating fruit in Queensland with eggs of a* Toxocara *roundworm.*

their larval stage. They roam the bloodstream and tissues, surrounded by white cells that cause inflammation wherever the larvae lodge.

Often toxocariasis is disguised by symptoms indicating a baffling variety of other diseases. It may show in skin rashes, for example, or persistent fevers, respiratory complaints, or attacks of vomiting. In many mild cases there are no noticeable signs — evidence of the disease turns up later in blood tests. If the larvae settle in deeper internal organs such as the liver and kidneys, they can do gradual tissue damage that may threaten the lives of untreated victims. And in a few cases the blood vessels of the eyes are infected, causing permanent blindness through damage to the retinas, the light-sensitive coatings behind the eyeballs.

The severity of toxocariasis increases with the number of viable eggs that are swallowed and the victim's exposure to repeated infections. Researchers have found that as few as 20 eggs can cause significant illness. Medical care is often needed, but many cases resolve themselves in a week or so if the source of infection is removed or avoided.

Publicity given to the risks of toxocariasis in Britain led to an outcry against the carelessness of some dog owners, and a stiffening of municipal by-laws against the fouling of public places. Tests indicated that 25 per cent of all dog droppings in parks and playgrounds contained eggs of *Toxocara*. The rate of human infection in urban populations was estimated at 2 per cent — but in children living near parks, it was found to be 4 per cent.☐

Bats were the key to Palm Island's puzzle

A mystery illness on a Queensland island turned out to be toxocariasis.

Fear swept the Aboriginal community on Great Palm Island, Qld, in 1979 when violent attacks of painful vomiting struck down 138 children and 10 adults in three weeks. All the victims were found to have enlarged livers. Many began bleeding internally or through the mucous membranes.

It took intravenous feeding and intensive care to save some of the children, who were ill for nearly a month. They were tested for every likely poison, bacterium or virus. Toxocariasis was not considered because it had never been responsible for such a sudden, explosive outbreak. Unable to find any causative organism, doctors called the illness 'the Palm Island mystery disease'.

Investigations continued and one common factor was found: every victim had been eating mangoes. Fruit collected on the island was examined and hundreds of eggs of a *Toxocara* roundworm were found on the rinds. They proved to be *Toxocara pteropodis* — a fruit bat parasite discovered in Vanuatu but never known to have existed in Australia.

Bats of the common black flying fox species *Pteropus alecto* were captured at Townsville, 65 km to the southwest. Three out of seven suckling juveniles had the worms and were excreting them in their faeces. More infected juvenile bats were detected in Brisbane and another species, *Pteropus poliocephalus*, was also implicated.

Mangoes were tested over a wider area and *Toxocara* contamination was found among backyard fruit at Yeppoon, near Rockhampton. Commercial supplies in markets were clean, however. A fungicide treatment, required in the packing process, had removed whatever bat deposits there may have been.

Fruit bats of many species range from the subtropical coast of Western Australia, across the far north and down the east coast to northern NSW — with a huge outlying colony based in the Sydney suburb of Gordon.

Enlargement of the liver, for no apparent reason, has long been noted as a common complaint of northern Aborigines. The solution to the Palm Island mystery suggests toxocariasis from bats, as well as from dogs, as a likely cause. It also indicates a possible risk to anyone eating unwashed fruit from trees that are visited by bats — especially to children who like peeling fruit with their teeth.

The worm eggs are excreted only by baby bats, clinging to their mothers and suckling while the adults feed on fruit or blossoms. After the first month or two of their lives the juveniles cease to excrete eggs. Since bats breed only once a year the dangerous period is November-December. But fruit ripening after that, if there has been no rain, could still have deposits of the long-living eggs. Washing the fruit — and the hands after picking it — will remove any risk.☐

Tetanus in wounds

Germs invading damaged flesh produce a poison that triggers choking spasms. Protection is easy — but far from permanent.

Immunisation of two generations of infants and schoolchildren has obscured the menace of tetanus. It no longer figures as a common consequence of injury. But in the unprotected, its effects are no less grave. People who fail to get booster injections at least every 10 years, and people over 50 who may never have been immunised, risk an affliction that still can kill.

Chances of tetanus infection bear no relation to the seriousness of wounds. Victims of major injuries are safer, in fact, because antitetanus shots are a routine part of any treatment. Trivial wounds, apparently not needing medical attention, give rise to most infections. Sometimes a patient is not aware of a wound at all. In such a case a doctor unfamiliar with tetanus can easily miss the diagnosis. Delayed injections can arrest the disease but they do not reverse symptoms that have already set in.

Tetanus is caused by a bacterium, the bacillus *Clostridium tetani*. It exists harmlessly in the intestines of many animals, including humans occasionally. Horses are the most prominent hosts — another reason why the disease has become less common. The germs are excreted, and on contact with the air they form spores. These can remain dormant for years, lying in soil, blowing about in dust or clinging to plants, buildings and implements, especially on farms. They can be touched — provided the skin is intact — inhaled or swallowed without danger.

The germs are reactivated only in animal tissue that is extremely low in oxygen. So the spores need to be completely embedded in tissue that has been killed — by injury usually, or by the action of other bacteria or chemical poisons. Many wounds escape infection. But in ideal conditions the spores incubate and the germs develop their rod-like form. Soon they start to multiply.

In the process they release a toxin of remarkable potency. It works like strychnine — but the amount needed to achieve the same effect is about 100 000 times less. In natural situations an infected animal is killed by the effects of the toxin, if it is not already dead. On the decomposition of its carcass, or its digestion by carrion-eating scavengers, more spores are spread.

Puncture wounds are the worst

Infection can occur in many ways. Spores can be rubbed into existing cuts and sores during contact with animals, soil or any contaminated surface. But more often the spores are forced into tissues at the time a wound is inflicted, because they are clinging to whatever causes the injury. The wounds most likely to be infected are punctures that readily close over, leaving deep-seated spores that cannot be dislodged by washing or reached by disinfectants.

Punctures by nails, tacks, wood splinters and plant thorns are likely starting points for tetanus. Claw wounds and bites inflicted by all kinds of animals can be infective, especially if they eat carrion. Even a bird's peck can be dangerous. Before child immunisation became the rule, a schoolboy attacked by magpies on the Queensland Gold Coast died of tetanus.

Incubation of tetanus spores can take hours or weeks. The sooner the symptoms show, the

TETANUS

EARLY SIGNS Stiffness of jaw; difficulty in swallowing; other muscle stiffness. Raised temperature, headache, sweating.

ACTION Obtain urgent medical attention.

graver the disease will be. Most often it sets in one to two weeks after an injury. A person who was once immunised but has gone too long without a booster injection may be resistant, though not immune — the symptoms may be delayed and relatively mild.

Tetanus infection causes no noticeable change to the wound itself, which may be inflamed for other reasons. The multiplying bacilli do no damage to the tissues that they occupy. Instead their toxin finds the nearest nerve and follows it to the spinal cord. There the chemical action of the toxin blocks the nerve cell restraints that allow control over muscle movements. Assailed by random impulses, the muscles twitch convulsively or tighten and fail to relax.

What happens to victims

Tetanus is often called lockjaw. A spasm that clenches the jaw is sometimes the first sign. But the affliction can start with any set of muscles. It progresses fairly quickly to involve all or most of a victim's normally voluntary movements — including breathing.

With contracted muscles pulling against one another, co-ordinated movement becomes impossible. In severe cases the victim collapses and may be in immediate danger of choking. The heart rate and blood pressure are unstable. Heavy sweating is common and there may be fever. In untreated cases — and some that are treated too late — death results from respiratory or heart failure, or from pneumonia.

Only a few metropolitan hospitals can undertake the full treatment of tetanus. It calls for specialised and virtually constant care. Often it takes weeks, and complications are frequent. Mechanical breathing aids, muscle-relaxant drugs and sedatives help, but dangerous new spasms can be triggered by sudden sensations — not just touch but also noise or bright light. A drug is often used deliberately to induce paralysis during the critical phase of the illness.

Any medical practitioner can give injections to prevent tetanus. People in risky occupations, such as farming and gardening, are advised to have antitetanus shots every five years. A frequency of 10 years, however, is considered safe enough for most people.□

Scratch and bite fevers

Teeth, beaks and claws are all highly likely to carry organisms that are harmful to humans.

The slightest wound inflicted by an animal should be thoroughly washed and disinfected. Chances are high that teeth, beaks and claws carry potentially harmful organisms. As these organisms vary widely, it is impossible to be specific about their effects. Some animals, however, can transmit their own diseases through bites and scratches.

Ratbite fevers are common in many parts of the world, and rare in Australia only because ratbites are rare. Rats and mice of many species frequently have bacterial or fungal infections. Passed to humans, they cause fevers that develop anything from a day to a month later, with a painful swelling of the lymph nodes or with joint pains, headache and vomiting. These fevers are not dangerous if antibiotics are used.

Cats can pass on a disease by biting or scratching. Fortunately rare, it affects all systems of the human body and can have long-lasting effects. The organism responsible is thought to be a species of *Chlamydia* — a microbe that has some of the characteristics of viruses and some of bacteria.

Cat-scratch fever starts about a week after the injury, showing only a small, pus-filled blister on the site of a wound that had apparently healed. A week to a month later the area

becomes painfully swollen. A mild fever is accompanied by headaches, chills and fatigue. Sometimes a rash breaks out on the hands and feet. With bed rest, a light diet and plenty of fluids, most victims recover of their own accord in about a month. But in some cases the lymph glands near the injury are infected. Painful inflammation occurs and serious abcesses develop. These have to be surgically drained. Recovery from this form of the disease can take months longer.

This little girl is hand-feeding a quokka, a wallaby found in the southeast of WA. Marsupials have been implicated in human infections through scratching.

Suspicion falls on marsupials

Kangaroo scratches have been implicated in one case of disseminated nocardiosis, an uncommon fungal infection of the internal organs. All the known occurrences of this infection have been fatal — most of them not diagnosed until after the victim's death.

A nine-year-old girl from Condobolin in western NSW was put into hospital in 1962 after a week of fever, coughing and chest pain. She had a small pustule on her abdomen and some curiously puckered scars from old sores that had started, according to her medical record, like mosquito bites. A fortnight before, and on many earlier occasions, she had played with a pet kangaroo.

In spite of every possible test and treatment, the child died a week later. By this time her lungs were overwhelmingly infected and her bloodstream starved of oxygen. In the meantime more pustules had broken out on her face and shoulder.

Lengthy laboratory work revealed that the skin lesions contained a microscopic fungus, *Nocardia asteroides*. Heavy growths of the same organism were found in the girl's lungs — where little natural tissue remained — and in the kidneys and brain. Damage to many other organs, including the heart, indicated other sites of long-term infection.

Nocardia is a fungus, found worldwide, that breaks down rotting plant matter. Infection of mammals is usually through breathing or eating its spores. The strange feature of the Condo-

bolin girl's case, reported in the *Australian Medical Journal*, was that she was apparently infected through her skin on repeated occasions over about six years.

Evidence against the kangaroo is not conclusive. But nocardiosis is well known as a chronic skin disease of marsupials and grazing livestock. And it seems to be communicable by scratching — captive mobs of kangaroos and wallabies in Brisbane in the 1950s suffered an epidemic.□

Children are often — rightly — encouraged to handle native animals. But parents would be well advised to consult a doctor about any scratch their child receives from a marsupial.

Psittacosis

Bird traders took Australia's own disease all round the world. Easily mistaken for pneumonia, it hits hardest at the elderly.

South American parrots took the blame when a wave of human psittacosis deaths swept three continents in 1930. Starting in Argentina, this previously little-known disease took more than 100 lives among bird fanciers in North America, Europe and Africa. It followed a travelling exhibition of caged birds.

Not until 1934 was it established that the infection had originated in exported Australian budgerigars. Researchers had discovered by then that the psittacosis germ is a natural parasite of all Australian parrot species, in the wild as well as among aviary birds and pets. The incidence varies seasonally and regionally, but sometimes well over half the members of wild flocks have the disease.

The organism responsible is *Chlamydia psittaci*. Like a virus it lives in the cells of its host. But like a bacterium it can be destroyed by some antibiotic drugs. The disease is usually

active only in juvenile birds. Adults, showing no symptoms, pass it to their young in the nest. Active infection of a bird in the first few months of its life shows in lethargy and apparent depression, often with eye inflammation, nasal discharges and diarrhoea.

Trading in caged birds has spread psittacosis

Psittacosis can be transmitted from birds to humans in even the briefest of associations.

worldwide, and to many species other than parrots. Finches and pigeons are commonly infected. Overseas the disease has caused problems in commercial turkey flocks.

How humans catch the disease

A solitary pet that survives its first year in good health is most unlikely to present a danger to its owner. People working with birds in large numbers are most at risk. But psittacosis can be contracted simply by handling an infected bird or being nibbled or pecked by it. The disease is usually caught, however, through breathing in the germs from dried droppings or from dust in feathers.

The onset of psittacosis is much like pneumonia — a sudden fever with chest pains and coughing — after one or two weeks of incubation. Cases are not commonly recorded but more are probably diagnosed, and successfully treated, as pneumonia. Deaths have been rare since tetracycline drugs became available. But elderly people with other health problems can succumb rapidly to psittacosis.☐

More than half of these wild galahs could harbour the psittacosis germ.

Allergy to pets

Distressing illness strikes even though no animal is around. Entering a strange room can be a nightmare for sufferers.

Horses produce the most allergenic of all common danders. Sufferers aware of their allergic reaction should avoid horses, donkeys and zebras.

Mammals and birds continually shed tiny flakes of skin. In domestic animals the dried particles are called dander — a word that shares its origins with the dandruff of humans. Easily inhaled, the fragments provoke sharp allergic reactions in a minority of people who are susceptible to various animal proteins.

The reaction is much like 'hay fever' — allergy to plant pollens. But it has no season. Soon after breathing dander, victims suffer an intense itching of the linings of the mouth, nose and throat, and fits of sneezing. Their noses stream and their eyes are red and swollen — the lids sometimes so puffy that they cannot be opened. If the victims are prone to asthma, a severe attack may be triggered.

Dander clings to all hair or fur that is shed by animals. It works loose as the hair ages. Long after a pet has died or been removed from a house, deposits of hair many years old can release clouds of active dander if they are disturbed. Similar allergens are contained in animal saliva and urine. In contact with the skin of sensitised people, they can cause disfiguring hives and rashes that itch maddeningly.

Feathers also are coated with particles that easily become airborne. Sensitivity to dander from feathers used in stuffing pillows and upholstery is an important cause of asthma and what people may think is hay fever. Pet birds less frequently cause problems. Workers in the poultry industry may carry home allergens in their clothing, affecting their families.

Sensitivity is acquired gradually, increasing with repeated exposures. Some people live with a pet for years before it starts to give them trouble. Allergy to any kind of animal is feasible, although cats, dogs and horses are the commonest offenders. Rabbits, mice and guinea pigs also are frequently responsible for illness, especially among laboratory workers.

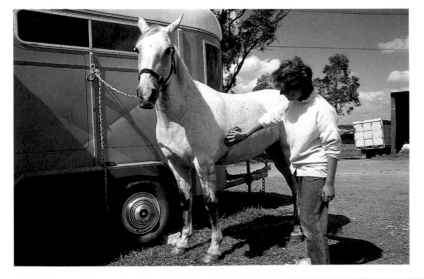

Avoidance is the only remedy

Antihistamines and possibly stronger drugs are prescribed to control severe cases of allergic reaction. But there is no cure. All that can be done is to break the contact between a victim and the animal or its shed allergens. Removal of a house pet must be followed by a thorough cleaning of furniture and floor coverings.

Allergy to cat dander is probably far more common than people realise. Children in particular may suffer from it through close contact with schoolmates who cuddle cats before they leave home. Visiting a house that has been inhabited by a cat — even if not in the past five or six weeks — can give an allergy sufferer distressing nasal and chest problems within seconds. Victims often seem to hate cats. But their primary emotion is fear of what a cat may unwittingly do to them.

Dog dander is not as strongly allergenic as the material from cats, but the closeness of many dogs to their owners makes allergy cases common. In milder cases the family dog may merely have to be taught to keep out of the house. The existence of the allergen in saliva and urine is especially relevant to problems with dogs.

Horses produce the most allergenic of all common danders. Their falling numbers since the introduction of motor cars makes it easy for susceptible people to avoid them. Most sufferers have worked directly with horses or ridden them regularly for recreation — though in some people allergy can be sparked merely by exposure to someone else's riding gear. People aware of their bad reactions to horses should be careful at a zoo or a circus, because donkeys and zebras are likely to have the same effect.□

WHEN PIGEON BREEDERS HAVE TO GIVE UP
A form of alveolitis forces some people to abandon their hobby.

Some people who breed and race pigeons are forced out of the sport by a chronic disease. Years of regularly inhaling fine particles from dried droppings turn their lungs into a permanent battle-ground between allergens and antibodies.

The main features of the illness, a form of alveolitis, are a dry cough, shortness of breath, weakness, attacks of shivering, and loss of weight. If the disease is neglected the victim becomes a complete invalid. Only a minority of pigeon fanciers develop the allergy. If it is detected early — a blood test is available — and the hobby is abandoned, the chances of recovery are good.

These colourful fan-tailed pigeons could transmit a form of alveolitis to their breeder.

House dust mite allergies

It took 40 years to track house dust mites down as a leading cause of asthma.
Yet we feed them by the thousand every day.

Household dust is the most common year-round cause of allergy. Many people exposed to it suffer an illness similar to hay fever, and some suffer severe attacks of asthma. In fact studies indicate that nearly 90 per cent of asthma patients are affected by house dust.

It is a complicated, ever-changing mixture of animal, vegetable and mineral materials, varying from season to season, house to house and room to room. It may contain well-known allergens such as plant pollens and the dander from the shed skin of pets. But the component that causes most of the trouble is an elusive animal, not half the size of the full stop at the end of this sentence.

Scientists starting hunting for the principal allergen in house dust in the early 1920s. The mite *Dermatophagoides pteronyssinus* was not implicated until 1964, when a Dutch researcher devised a way of counting its population in dust samples. He found up to 490 mites per gram of dust in the bedrooms of allergy sufferers, but hardly any in dust from Swiss houses where they stayed to get relief. The case was proved in tests using dust extracts with a varying content of mites. Further research showed that the main allergen is excreted in the mites' faeces. Accumulations of dried faeces become airborne with any disturbance. The body tissues of dead mites are also allergenic.

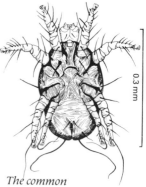

0.3 mm

The common house dust mite Dermatophagoides pteronyssinus.

Sydney: mite capital of the world?

Specialists in allergies and asthma have suggested that Sydney and the NSW central coastal region have the world's highest concentrations of house dust mites. The climate is ideal: the mites thrive best within a temperature range of 18-28°C and in relative humidity that does not fall below about 70 per cent. Then they can breed in almost any darkened part of a house, instead of being restricted to bedding as they are in much of the world.

A sample of vacuum cleaner dust from a Sydney house, tested in 1983, held 534 of the common house dust mites per gram, plus 30 *Dermatophagoides farinae* — the first identified in Australia. This species is the dominant cause of allergy in North America, but the European type is of much more importance here.

Humans supply the main food of *Dermatophagoides pteronyssinus*. The mites eat the tiny flakes of skin that we shed all the time. But they can also feed on pet dander, carpet wool, clothing and furnishing fibres, feathers,

kapok and so on. The American type mainly eats milled cereals.

Hospitals have the right idea

Dust mite counts in most hospitals are extremely low — less than 2 per cent of the density in a normal house. That is because floors are smooth-surfaced instead of carpeted, and frequently scrubbed and polished. All bedding, furniture and soft furnishings are washed regularly. Hospital practices were aimed at discouraging bacteria, long before dust mites were thought of.

Ideally the allergy sufferer should copy hospital procedure. Short of that, some benefit is gained from stepping up the frequency of vacuum cleaning and getting rid of quilts, rugs and so on that are not readily washable. Pets should not be allowed on beds — their hair and dander provide extra food for the mites.

Insecticides are available that will kill house dust mites, but they are generally unsuitable for the frequent use required. The only way to eliminate mites is by complete fumigation. The smell makes a house unliveable for a few days — by which time the mites are starting to re-establish themselves.

Mild allergy sufferers should need no medical treatment if the dust in their environment is reduced. People who still experience some symptoms usually get over them by taking oral antihistamines for a few days. An unfortunate minority who continue to react severely may get relief from some newly available prescription drugs. Failing that, a doctor may recommend a course of desensitisation with minute injections of a mite extract, increasing over several months. In many cases the allergy is dampened down.☐

THE ANSWER COULD BE IN YOUR TEACUP

A solution of tannic acid could solve allergy problems.

Tannic acid, which gives tea its flavour, has the ability to cross-link proteins and change their characteristics. It does this when animal hides are tanned and turned into leather.

Mr Wes Green, a medical researcher at the University of Sydney, has shown that a 1 per cent solution of tannic acid in water destroys the allergy-producing properties of dead dust mites and mite faeces. It could be used for washing floors and bedding.

Development of an aerosol spray is under investigation. It would be non-toxic and non-irritant, smelling something like tea leaves. In time it may be combined with a suitable pesticide so that living mites can be killed at the same time.

A case against cockroaches

These household pests head a list of insects that spell trouble for allergic people.

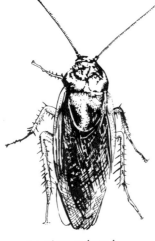

American cockroach
Periplaneta americana.

Though they often carry germs on their legs and bodies, cockroaches are free of any dangerous diseases of their own. But they head a virtually endless list of insects (excluding dust mites, which are arachnids) that can make breathing a hazard for allergic people.

Any insect matter can provoke allergic reactions. Inside houses, most trouble is caused by dried droppings and fine particles rubbed off body scales. Cockroaches add to the problem by softening their food with copious amounts of regurgitated salivary fluid. Dried traces of this add to the allergenic material in the air. Overseas studies indicate that in heavily infested houses, asthmatic children react as least as

much to cockroaches as they do to dust mites.

Locusts are a well-known cause of allergic reactions — something to be borne in mind when their populations occasionally explode to plague proportions in agricultural areas of southeastern Australia.

Inhaled insect allergy is often an occupational disease. People working with foodstuffs can become allergic to material from weevils. Laboratory staff dealing with insects are frequently susceptible. Among the more curious allergies recorded are those of pilots who flew consignments of flies for scientific experiments, and those of people who made their livings raising maggots and moth larvae for fishing bait.☐

The price of rabies

People and domestic animals could be protected if the rabies virus took hold in Australia. But part of the cost would be a wildlife massacre.

Keeping Australia free of rabies is worth any effort and expense. The strictest quarantine rules, the tightest coastal surveillance and the sternest penalties for animal smuggling are warranted. For if the virus ever established itself here, the measures that would be needed to eradicate it are ugly to contemplate.

Rabies can infect any mammal, turning placid and timid creatures into mindless aggressors that bite whatever they see. For their human victims, once the virus has incubated and the symptoms show, there is no hope. In medieval times their suffering was ended by smothering them with pillows. Of all the world's victims of a disease that goes back at least 4000 years, only three people are known to have recovered.

Mass killings could not be avoided

Dogs present the greatest direct danger to humans and livestock because we keep so many close by us and employ them in farming. People at risk in an Australian rabies outbreak could be vaccinated. So could their pets and working dogs. Livestock could be protected too, given time and a colossal diversion of funds. But the problem would not disappear.

In outback Australia, rabies would go wild. Carriers would include dingoes, feral cats, foxes and marsupials. Bush rats, exploding across the inland in their periodical plagues, would take death far and wide. Worst of all, our millions of bats would become mobile reservoirs of the virus. Migrating seasonally and using breeding

Australia has millions of bats like these flying foxes — and bats would spread rabies throughout the land if the disease broke out here.

The red fox is the main reservoir and spreader of rabies in western Europe.

sites shared by many species, they would spread it throughout the continent and keep bringing it back to settled areas.

Of Australia's 50-odd species, only the ghost bat is a flesh-eater. But bats that normally live peaceably on insects or fruit will attack anything if they are rabid. Insect-eating bats much smaller than Australia's flying foxes have killed several people in the United States.

In western Europe the red fox is the principal reservoir and spreader of rabies. Where their locations are known, some fox families have been protected from infection with an oral vaccine. Treated food is laid around their feeding grounds. The method could also work well in North America, where skunks and racoons frequently harbour rabies.

No such technique aimed at saving animals could be effective in the vast unpopulated spaces of inland Australia. Nor would it help much even in towns, where bats often roost in the ceilings of buildings. Possums and bandicoots, for example, would quickly become threats. The forced response to rabies here would be campaigns of wildlife extermination.

Illegal activities increase the risk

Rabies has not far to come. It is entrenched on many Indonesian islands and frequently kills people. Thirty-six deaths in the first six weeks of 1986 prompted a renewed crackdown on Jakarta restaurants serving dogmeat. The prac-

Dogs entering Australia must spend up to nine months in quarantine stations like this one at Kemps Creek, NSW. Such measures are essential to prevent the introduction of rabies into this country.

tice fosters a dangerous, illicit trade in dogs.

Quarantine laws control the movement of dogs and livestock between Indonesian territories, but nothing stops that country's bats. If rabies ever spreads to Irian Jaya and across New Guinea, other bats could well be the vehicles for its entry to Australia. Many migrate across Torres Strait.

More urgent worries are caused by illegal activities. Health authorities believe that landings by smugglers and poaching foreign fishermen are fairly frequent on remote parts of the northern and northwestern coasts. Dogs accompanying the intruders could have been exposed to rabies at earlier ports of call. And one runaway rabid dog could distribute the virus among dingoes and marsupials with amazing speed. In a French village in 1983, in the space of just a few hours, a frenzied cocker spaniel bitch bit her owner, his daughter, five other people, 13 dogs and cats and five sheep.

Any dog brought into Australia must spend nine months in quarantine. Unless it comes from a declared rabies-free country such as Britain or New Zealand, it must also have spent the previous six months in quarantine overseas. Since these measures were imposed, Australia has never had a proved case of rabies. The only outbreak, which killed one or two people, was in Tasmania last century.

Some people risk heavy fines by attempting to evade the quarantine process. The dogs they smuggle are usually from rabies-infected countries. Others in search of profit try to import and breed novelty pets such as hamsters and gerbils — prohibited altogether in Australia. Hundreds of hamsters were seized in southeastern Queensland in 1981. They were the progeny of only four pairs, illegally imported from Asia a year before.

What the virus does to animals
The cause of rabies — also known as hydrophobia or lyssa — is called *Lyssavirus*. It attacks the nervous systems of mammals and multiplies in their brain cells. Recent research shows that the virus exists in at least four strains, differing slightly according to the types of animals that act as wild reservoirs. This probably accounts for wide differences in the symptoms of rabies. As well as a 'furious' form, in which infected animals attack everything in sight, there is a 'dumb' form, in which the main effect is paralysis. And there are variations in which both forms seem to be mixed.

The first sign of rabies, usually a few weeks after an animal has been bitten, is a change in its disposition. Normally friendly pets become furtive and surly. Shy or aggressive pets, on the other hand, become friendly. Wild animals lose all their fear of humans. Docile farm livestock become excitable. Hoarseness may be noticed in the voices of dogs or cats. Their eyes may appear oddly glazed and vacant, with the pupils unevenly dilated.

A dog with 'furious' rabies is likely to run from its home and trot for long distances, holding its head low and drooling saliva, and biting at everything it meets. A dog that is chained or shut up may injure itself severely in its attempts to get away. Rabid cats seldom run so far. But sudden movements and noises give them muscle spasms that cause them to leap about and bite.

In the final phase of the illness an animal starts to stagger and paralysis sets in. It may lose consciousness or suffer a violent convulsion. Death usually takes place four days to a week after the first signs appeared. In 'dumb' rabies the phase of excitement and savagery is very brief or it may not occur at all. Paralysis sets in early, usually starting in the throat. Saliva drips from the animal's part-open mouth. It cannot swallow, and may look as if it is choking.

Madness at the sight of water
More than 90 per cent of human infections come from the bites of dogs in the 'furious' phase. But animals become infectious several days before symptoms are obvious. Contagion is possible without a bite, through saliva or particles of infected tissue from a slaughtered animal getting into an existing wound. Dried saliva is harmless, however — the virus dies quickly in air.

Lyssavirus has an extraordinarily varied incubation period. The average time for symptoms to appear in humans is about two months — but it can be as little as 10 days or well over a year. People bitten in a rabies outbreak can be given courses of injections that should prevent the disease. But it has been suggested that in the case of bites by bats, the anti-rabies serum must be administered within two days.

A flu-like fever and severe headache mark the onset of human rabies. Within one to four days, as brain cells come under increasing attack, the victim grows agitated and mentally confused, and may hallucinate. Periods of terror alternate with violent rage or listless apathy. Muscle spasms start. A sound, a flicker of light or even a puff of air may trigger convulsions.

Hydrophobia — fear of water — is an apt alternative name for the disease. Any attempt to swallow provokes agonising, strangling throat spasms. Foaming at the mouth from an excess of saliva, the victim grows desperately thirsty but dare not drink. Finally the sight of water, or the mere mention of it, provokes a maniacal fury. Death from heart or respiratory failure ends three or four days of torment.□

WHAT TO DO IF THE WORST SHOULD HAPPEN

Swift and sensible action by the public would give authorities a chance to control a rabies outbreak.

- If a pet dog turns unaccountably savage and drools saliva, confine it and call a veterinarian.
- If you see a dog on the loose, trotting with its head down and dripping saliva, call the police.
- Don't panic if you are bitten — there is time to get injections. Seek medical attention.
- If a wild animal of a normally timid type wanders about by day, showing no fear and allowing you to approach it, stay away. Tell the nearest wildlife service office.
- If a wild animal invades a campsite or farmland and acts viciously, confine it if you can do so safely and call the police or a wildlife officer. If it must be destroyed, don't shoot it in the head. Rabies will be proved or disproved by examination of brain tissues.
- If you know of animals brought from overseas without quarantine, give the facts to the police or public health authorities.

Beware the air in bat caves

Infectious organisms make bat breeding caves dangerous places for humans.

Considering the high potential of bats for spreading disease, they let Australians off very lightly. But evidence is mounting that bat breeding caves are places to avoid. Infectious organisms thrive in droppings that may have accumulated for centuries.

Overseas, the most common bat-carried disease is histoplasmosis — a fungal infection of the lungs and sometimes of vital internal organs. It hits hundreds of thousands of people a year in the Americas, usually with mild and easily curable effects, but it is occasionally fatal. In Australia, even allowing the possibility that some cases are wrongly diagnosed, it is unaccountably rare.

The microscopic fungus *Histoplasma capsulatum* grows in the droppings of bats and birds. It multiplies by forming spores that are breathed in by the victims of histoplasmosis. In the United States the disease is frequently associated with domestic chicken flocks and with starlings that roost near houses. But the birds do not carry the fungus. It infects only mammals, and is spread about by bats in their seasonal migrations.

Occurring as an acute lung infection — two or three weeks after inhaling the spores — histoplasmosis causes fever and night sweating, pain around the lungs, shortness of breath and dry coughing. Fungicidal drugs can be prescribed, but many cases clear up of their own accord within a month or so. Chronic infections are more dangerous because they may show no symptoms until internal organs are damaged.

Until the 1980s, the fungus in Australia had been identified only in a few chronic histoplasmosis patients, and once in the soil of a chicken coop. It had never been detected in acute infections, or isolated from any environment that established a link with our bats.

The trouble at Wee Jasper
Bats were under strong suspicion before then,

Bats crowding together in breeding caves like this one produce droppings which, accumulating over the centuries in deep layers on the cave floor, harbour spores of the fungus Histoplasma capsulatum.

The bentwing bat Miniopteris schrei-bersii *breeds in Church Cave, NSW.*

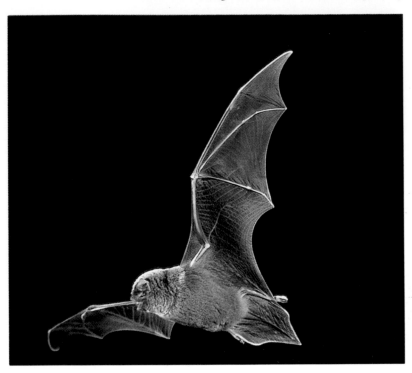

however. In 1976, scientists learned of an epidemic of histoplasmosis among members of a caving club. Thirteen fit young men had contracted lung infections after exploring Church Cave at Wee Jasper, near Yass, NSW. The cave is a breeding site for the bentwing bat *Miniopterus schreibersii*.

Applied science lecturers and students at the Riverina College of Advanced Education, Wagga Wagga, launched a spare-time programme of investigation. From 1977 to 1983 they made many visits to Church Cave, trying various methods to isolate the fungus in the air or in the deep layer of old bat droppings — guano — on the cave floor.

The researchers spent an hour or two in the cave on each visit, wearing the respirators that farmers use when spraying dangerous pesticides. In spite of this precaution, three out of eight people caught histoplasmosis. But their efforts to find the fungus bore no fruit until the end of 1983. Then it was grown in laboratory cultures from guano and soil samples and respirator filters, and from phlegm taken from the last of the histoplasmosis sufferers.

Bentwing bats breed in at least 10 other localities in eastern Australia. Some are believed to change from one cave to another, and to share quarters with other species. Many of those are the same as the species that are known to carry the disease in other countries. Scientists will be surprised if the cave at Wee Jasper proves to be the only trouble spot.☐

Hot pool hazard

When water lies too long and gets too warm, a deadly amoeba may flourish. The more the fun, the greater the risk.

Tragedy among rural South Australian families in the 1960s opened a new field of worldwide medical concern. Investigations into the deaths of six children proved that free-living amoebas — not just the parasites that cause dysentery — can be killers.

The children, all healthy and active, died rapidly of what became known as primary amoebic meningoencephalitis, or PAME. Years of laboratory work established a previously unknown amoeba, *Naegleria fowleri*, as the cause. Meanwhile multiple deaths from PAME were reported from New Zealand, the United States and Czechoslovakia.

PAME was obviously not a new disease — simply one that had been in disguise. Viruses would have been blamed in the past. Since pathologists have known what to look for, well over 100 PAME deaths have been reported, including nine more in Australia.

The amoeba is probably widespread, but it needs exceptional conditions to grow and outstrip other organisms that would normally eat it or starve it. The common factor in PAME cases is water with a low salt content, lying stagnant and becoming unusually hot. Infections in New Zealand and California, for example, came from pools fed by geothermal springs. Deaths in Czechoslovakia were traced to a heated indoor swimming pool, where presumably the water had not been changed.

Temperatures soar in overland pipelines

Most Australian cases have been in hot, arid districts where water has to be piped overland from sources that are often depleted by drought. Two of the most recent deaths occurred in the same week in Western Australian wheat belt towns that are 150 kilometres apart but share the one overland water supply. In a January heatwave, the child victims used backyard playing pools.

Naegleria fowleri seems to present no danger to people who drink the water it contaminates, or who bathe in it quietly. The route of brain infection is rapid and direct, through the upper nasal cavities. To force water into these cavities requires vigorous action — the sort that child-

ren perform when they jump feet-first into pools or are 'ducked' by having their heads shoved under. In some South Australian cases the victims had not been swimming at all. But they had frolicked in household baths that were filled from a contaminated water supply.

PAME has a sudden onset, within a few days of receiving the amoebas. A severe, persistent headache is quickly followed by fever, sometimes with a sore throat and nasal inflammation. The neck stiffens and there is frequent vomiting. Fits resembling epileptic attacks occur before the victim lapses into coma. The victim dies four or five days after the first signs of illness.

There is some evidence, from old cases in

Vigorous activity in any still water that has been excessively heated by the sun brings the risk of a deadly amoebic infection.

Britain, that PAME may be treatable with a drug. But the disease is rare, compared with many other forms of meningitis and encephalitis with similar symptoms. It is hardly ever diagnosed until after the victim's death.

In Australian conditions, any still water that is excessively heated by the sun should be regarded with suspicion. Children's playing pools should be shaded in summer and their water chlorinated and renewed frequently.

A similar disease but a slower death
Another free-living amoeba, not positively identified but thought to be a species of *Acanthamoeba*, has on rare occasions caused a form of meningoencephalitis that is just as deadly but takes a slower and more subtle course. Most victims are already suffering another disease. Infection apparently starts elsewhere in the body and reaches the central nervous system through the bloodstream. No background link such as water contamination has been found. Symptoms fluctuate and death can take weeks or months. It has also been the cause of blindness — after eye wounds were invaded — and respiratory complaints. □

Bather's itch: just a fluke

A case of mistaken identity results in a maddening rash.

Tiny worm larvae, drifting near the surface of tidal lagoons and shallow coastal inlets, mistake humans for seagulls. It sounds comical — but it is no joke for people who are disfigured for days by rashes of pimples, or deprived of sleep by a maddening itch.

The complaint is hardly dangerous. Many a holiday has been ruined, however, and many a child made frightened of the water.

Schistosome dermatitis or bather's itch — sometimes blamed on biting sea lice — is caused by skin invasion by the larvae of a flatworm, *Austrobilharzia terrigalensis*. The adult worm was first identified in a seagull at Terrigal on the central coast of NSW. It belongs in a class of parasitic flatworms called flukes. A closely related blood fluke in African rivers causes a dreaded wasting disease, bilharzia. Another in the class, the liver fluke, often kills livestock. On rare occasions it infects people who accidentally eat its larvae on wild watercress.

Austrobilharzia has a complicated and insecure life cycle. It requires a seabird such as a gull or a pelican as its adult host, in which it lays eggs. They are excreted and hatch in water, but the larvae cannot develop beyond a juvenile stage unless they find a certain marine snail, the whelk *Velacumantus australis*.

In the snail they reproduce, and their descendants reproduce again. The third generation takes a different form, called a schistosome cercaria, and swims out to find the final host. Its only guide to a suitable bird is body warmth — and that is where paddling children and wading fishermen come in.

Anchored by a sucker, the cercaria secretes an enzyme that dissolves tissue and allows it to penetrate the skin. Scarcely a pinprick is felt. If humans were its proper hosts it would enter the bloodstream. Instead it stays under the skin, provoking the body's natural reaction against an invader. A day or so later a pink spot appears where each cercaria has entered. The spots rise, redden and enlarge to about 3 mm in diameter. Clothing rubbing against them sets up a fierce itch. There is a risk of bacterial infection if they are scratched. Otherwise they darken and recede in two or three weeks. Marks like freckles may remain for months longer.

People's reactions differ in severity, and can become more intense with repeated infections. Antihistamine creams are usually effective against the itching. Repellants are available but they do not last long in water. Perhaps the best defence is petroleum jelly, smeared on exposed parts of the body that will be under the water.

Bather's itch occurs mostly on the NSW coast. It is regarded as a summertime hazard, but only because that is when most people suffer from it. The worms are active at any time of the year. □

The mudwhelk Velacumantus australis *is the host of the parasitic flatworm* Austrobilharzia terrigalensis. *The larvae of this worm can penetrate human skin and provoke an irritating reaction.*

Free oysters may be no bargain

Think twice before helping yourself to 'wild' oysters.

At low tide, hundreds of rock oysters are temptingly exposed along some shorelines.

Shellfish do not digest all the organisms that they filter from water. Some bacteria and viruses collect in their tissues, still alive. Infection is likely from shellfish that grow in impure water. Shellfish growing in heavily polluted water are capable of passing on the germs that cause typhoid fever.

In 1978 about 2000 people in four states suffered gastroenteritis from oysters farmed in Sydney's Georges River. They were infected with a *Vibrio* — a bacterium closely related to the one that causes cholera. And the cholera organism itself was reported in 1986 to be living in at least 13 Qld and NSW rivers.

Since the 1978 scare, commercially farmed oysters have been subject to regular testing and are given time to cleanse themselves in purification tanks of sea water, under ultraviolet light, before they are marketed.

People harvesting 'wild' oysters or mussels — or helping themselves from oyster leases — run a risk unless they are prepared to cook them thoroughly. Those who insist on their oysters raw should be sure that they come from an authorised source.□

Trouble from handling crayfish

Ungloved food handlers risk infection from bacteria in fish slime and soil.

People who constantly handle crustaceans or fish in their work are wise to wear heavy gauntlets. Otherwise they risk not only cuts and punctures but also a high rate of bacterial infection.

A form of erysipelas, peculiar to food handlers, is particularly unpleasant. A purplish rash, spreading across the fingers from any trivial wound, turns into a series of pus-filled ulcers. Meanwhile victims suffer up to 10 days of fever, headache, joint pains, lymph node tenderness, weakness and weight loss. Full recovery can take a month or more.

The cause is *Erysipelothrix indiosa*, a bacterium of fish slime that is also found in soil. It can be picked up in the slaughter of livestock and poultry and in handling their meat — one of the many common names for the disease overseas is 'pork finger'. But the people most likely to be infected are those whose work also carries a risk of skin injuries.

The most detailed study of the disease in Australia was among crayfishermen based on the Houtman Abrolhos Islands off Geraldton, WA. Before they took to wearing gauntlets they suffered badly from what they called coral poisoning. The crayfishing grounds are fringed by coral reefs. It is possible that traces of toxins from corals contributed to the severity of their symptoms, and perhaps increased their sensitivity season by season.□

For full protection against a painful infection carried in fish slime, this crayfisherman should be wearing thick gauntlets on both hands.

HUMAN CARELESSNESS FEEDS FLIES

Few species of flies — usually three or four in any place, out of nearly 7000 species in Australia — seek a permanent association with people. That is just as well, because those that have adapted to exploit human habits do it with great efficiency. Ideally suited by the Australian climate, they breed in enormous numbers.

No disease relies on flies, in the way that malaria relies on mosquitoes. But houseflies and bushflies, compulsively drawn to our food and our faeces, are the animals most likely to spread illness. What actual harm they do depends on us — on our standards of hygiene. Our neglect can turn flies into the most dangerous creatures on earth.

Where houseflies originated is beyond guessing. They have gone everywhere in the world where people have settled. Ours came from England in every colonising ship, breeding in rotting food. The main species of bushfly — misleadingly named because it pesters most people in cities — came thousands of years before with the Aborigines. Blowflies, with a less deliberate interest in humans, are mostly local species. Their diet of wildlife carrion has broadened only recently to include livestock and the Sunday joint.

Wings Houseflies belong to the order Diptera, the two-winged or 'true' flies. Adult flies have only one pair of functional wings, the forewings. The hind wings are reduced to tiny 'halteres' which act as balancing organs when the fly is in flight.

Antennae All flies have antennae. The housefly has short, broad antennae.

Eyes The enormous compound eyes of houseflies take up most of the head. Each eye has 4000 facets.

Housefly Musca domestica

HOUSEFLIES AND FILTH

Houseflies are drawn to any decomposing animal or vegetable matter. Rotting food, household garbage and human or pet faeces are sought as egg-laying sites and as nutriment for the larvae. But the adults cannot live long without a rounded diet of sugars, proteins and water — so they seek out fresh sources as well.

Flies in this group have no biting or chewing ability — their mouth is a spongy pad on the end of a sucking tube. Everything they eat must be liquefied. This is achieved by enzymes in their gut contents, which they regurgitate and dribble onto food. On the feet of flies or in their guts, any number of different bacteria, viruses and small parasitic organisms can be transferred from faeces and rotting matter to fresh food, water and open wounds. Many so-called food poisonings and secondary infections of sores are caused this way. If more serious diseases are about in a community that has poor sanitation, flies could be an important factor in their spreading.

A new generation in 14 days

The most familiar pest and potential menace is the common housefly *Musca domestica*, which tends to rest quietly on walls and ceilings when it is not feeding. The lesser housefly *Fannia canicularis* circles restlessly, especially under suspended objects such as light fittings.

Females of *Musca domestica* lay up to 2000 eggs during a normal life of about three months, in batches of more than 100 at a time. At moderate room temperatures the maggots hatch out in 10-14 hours — sooner in very warm weather. They feed for at least three days, growing to about the length of a fingernail before the pupal stage of transformation starts. Adults emerge from their pupal cases three to six days later, full-sized and ready to fly.

In warm conditions a female can lay eggs within a fortnight of having been one herself. Anywhere in Australia, many generations of houseflies are reproduced each year.

KEEPING YOUR HOME FREE OF FLIES

Screening of doors and windows is the best way of keeping out flies. If that is not possible, many types of space or surface sprays will give you some temporary control — but warnings and directions on the containers must be read and followed carefully.
To avoid disease and discourage the breeding of flies:
- Keep all food and drink covered.
- Wrap uneaten food and other kitchen wastes immediately.
- Make sure that garbage bins have tight-fitting lids.
- Wrap and remove animal droppings from gardens and yards.
- If you make garden compost, do it in a closed container.

A SMALL MERCY: MAGGOTS ARE CANNIBALS

Flies may lay their eggs communally when they find a good site. The first to start laying seems to stimulate others to take positions nearby. The object may be to ensure a food supply for at least some of the young — for maggots are cannibals that readily eat their own kind.

Maggots are also preyed on by wasps, and parasitised by some other insects and worms that lay eggs in them, eventually killing them. And although the flies' eggs need dampness to hatch, maggots drown in a saturated breeding site. But in suitable conditions they are prolific: scientists have raised more than 100 bushfly maggots at once in one human stool.

A maggot is legless and virtually headless, with a breathing system that opens at both ends of the body. At the narrower end, a feeding siphon is accompanied by one or two hook-like appendages that break up food and can be dug in to help the maggot crawl along.

Moulting their skins twice as they grow, maggots work their way out to a drier position near the surface or the edge of their breeding site. The second skin, though loose, is not shed. It hardens to make a barrel-shaped case. Inside it the maggot becomes a pupa, quiescent during the metamorphosis that turns it into a fully developed, adult fly.

Without claws or teeth to break out of the case, the fly uses a balloon-like sac that develops in front of its head, late in the pupal phase. Blowing into the sac, it inflates it so that the case is forced open.

Eggs can hatch into maggots within hours. Maggots feed on decaying organic matter, particularly garbage and faeces. The maggots of some blowflies also attack living flesh.

Scanning electron micrographs of the mouthparts (right) and foot (below) of a housefly. Regurgitated saliva and stomach juices liquify food before it can be sucked up by the spongy pad on the end of the sucking tube. 'Fly-spots' seen on light-coloured food are not flies' excreta but dried patches of material from their guts. The feet of the housefly have two tarsal claws above a pair of pad-like pulvilli and a projecting central lobe, the empodium. Houseflies secrete an adhesive substance onto the pads that enables them to cling to smooth surfaces.

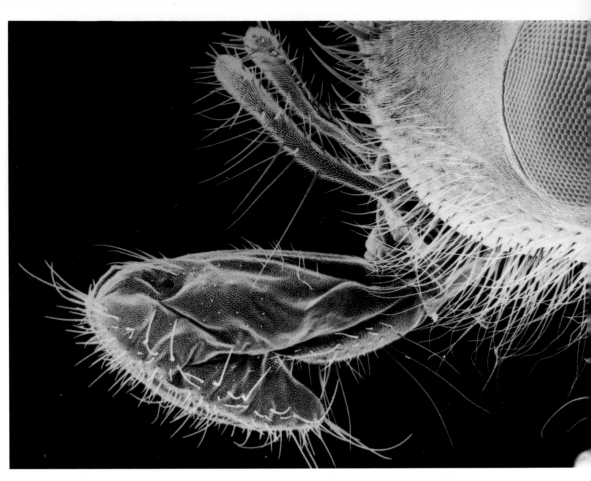

SPOILER OF THE GREAT OUTDOORS

Bushflies have never had it so good. In the country, agriculture gives them ample breeding sites in livestock dung. In town, a suburban fondness for outdoor eating rewards their summertime invasions with liberal supplies of accessible food. They rarely enter buildings, but do not have to. The ever-increasing bushfly population, coupled with generally relaxed standards of food hygiene at picnics and barbecues, presents a mounting health risk.

Musca vetustissima is exclusively a dung-breeder. It could not have thrived on the hard little pellets dropped by marsupials — it shuns them. So there is little doubt that its ancestors arrived with early human immigrants, relying on their faeces. Later the droppings of dingoes would have helped. But the explosion of bushfly numbers came with European pastoral agriculture. In the absence of specialised dung beetles, the flies have first rights to millions of tonnes of sheep and cattle pats a year.

The bushfly is closely related to the common housefly. Though slightly smaller and marked with light patches, it is essentially very little different. But because bushflies do not shelter in buildings there is no wintering-over in cooler parts of the country — they vanish for up to six months of the year.

Great migrations from the north

For generations in southern Australia, the mystery was not where flies went in winter but where all the extra ones came from in summer. No bushfly breeding sites with dormant maggots could be found. Nor were the flies ever seen emerging in spring. Towards the year's end they just arrived, in numbers sufficient to blacken the backs of pedestrians on the street corners of Melbourne, Adelaide or Canberra.

A long study by a CSIRO team in the late 1960s found the answer. Bushflies migrate, borne by hot winds from the north. One route was traced all the way from Cunnamulla, Qld, to Canberra — nearly 900 kilometres. A few flies make the whole journey, taking about seven weeks and arriving in spring. But eggs are laid on the way. It is the progeny that mounts the summertime onslaught.

Why bushflies cling to people

The bushfly's habit of resting and riding on people's backs is sometimes put down to a thirst for sweat. While it is true that flies are attracted to body secretions, it is unlikely that sweat offers anything like the refreshment and nourishment that they could get from many other sources. Researchers believe that bushflies are simply conditioned to stay with humans —

the original source of faeces for breeding — and they make for the place where movements are least disturbing.

When people are not moving, bushflies tend to cluster more round their eyes, nostrils and mouths. Again there is not normally much nourishment to be had — unless they find sores or wounds. Ridges on their lapping pads can be used to rasp at scabs, opening them and promoting an ooze of blood or serum.

The bushfly's potential for causing and spreading disease, especially among infants and people who are already in poor health, is enormous. It is a leading contributor to eye infections — the worst of which, trachoma, causes blindness in hundreds of outback Aborigines. The germ, *Chlamydia trachomatis*, is chiefly spread in water and by touch, but flies undoubtedly play their part.

Clusters of bushflies.

Bushfly Musca vetustissima

331

THE BLOW-IN BLOWFLIES

Most blowflies breed naturally in the flesh of dead animals, or pester livestock. But if winds blow them near settled areas, some may invade buildings. These casual visitors have a stouter build and brighter colours than houseflies. The best-known are species of *Calliphora*, in particular the blue-bodied *Calliphora auger* — sometimes called a bluebottle.

The usual target of the semi-domesticated blowflies is raw meat, though some will 'blow' woollen fabrics with their eggs. Some comparatively innocent species, known as cluster flies, simply seek shelter and warmth in buildings. Groups of them gather in darkened areas — often the same places every autumn. Cluster flies lay their eggs in earthworms.

The threat to livestock

Calliphora species, along with the introduced sheep blowfly *Lucilia cuprina*, distinguished by its bronze colouring, are more important to the agricultural economy than they are as agents of human disease. They initiate the 'strike' against livestock with damaged skins, which is followed up by other blowflies and may lead to the death of maggot-infested animals.

Blowflies of the Calliphora *species (above and below right) and a greenbottle of the* Lucilia *species (below left).*

Blowfly eggs are fast-hatching. In hot weather they may even hatch prematurely in the female's body, so that live maggots are laid. A closely related group, the fleshflies, always bear their maggots live. Some blowflies and fleshflies, in full flight, can 'bomb' meat with newborn maggots.□

Food infections spread by flies

Organisms responsible for dozens of viral, bacterial and parasitic diseases can be carried by flies. They include some fearsome afflictions.

Some of the diseases carried by flies are unknown in Australia. Others are controlled, if not entirely stamped out, by routine immunisation and good sanitation.

Most of the persistent problems stem from food and water contamination. They range in severity from typhoid fever and infectious hepatitis to mild gastroenteritis, and they include the bacterial infections of the digestive system that are commonly called food poisonings. Outbreaks are usually the result of individual negligence. But if general standards of hygiene were to slip, flies would become the agents of alarming epidemics.

Typhoid and the unknowing carriers
Typhoid fever was the principal killer of miners during the goldrush era, and it continued to take hundreds of lives a year in bush camps and city slums until well into this century. About 20 cases are reported in most years now. Nearly all are successfully treated with antibiotic drugs. Vaccination, effective for about three years, is easily obtained.

The bacillus *Salmonella typhi* is active only in humans. Germs multiply in the intestines and are excreted in faeces and urine. They can be passed on in contaminated drinking water and sometimes in shellfish that grow in heavily polluted water. And they are easily carried from excreta to food by flies, or by infected people with dirty fingers.

Typhoid fever, setting in after 10-14 days of incubation, is marked by abdominal pains but an absence of vomiting. Patients have constipation at first, rather than diarrhoea. In the second week a rash of pink spots appears. Recovery may be rapid in mild cases, but the fever can last for up to two months. The intestinal wall may be perforated or blood vessels ruptured. Untreated, more than 10 per cent of victims die.

A sinister feature of typhoid — especially in mild cases that are not diagnosed — is that about 3 per cent of recovered patients keep the bacteria. They harbour them in their gall bladders and go on excreting them. If they are careless of personal hygiene and handle food for other people, they can start epidemics while in perfect health. Paratyphoid, transmitted in the same ways as typhoid and producing similar symptoms, is less severe in its effects but still a serious disease. The bacillus responsible, *Salmonella paratyphi*, is less likely to be carried unknowingly by people who recover.

Other species of *Salmonella*, along with various bacteria of the *Shigella*, *Camplyobacter* and *Staphylococcus* groups, can cause less feverish gastroenteric illnesses marked by persistent vomiting, diarrhoea and abdominal cramps. While flies and unhygienic food handling are largely to blame, some of these bacteria occur naturally in animals. People can be infected by eating undercooked food. Severe gastroenteritis

continuing for more than three days should have medical attention — it can lead to dangerous dehydration and shock.

Amoebas at home in our bodies
Flies and poor hygiene inflict a parasitic amoeba on a surprising number of Australians. Studies suggest that as many as 4 per cent of the population could be infected. Most are unharmed, because *Entamoeba hystolica* finds its natural home in the large intestine, usually eating food particles and bacteria.

But in consistently unsanitary conditions, especially in tropical regions, people who are heavily and repeatedly infected contract a severely debilitating disease called amoebiasis. The amoebas attack intestinal tissue, producing ulcers, and some attack the liver, causing painful abcesses and fever. The most prominent symptom is persistent diarrhoea including blood and mucus — amoebic dysentery.

Another single-celled parasite, *Giardia lamblia*, causes diarrhoea that is less severe but may last for weeks if not treated. It is accompanied by nausea and an uncomfortable distension of the stomach. Giardiasis, amoebiasis and the bacillary form of dysentery — caused by *Shigella* bacteria — are all treatable with drugs, and all preventable by good sanitation and by keeping flies off food. □

Flies are dangerous because they move from rotting garbage (top) to food (above), to faeces and to people indiscriminately, transferring infective organisms as they go. Houseflies and blowflies also lay their eggs in meat or decaying matter wherever they find it.

TYPHOID FEVER

EARLY SIGNS Rising fever. Headache, mental confusion, abdominal pain.

ACTION Isolate patient and take special care in disposing of body wastes. Obtain medical attention.

Cunjevoi *Alocasia macrorrhiza*.

Part 6

PLANTS WITH
HARMFUL PROPERTIES

*The kingdom of plants is a peaceful one —
but it relies on an ancient chemistry that
is often at odds with our own. Safety, in the
main, is simply a matter of recognition.*

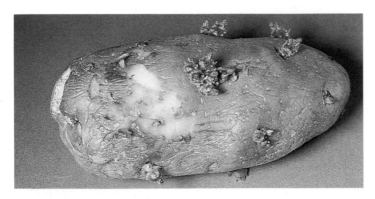

Above and left **POTATO** *Solanum tuberosum*. Any greenness indicates solanine, a poisonous alkaloid. *Below* **RHUBARB** *Rheum rhabarbarum*. Leaves are poisonous.

PROBLEMS IN THE PANTRY

Some common household foods contain poisonous substances.

Among commonly eaten vegetables and fruit, potatoes pose the most frequent possibility of ill effects. Any greenness in or under the skin indicates the presence of solanine, a poisonous alkaloid. But there is no need to throw a whole potato away. If the green part is cut off — along with any sprouting shoots — the rest of the tuber is safe to cook. Stems, leaves and berries of the potato plant, however, should never be eaten. They always contain solanine.

Rhubarb stems get their delicious tartness from two harmless acids. But a third, the potentially deadly oxalic acid, is concentrated in the leaves. The leafy tops of beetroot must be discarded for the same reason. All members of the brassica family — cabbages, cauliflower, broccoli and so on — have an irritant oil in their seeds. The leaves, if regularly eaten raw in substantial quantities, are capable of causing stomach pains and blood disorders.

Kernels of stone fruits such as peaches, plums, apricots and cherries may contain a little hydrocyanic acid — often called prussic acid. Swallowing the odd one does no harm, but children should be discouraged from making it a habit, and from breaking open stones to eat the kernels. Great care should be taken if almonds are picked from a tree. One variety is grown for its edible kernels. The other has kernels that are noticeably bitter — and extremely toxic.

Bulbs of onions and their relatives such as garlic and chives are delicious — but the bulbs and corms of all other garden plants should be treated with suspicion. They should never be left lying where children could find them and mistake them for edible species.

Food or foe?

Plants are the mainstay of human survival and progress. But many should never be eaten — and some not even touched.

Some 2000 plant species growing in Australia are capable of having harmful toxic effects on us. Awareness of most of them came out of misfortunes in livestock grazing. Relatively few are normally encountered by people, and fewer still are likely to suggest themselves as food.

Serious poisonings are in fact rare in this country. They do not rank in importance with the allergic torments of hay fever and contact dermatitis victims, to whom a whole range of plants are enemies. And cases of plant poisoning do not occupy doctors nearly as much as the physical injuries suffered by people running into thorns or sharp branches, entangling themselves in vines and falling out of trees — or having trees fall on them.

In the main, the most dangerously toxic plants are too corrosive or taste too unwholesome to be eaten. But there are significant exceptions, and this section is mainly concerned with them. It offers a representative range of the more likely risks and the situations in which they arise. Given that people can recognise them, most are avoidable.

Many edible, palatable plants contain poisons in amounts so small that eating a normal quantity is harmless. Celery, for example, is toxic — but you would have to gorge yourself on kilograms at a time to be aware of it. Dangers can arise from obsessive 'health' diets and quack cures, however. People have died from persistently overeating the herb comfrey — now banned from sale. An American trying a cancer cure ate a whole cupful of apple pips at a sitting and was killed by hydrocyanic acid.

Some children acquire a habit of nibbling small parts of the same plant day after day. Toxins too unpalatable to be taken in dangerous quantities all at once can be accumulated in this way, perhaps to damaging levels. Very often the plant involved is a home garden species. So if parents notice such behaviour and are worried, the remedy may be in their own hands.

Plant chemistry still holds mysteries

There are no shared characteristics of taste,

smell or appearance that indicate plant toxicity, and there is no easy way of testing an unknown plant with certainty. Hundreds of different toxic substances have been found, but in many cases the exact nature of the poisonous principle remains unexplained.

The properties of plants are made even more diverse and unpredictable by variations in soil conditions or climate. Some species can be more or less poisonous in particular places, in certain weather or at different stages of growth. Breeding techniques and fertilisers can change the chemistry of plants. Industrial wastes, radioactive materials and some weedkillers and pesticides, applied carelessly or illegally, leave soil residues that can poison crops and may prove genetically damaging to humans.

Mineral elements such as copper, molybdenum and selenium, naturally present in some soils, can alone make plants toxic if they occur in unbalanced proportions. Many other minerals form poisonous inorganic compounds, simple in structure and nearly always nitrogen-based. Nitrates are not especially toxic but our systems reduce them to nitrites, which are.

The more significant plant toxins are highly complex organic compounds, always containing carbon. They include sugars, alcohols, resins, peptides, phenols and acids. Again the plant material may not be harmful in its natural state, before our digestive enzymes break it down. Deadly hydrocyanic ('prussic') acid, for example, comes from a sugar.

Foremost among the organic compounds, outnumbering all the others, is a bewildering range of different alkaloids. These in particular, along with their capacity for harm, have taken on a vital role in the preparation of many medicinal drugs.

Tropane alkaloids and steroids are the best known, not only for their medicinal uses but also unfortunately for their abuses. Both kinds come mainly, though not exclusively, from the big plant family called *Solanaceae* — potatoes, tomatoes, tobaccoes, nightshades and scores of other wild and cultivated shrubs.

A striking fact about most of the solanum alkaloids is that they have greater impact on our systems than they do on livestock or pets. Short of directly killing people, some of them can induce a wild delirium. Several Australians have died through misadventures while their minds were deranged after drinking plant potions. And there is evidence that alkaloids in related plants, if deliberately taken in a similar way, would cause foetal deformities. □

Pollen from grass is the commonest cause of allergic rhinitis in summer.

BLOWING IN THE WIND

Legions of people suffer the torments of pollen-caused allergies.

Powdery pollens, borne on the wind, enable plants to cross-fertilise without the aid of insects. But they are also the leading cause of plant-related illness, surpassing all kinds of poisoning. In allergic people they provoke what is called hay fever — though it has nothing to do with hay and does not involve a fever. The formal name of the disease is allergic rhinitis. Victims suffer hours of red-eyed, runny-nosed misery, with sore throats, headaches and fits of sneezing. In severe cases the skin may erupt, and the worst allergies are linked with dangerous attacks of asthma.

The direct cause of the symptoms is histamine, which the body produces in reaction to pollen. They can be eased with anti-histamine preparations, and some cases are cured by long courses of desensitising injections. The problem is seasonal for most sufferers, who usually acquire the allergy in their teens and if they are lucky lose it in their twenties. Virtually any flowering plant can be responsible — even freshly cut flowers in a house. But grasses are the leading cause in summer, with trees causing more trouble in spring. Newspapers in some cities publish daily counts of the amount of pollen in the air, aiding allergic people to reduce the risk by keeping indoors at peak times.

Below are scanning electron micrographs of individual pollen grains. Each type of pollen has a distinctive shape, and the material of the protective spore coat is so durable that botanists have been able to use fossilised pollen grains found in ancient rocks to identify plant species that grew millions of years ago.

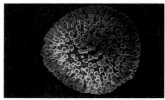
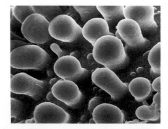

Above Franklandia triaristata. Detail of grain is shown right.

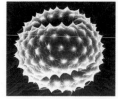
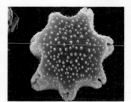

Calomeria amaranthoides.   *Nothofagus falcata (fossil).*   *Athertonia diversifolia.*

Left **ANGEL OF DEATH** or **DEATH CAP**
Amanita phalloides. This mushroom can reach
a height of 15 cm, with a flat or dish-shaped
cap of similar width. Its tapering stem is
enclosed at the base by a cup that projects
noticeably from the soil. The delicious
common field mushroom *Psalliota campestris*
(above) looks much the same. But mature
angels of death have a less pleasant smell,
suggestive of raw potatoes.

The angel of death

Blessedly rare so far, this newcomer accounts for more than 90 per cent of the world's fatal mushroom poisonings.

Australia was spared the world's most toxic mushroom until 1967, when *Amanita phalloides* turned up in Canberra. Known as the angel of death or death cap, it was probably imported from its native Europe in spore form, with tree seedlings. Since its first appearance here, more of the fungus has been found in Melbourne and the ACT.

The angel of death has none of the lurid colouring that some people take as the danger sign in mushrooms. In its general appearance it is unfortunately similar to the field mushroom *Psalliota campestris*. Sometimes the cap has a tinge of green, but more often it is white.

Overseas, this plant is blamed for up to 95 per cent of all fatal mushroom poisonings. A single full-grown specimen is enough to kill a person. Hours after a meal there is an onslaught of stomach pain with bloodied vomiting and diarrhoea. Even if these symptoms ease, in all but the mildest cases irreversible damage to the liver, kidneys and nervous system has already set in. Death may follow in two days, or take more than a week.

Amanita phalloides grows singly or in groups among tree roots in cooler areas. In Australia it is likely to be associated with introduced trees — especially oaks. In the interests of public safety, sightings should be reported to health authorities or to officials of a nearby botanic gardens or parks and reserves department.

Deadly dining — for houseflies

Before the angel of death was detected, Australia's most widely feared mushroom was the related *Amanita muscaria* — fly agaric. Its name came from its use in Europe to kill houseflies. Plants were broken up and the pieces left out in saucers of milk. Fly agaric poisoning is unlikely to threaten human life, but it causes severe gastro-intestinal upsets. Effects on the nervous system suggest wild drunkenness. This fungus is usually found in association with pine trees.

A third member of the group, *Amanita preissii*, has caused severe poisoning and collapse in Perth, where it grows widely in King's Park. Like the other two it was introduced. Australia also has native species of *Amanita* that are best avoided, though their toxicity is unproven.

The native mushroom that seems likeliest to

FLY AGARIC
Amanita muscaria.

CUPS ARE A WARNING SIGN

Any mushroom that is surrounded by a cup at the base of its stalk is likely to be toxic. The likelihood is greater if there are rings above the cup. Redness anywhere on a mushroom is also a warning of possible danger. But that does NOT mean that a mushroom showing no redness is safe.

If you pick a mushroom that has a cup at the base of its stalk like any of these, don't eat it.

cause violent illness, at least if it is eaten raw, is *Lepiota molybdites*. It is found in the east from tropical Queensland to northern NSW. *Psilocybe cubensis*, the so-called mad mushroom, grows in the same regions. Its effects resemble hilarious drunkenness but vomiting often follows. Possession and use of this mushroom as a drug are criminal offences throughout Australia.

Little can be said usefully about hundreds of other mushrooms in Australia because little is reliably known. In many cases, what is known of them here conflicts with their forbidding reputations overseas. Some may be far less poisonous than was supposed. But the risk of unpleasant illness from eating unknown types remains high. Anyone who is not absolutely sure that a species is harmless should refrain.

Mushroom or toadstool?

Some people use the word 'mushroom' only for species of *Psalliota* —the edible field mushroom and its close relatives. They call all others 'toadstools', whether they are edible or not. The distinction has no botanical point, but it can serve a purpose by instilling the idea — especially in children — that the less familiar fungi are different, and therefore risky.□

Right Amanita preissii. This imported mushroom has been found growing around Perth. *Below* **MAD MUSHROOM** or **GOLDEN TOP** *Psilocybe cubensis.* A native, this species grows mostly in groups and is found in coastal Qld and north-east NSW. It reaches 10 cm in height and has a ring of torn tissue around it just below the cap.

The pretty poisoners

Many toxic plants are cultivated for their flowers. A nibble won't hurt — but none of these should be eaten in quantity.

Although it is not 'dangerous' in the strictest sense of the word to grow these familiar, attractive flowers in your garden, it makes sense for parents to be aware of their harmful properties. Their very familiarity and attractiveness makes them all the more likely to be used as children's playthings, with the possibility of accidental or deliberate ingestion. Children should be discouraged from sucking nectar from flowers and from making 'tea' with leaves or flowers.□

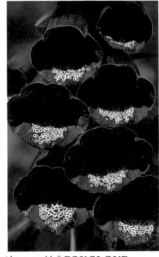

Above and left **FOXGLOVE** *Digitalis purpurea.* All parts, including the nectar, can cause stomach pain, diarrhoea and disturbance of the heart rate.

Above and below **SWEET PEA** *Lathyrus odoratus.* Seeds relied on as food can produce muscle wastage and bone abnormalities. So can the seeds of related *Lathyrus* species such as the everlasting pea and chickling vetch.

Above **GLORY LILY** *Gloriosa superba.* Eating any pa[rt] invites neurological disorders as well as dangerous gastric attacks. Grown in tropical Queensland.

Above and right **ARUM LILY** *Zantedeschia aethiopica.* The flower spike and surrounding white spathe cause dangerous stomach inflammation.

Above and right **CHRISTMAS ROSE** *Helleborus niger*. All parts can cause severe gastro-enteritis.

Right and above **MONKSHOOD** or **WOLFSBANE** *Aconitum napellus*. Dangerous respiratory distress will follow if much of any part is eaten. Toxins can also enter the bloodstream through cuts if the plant is handled.

Below and right **LARKSPUR** *Consolida ambigua*. Seeds and young plants of many delphinium species can cause gastro-enteritis and convulsions.

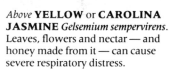

Above **YELLOW** or **CAROLINA JASMINE** *Gelsemium sempervirens*. Leaves, flowers and nectar — and honey made from it — can cause severe respiratory distress.

Above and right **LILY OF THE VALLEY** *Convallaria majalis*. All parts can cause severe vomiting and diarrhoea.

Above and right **GOLDEN DEWDROP** *Duranta repens*. Berries cause fever, accelerated heart rate and convulsions.

Above **COMMON** or **ENGLISH YEW** *Taxus baccata*. Seeds and leaves cause gastro-enteritis and respiratory distress.

DAPHNE *Daphne odora*. Berries cause swelling of the lips and tongue, vomiting and collapse.

Toxic ornamentals

Introduced trees and shrubs, prized for their distinctive forms and colours, bring hazards to the home garden.

Children are most at risk from these attractive trees and shrubs. Youngsters find their bright berries and seeds very tempting, and may try to eat them. Adults are far less likely to experiment.

Although you don't want to make your children fearful of nature, making them familiar with the poisonous properties of these plants could help to save their lives.☐

Above **PHYSIC NUT** *Jatropha curcas*. Seeds (below, left and right) cause severe gastro-enteritis. Related trees, often called coral plants, carry a similar risk.

Above **ANGEL'S TRUMPET** *Brugmansia candida*. Seeds, leaves or flowers can induce delirium, convulsions and violent behaviour.

STRIPED DUMBCANE
Dieffenbachia amoena.

SPOTTED DUMBCANE
Dieffenbachia maculata.

HOUSE PLANT WAS USED FOR TORTURE

Chewing the leaves or stems of this plant produces severe, burning pain in the mouth.

The popular indoor plant dieffenbachia, two varieties of which are pictured left, should be kept out of the reach of crawling babies. It earned its common name, dumbcane, when it was used to punish slaves in the West Indies. Forced to chew the leaves, they suffered an intense burning and swelling of the lips, mouth and tongue and were unable to speak for days.

Cats with a habit of nibbling dieffenbachia leaves have been killed. A similar problem among cats — for some reason they are usually Siamese — is often caused by philodendrons.

So many ways to kill

Oleanders, lining streets and decorating gardens, are by far our most dangerous plants. Even burning them is hazardous.

The two kinds of trees or shrubs that are called oleanders are not related. They grow in noticeably different ways. What they have in common are toxins that not only produce violent stomach reactions but also attack the central nervous system and the heart. Both are versatile in their ways of poisoning.

Children have died after swallowing leaves, flowers or seeds. But oleanders do not have to be eaten. Leaves dropping into drinking water, or the fumes of the plants being burnt, or twigs being used as drink stirrers or toothpicks can cause severe illness.

Nerium oleander, the more widespread type in Australia, is from the eastern Mediterranean and has blossoms that range in colour from white to reds and pinks. It bears narrow pods full of small seeds. It takes a shrubby form with slim stems branching out from near the ground.

Above **YELLOW OLEANDER** *Thevetia peruviana.* Flowers are always bright yellow. Sometimes it forms a rounded shrub about 3 metres tall but trees reaching 10 metres are not uncommon.

Right Samantha Oakley, a Brisbane 5-year-old, ate some yellow oleander flowers 'because they were pretty'. After two days in intensive care, she made a full recovery.

The yellow-flowered *Thevetia peruviana* from tropical America — called yellow oleander or daffodil tree — likes tropical or subtropical conditions. It has a fleshy fruit containing a stone with twin kernels. Fruits are usually green but blacken when they are fully ripe — and most dangerous.

Both kinds thrive in virtually any soil and resist drought. Attractive and so easily grown almost anywhere in Australia, oleanders are made more dangerous by their popularity. They line many urban streets and brighten countless home gardens and public parks. Parents may consider it wise to get rid of them from their own gardens. But risks arise when children go elsewhere to play. They should be taught at an early age to recognise oleanders and shun them.

Symptoms of oleander poisoning can take some hours to develop. Severe gastric effects are

OLEANDER *Nerium oleander.* Flowers vary from white through pink to dark red. The shrub seldom grows to more than 2 metres high but it can grow to 6 metres.

seen at first, but the major harm is to the functioning of the heart muscle. During the early part of the illness the heart rate is slowed. Later it becomes abnormally fast, with convulsions and breathing distress.

Tracking the poisoned battalion

Many books on plants and poisons include a mention of 300 soldiers who ate meat roasted on oleander skewers. All were poisoned and some were killed. Newspapers and magazines seize on the tale as if the incident were recent. In fact it goes back through a succession of old texts and medical treatises to Pliny the Elder, a Roman historian of the 1st century AD. And he was relying on an even earlier account — probably from a campaign of Alexander the Great, some 330 years BC.

Australia has no need of ancient history to highlight the danger of oleanders. It has stories enough of its own, some of them tragic. In a recent Brisbane case a three-year-old girl was sent to hospital after a prolonged bout of vomiting and stomach pain. Her heart failed in the ambulance and she could not be revived. Doctors later found a concentration of oleander toxin in her heart tissue that they calculated could have come from eating two leaves. Earlier that day the little girl had been playing under a yellow oleander in the backyard of her home.

Victims treated in time, however, stand an excellent chance of recovery with modern techniques of intensive care. Aberrations of the heartbeat can be corrected with drugs, the chief one of which is atropine — another plant poison, extracted from the deadly nightshade *Atropa belladonna*. □

Oleanders line streets and park pathways all over Australia. Their popularity makes them all the more dangerous.

Kitchen heater was a knockout

Beating the cold on a winter night almost cost a Sydney family their lives.

One winter night in 1983 an elderly Sydney couple, giving dinner to their daughter-in-law and grandson, brought a homemade metal firebox into the kitchen. They lit a heap of lopped oleander branches, collected from a nearby cemetery.

Within half an hour the daughter-in-law and her son were ill, vomiting and writhing on the floor with stomach pains. They seemed to recover. But after a further hour or so all four were stricken. The grandfather managed to attract the attention of neighbouring relatives before collapsing.

A doctor found the four unconscious in different rooms, with the oleander branches still smouldering and filling the house with their toxic fumes. The family were resuscitated in hospital — where one of the policemen who were called to their aid also required treatment.

A year previously, in another Sydney suburb, a four-year-old boy was dared by an older playmate to smoke a 'cigar' of rolled-up leaves. Their street was lined with oleanders. The boy took a few puffs. That night he suddenly screamed, clutched his stomach and began rolling in agony on the floor of his home. When the pain eased his parents questioned him about what he had eaten. Not until the following day did they discover the true cause and rush him to hospital. He made a full recovery. □

OLEANDER POISONING

EARLY SIGNS
Vomiting, stomach pains, trembling. Dilated pupils; slow pulse; perhaps redness around mouth.

ACTION Seek urgent medical attention.

Right **GREVILLEA**
'Robyn Gordon'.

SCARLET RHUS *Toxicodendron succedanea*, also known as *Rhus succedanea*.

MANGO *Mangifera indica.*

A blistering attack

Just touching certain plants inflicts an ugly torment on sensitised people — and the list of hazards is growing.

Minutes after clambering into the prettiest tree in his backyard, a Sydney four-year-old was covered from head to foot with an itching rash. Soon his whole body was swollen and blistered. A week later he was still pitifully sore in bed, swaddled in damp dressings and dosed with steroid drugs.

The tree is gone now, and the little boy has learned to be careful when he plays elsewhere. He was yet another urban victim of plant-contact dermatitis. The cause was the scarlet rhus, a favourite of southern gardeners because of its richly coloured autumn foliage. It seems to be giving more trouble every year — a reflection of increasing suburban populations, and the fact that some people's sensitivity is developed over several seasons.

The scarlet rhus *Toxicodendron succedaneum*, also known as the Japanese wax tree or sumach, is a close relative of poison ivy *Toxicodendron radicans*. It exudes resinous substances potent enough to cause severe skin eruptions even at second hand — from particles clinging to washing that blows against a tree, for example, or in the fur of pets that rub against it.

Scarlet rhus belongs in a varied plant family called the anacardiaciae, many of which are notorious for causing dermatitis. Other introduced trees that are commonly cultivated here and often give rise to skin problems are the mango *Mangifera indica* and the cashew *Anacardium occidentale*.

A native member of the family, the tar tree or marking nut *Semecarpus australiensis*, grows in far northern Queensland and is sometimes planted as a coastal shade tree — but visitors should be wary of its black, dripping resin and not choose it to camp or picnic under.

Many otherwise innocent garden plants — especially if they touch the face — raise blisters on some unfortunate people. Offenders include chrysanthemums, geraniums, primulas and tulips in the flower bed, and carrots, celery, cucumbers, lettuce, onions and tomatoes in the vegetable patch.

Horticultural science, developing new plant varieties and devising hybrids that do not occur in nature, adds to the risks. A recent example is the artificially hybridised grevillea 'Robyn Gordon', flowering year-round and enormously popular since the 1970s in home gardens and public parks. It is proving extremely troublesome — though neither of its parent species was. Doctors are especially worried because once a sensitivity to a plant is established it extends to the parent plants and other relatives.

If a particular cause of plant dermatitis is recognised, and not confused with a food allergy or irritation by a substance such as washing powder, contact can be avoided. Should there be a mishap, the effects can be alleviated by antihistamine creams or tablets. □

CASHEW NUT *Anacardium occidentale.*

TAR TREE or **MARKING NUT**
Semecarpus australiensis.

POISON IVY *Toxicodendron radicans.*

CORROSIVE SAP ATTACKS THE EYES

Intense pain and temporary blindness, lasting for days in severe cases, can result if the sap of some plants gets into the eyes. Delicate membranes are damaged by corrosive chemicals. Among common plants with which care must be taken are many species of euphorbia, including the poinsettia *Euphorbia pulcherrima* and naked lady *Euphorbia tirucalli*. The sap does not have to be in liquid form — dried traces rubbed into the eyes with the fingers can also be harmful.

Along tidal streams throughout the subtropical and tropical regions, the milky mangrove *Exoecaria agallocha* has a bad reputation for the same reason. The bushman's name for it is 'blind-your-eye'. A widespread shrubby scrambler, the caustic bush or vine *Sarcostemma australe*, has a milky sap that can irritate the skin and cause blisters.

These and other plants with corrosive saps also have violent gastric effects if they are eaten — but they cause a burning sensation in the mouth so quickly that they are most unlikely to be swallowed.

CAUSTIC BUSH or **CAUSTIC VINE** *Sarcostemma australe.* Flower and milky sap (above); whole plant (below).

NAKED LADY *Euphorbia tirucalli.* Sometimes called **RUBBER PLANT** or **PENCIL BUSH.**

POINSETTIA *Euphorbia pulcherrima.*

MILKY MANGROVE *Excoecaria agallocha.* Whole tree (above); leaves and flowers (right).

Right and far right **DEADLY NIGHTSHADE** *Atropa belladonna*. Berries and other parts cause nausea, blurred vision and muscle weakness. Bigger doses can cause breathing distress and death from heart failure.

Above **COW'S UDDER PLANT** *Solanum mammosum*. Unripe fruits and leaves cause gastro-enteritis.

Below and right **APPLE OF SODOM** *Solanum hermannii*. Fruits cause headache, dizziness and gastro-enteritis. Many other *Solanum* species are poisonous.

Above and below **CASTOR OIL PLANT** *Ricinus communis*. A few seeds (below) can cause severe gastro-enteritis, convulsions and collapse. Only one or two seeds need be eaten to make a child severely ill.

Above and right **THORNAPPLE** *Datura stramonium*. All parts can cause delirium and convulsions. *Left* **HAIRY THORNAPPLE** or **HAIRY ANGEL'S TRUMPET** *Datura metel*. Has purple or lavender flowers and is equally dangerous.

Weed worries

More than just nuisances, some weeds are dangerously poisonous. These should rank high on any gardener's hit list.

Gardeners are likely to know about any cultivated plants with harmful properties growing in their garden. They are less likely to be aware of similarly poisonous weeds, springing up unbidden in neglected corners. And all parents should know about the kinds of toxic weeds children may come across on waste land, in paddocks and along verges.□

Above **LANTANA** *Lantana camara*. Unripe fruits may cause gastro-enteritis, weakness and breathing difficulty. Flowers differ in colour and form; all varieties should be treated with suspicion.

Above **HEMLOCK** or **CARROT FERN** *Conium maculatum*. All parts can cause muscle weakness, breathing distress and stupor. Often mistaken for parsley, which it resembles. *Inset:* Hemlock has a hollow stem (right), parsley a solid one (left). Children have been poisoned by using the hollow stems of hemlock as whistles or pea-shooters.

Above **GREEN POISONBERRY** *Cestrum parqui*. Fruits (glossy black when ripe) cause gastro-enteritis and fever, sometimes leading to paralysis. The orange-flowered poisonberry *Cestrum auranticum* is also toxic.

Bush temptations

Some native trees bear enticing fruits that are strictly for the birds. Our systems can't handle their potent toxins.

If you are tempted to sample an attractive-looking fruit or berry while out bushwalking or camping, DON'T — unless you know without doubt that the plant is harmless. The range of poisonous and non-poisonous plants is vast, and it is difficult for anyone who is not an expert to remember which species is poisonous and which isn't, or even to differentiate between species. Accidental poisoning of adults is usually due to wild plants being mistaken for edible species.□

WHITE CEDAR *Melia azedarach australasica*. Fruits cause gastro-enteritis, drowsiness and convulsions. Rainforests, northern NSW and Qld. Grown as a shade tree in WA and SA, under the name Cape lilac.

SICKENING LESSON FOR EXPLORERS

Cycad seeds look edible but most are highly poisonous.

Cycad seeds and stems were a rich source of nutrition for many Aboriginal tribes. Early European visitors followed their example without realising that the plants needed careful preparation and cooking. Violent illness struck Vlamingh's Dutch party at the Swan River in 1698, Cook's men at the Endeavour River in 1770, and French sailors under La Perouse at Botany Bay in 1788.

Australia has about 30 species of cycads — cone-bearing plants that shared dominance with pines and other conifers for more than 100 million years before flowering plants evolved. Most have been shown to be highly toxic and all are suspect. The three main groups, *Cycas*, *Lepidozamia* and *Macrozamia*, have a fairly similar palm-like form with prominent seed cones that sometimes look like pineapples. *Bowenia* cycads, found in limited areas of tropical Queensland, have foliage like a coarse maidenhair fern.

COMMON BURRAWANG *Macrozamia communis*.

SAGO PALM *Cycas revoluta*.

GIANT BURRAWANG *Lepidozamia peroffskyana*.

FINGER CHERRY or **NATIVE LOQUAT** *Rhodomyrtus macrocarpa*. Toxin in fruit attacks the optic nerve and has been blamed for many cases of permanent blindness. Rainforests, northern Qld.

RED NUT *Triunia youngiana*. One bite of a seed can cause severe gastro-enteritis and lowered heart rate. Rainforests, northern NSW and Qld.

SURVIVAL ON UNKNOWN PLANTS

Anyone forced to choose between strange plants and starvation can take some precautions against poisoning.

If you are stranded in remote country, you may be forced to live off the land, and that will mean trying to survive mostly by eating plants. (Some advice about eating wild animals is given in Part 4 of this book.)

In an emergency, when the risk of making yourself ill is outweighed by the risk of starving, follow these simple rules and techniques about eating unknown plants:

- DON'T be guided by what native birds eat. Their systems cope with poisons that are deadly to us.
- Crush the soft parts of plants growing plentifully around you and smell each of them. Choose the least pungent.
- Rub a little on the tender skin inside your upper arm.
- If the skin doesn't itch or redden, try a tiny amount on the tip of your tongue. Wait for a few minutes. If you detect any stinging, burning sensation or a strong bitter taste, discard the plant.
- If you detect none of these, chew a slightly larger piece of the plant, gargle, and spit it out.
- If your mouth and throat remain comfortable, eat the least amount — cooked if possible — that is needed to survive. Wait six hours to make sure there are no severe gastric effects before having any more.

TIE BUSH *Wikstroemia indica*. Fruits cause severe gastro-enteritis and have been blamed for the deaths of two children. Open forests, northern NSW to NT.

CRAB'S EYE, GIDEE-GIDEE or **JEQUIRITY BEAN** *Abrus precatorius*. Chewing one seed can cause severe gastro-enteritis, muscle weakness and accelerated heart rate, leading to circulatory failure. Deaths have been reported. Coastal heathlands, tropical WA, Qld and NT.

CUNJEVOI *Alocasia macrorrhiza*. All parts of the plant cause pain and swelling of the mouth and throat if chewed. Rainforest gullies, NSW and Qld.

Trees that sting

They mount a fearsome guard on our eastern rainforests. A brush with the leaf hairs brings agony in seconds.

Above **STINGING** or **SCRUB NETTLE** *Urtica incisa.*

Below **SHINING-LEAFED** or **MULBERRY-LEAFED STINGING TREE** *Dendrocnide photinophylla.*

Bushwalkers who blunder into stinging trees never forget the experience. Stabbing pains of maddening intensity strike within seconds. If the stinger is Queensland's notorious gympie bush, a more general pain lasts for days. Even a month or two later, washing can be an agonising ordeal.

Stinging trees of at least four related species grow in and around the rainforests of NSW and Queensland. Young plants — the most dangerous — are common as regrowth where forests have been cleared, or damaged by storms. They are the bane of forestry: Queensland workers don gloves and respirators and carry antihistamine tablets if their tasks take them among stingers.

The oversized leaves of young stinging trees bristle with rigid, hollow hairs. They have brittle tips of silica that snap off if they are touched, lodging in the skin and releasing a potent combination of toxins. Chemists have identified many components, but they are not sure which are responsible for the drastic sequence of reactions.

A faint itch is quickly followed by a prickling sensation mixed with jabs of excruciating pain, radiating from the site of contact. In a gympie bush stinging the most intense pains are replaced before long by a sharp, widespread tingling and localised soreness, well away from the point of stinging. If a hand is stung, for example, the other arm or the throat or face may become painful — especially if touched — within five minutes.

Pain comes back with cold

At the same time tiny red spots appear at the injury site, merging to form an angry swelling that persists for hours and sweats profusely. Pain often starts in the armpits and groins, going on for days after the other symptoms have disappeared. But rubbing, contact with cold water or the onset of cold weather can bring a return of severe pain to the stung area at any time in the following six or eight weeks.

Antihistamines ease some of the symptoms of a bad stinging but no antidote is known. Cures suggested in bush lore help some people; others find them useless — or worse than the sting. Victims of a limited stinging can expect to make a complete recovery. The outlook for people who are extensively stung — by falling into a tree, perhaps — is less certain. Pioneer bushmen dreaded the possibility of being

Above **GYMPIE BUSH** *Dendrocnide moroides.*

Above and right **GIANT STINGING TREE** *Dendrocnide excelsa*. Mature trees (right) can reach 35 metres; young trees (above) are more commonly seen as dense shrubs.

thrown from their horses in stinger country. Deaths are mentioned in some old stories, and two soldiers are said to have been killed in training during World War II. In a more recent case a man collapsed unconscious after falling into a stinger.

The gympie bush *Dendrocnide moroides* grows mainly in Queensland, north from the town of Gympie opposite Fraser Island. Some are found in the Tweed Valley–Lismore region of northern NSW. Most specimens are fairly spindly shrubs, though they can exceed 4 metres in height. *Dendrocnide cordata*, of similar appearance and probably just as toxic, has been found only on the Atherton Tableland, near Cairns. The other two stingers are more imposing as trees, but their painful effects are less severe and prolonged. The giant stinging tree or fibrewood *Dendrocnide excelsa* has a range from southern Queensland through NSW to the Victorian border. Mature trees reach 35 metres, but it is younger plants growing as dense shrubs that cause trouble. *Dendrocnide photinophylla*, the shining-leafed or mulberry-leafed stinging tree, has a similar stature. It is found from southern NSW to far northern Queensland.

Stinging nettles: sensitivity varies

Cooler regions of Australia have a native nettle, *Urtica incisa*, as well as the common weed *Urtica urens*. Both yield chemicals that are apparently identical with the toxic principles in the *Dendrocnide* species. Their stings cause itching, swelling and a burning sensation but for most people these effects are short-lived. A minority of victims seem to have a special sensitivity and may suffer intense discomfort for a day or two. Antihistamines ease the symptoms.☐

Part 7

FIRST AID IN EMERGENCIES

Could you help? Or would you panic? Major woundings, venomous attacks and poisonings are sudden and frightening — and often far from the reach of medical services. Presence of mind and a little knowledge can buy a lot of time.

Bushwalkers in leech-infested rainforest.

Poisonings

Few are life-threatening, but knowing what to do if a poisoning occurs can prevent needless suffering.

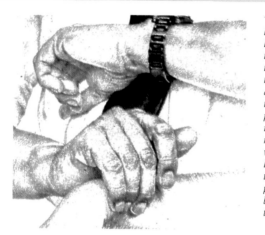

The wrist pulse (left) can be felt about 25 mm below the base of the thumb and 10 mm from the edge of the arm. Place three fingers on the pulse and press slightly. Time the beats. The carotid pulse (below) can be felt in the groove between the neck muscle and the windpipe. Press on it lightly with the flats of two or three fingers. The pulse beats on both sides, but feel only on one — the side farthest from you.

Monitoring the pulse rate is essential in cases of bites, stings and poisonings.

Taking a pulse is a simple way of checking the rate of heartbeat. A pulse that is abnormally fast or slow is an indication that a venomous bite or sting or a poisoning is having a serious effect. Keep a frequent check on it, even if you think a victim is recovering. If the pulse stops, resuscitation by external cardiac compression (page 357) must be started immediately.

The normal pulse rate for adults at rest is 60-80 beats a minute — just over 70 on average. It is slower in old people and faster in children. A young child's normal rate could be 90-100 beats a minute, and a baby's as high as 140.

Using a watch if possible, or estimating one-second intervals if you have to, feel the pulse for 30 seconds and double the number of beats that you detect.

Few poisonings from eating animal or plant substances are dangerous. A brief bout of vomiting and diarrhoea is seldom any cause for concern. But if a gastric attack goes on for days, or if violent illness leads to fainting or loss of muscular control, call a doctor.

Make sure that an unconscious victim's breathing passage is not blocked by the tongue or by vomit. Be prepared to use resuscitation techniques (pages 356-357) if breathing stops or the pulse fails.

Do NOT try to induce an unconscious person to vomit. A fully conscious patient can be helped to vomit by putting two or three fingers over the tongue and pushing them well down the throat — but only if you are certain that the poison was not corrosive. Some plant poisons are, and their return can do further damage.

Victims of corrosive poisons usually show signs of burning in and around the mouth. Provided they are conscious they can be given sips of water or milk to dilute the poison in their stomachs. The pain of a burnt mouth can be eased by rubbing butter or vegetable oil on the lips, tongue and the inside of the cheeks, perhaps following it up with some icecream.

If you find a child eating something that you know is poisonous, but not corrosive, induce the child to vomit as a precaution and consult a doctor. Nearly all poisonous plant matter is unpalatable and children rarely eat enough of it to suffer serious effects. But urgent medical attention should be summoned if a child's pulse or breathing show any marked change.

If you know what was eaten in any case of serious poisoning — especially if it was food that is normally safe to eat — put whatever remains of it in a closed container. It may be required for testing to identify the toxin and decide the best treatment.

Poisons Information Centres

Anyone who cannot find a doctor in a poisoning emergency, or who fears that such an emergency is about to arise because something is acidentally swallowed, can get advice at any time from a Poisons Information Centre. Each capital city has a designated hospital where experts are available round the clock to help in all cases of poisoning, including encounters with venomous animals and mishaps with drugs and other manufactured chemicals. Doctors themselves use the service frequently.

- ADELAIDE: (08) 267 4999.
- BRISBANE: (07) 253 8233.
- CANBERRA: (062) 43 2154.
- DARWIN: (089) 27 4777.
- HOBART: (002) 38 8485.
- MELBOURNE: (03) 345 5678.
- PERTH: (09) 381 1177.
- SYDNEY: (02) 51 0466.

Techniques to save a life

A person who has stopped breathing will die within minutes. So it is essential to get air into the lungs as quickly as possible by expired air resuscitation. If the casualty's heart stops beating, additional techniques will be needed.

Breathing may have stopped because the airway has become blocked. Your first step, therefore, is to clear and open the airway. With the casualty lying in the lateral recovery position (see bottom of opposite page), tilt the head backwards, push the chin up and clear out any obstruction in the mouth with your fingers.

If the casualty does not begin to breathe when the airway is opened, turn her on her back in preparation for expired air resuscitation. Instructions for applying this technique to an adult are given below; instructions for babies and small children are given at the right.

Occasionally it is not possible to use a casualty's mouth for expired air resuscitation — perhaps because of mouth injuries. In such a case breathe into the nose, holding the mouth shut with your thumb as you do so. Open the casualty's lips after each breath to allow the air to escape.

Infants and small children *Clear the airway and hold the child with his head horizontal. Cover his mouth and nose with your mouth and puff gently into his lungs, making his chest rise. Remove your mouth and turn your head to watch his chest fall. Give one shallow puff every three seconds. Repeat until the child is breathing regularly, then put him in the lateral recovery position.*

EXPIRED AIR RESUSCITATION

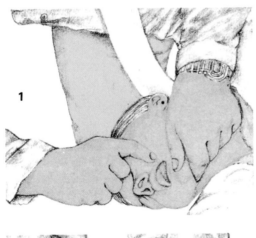

1 *With the casualty in the lateral recovery position, clear her mouth. Tilt her head back and support her jaw, keeping her face turned slightly downwards.*

2 *If not breathing, turn the casualty onto her back. Tilt her head back with one hand on her forehead and support her jaw in the 'pistol grip', with your fingers clear of her neck.*

3 *Place your lips around the casualty's open mouth, sealing her nose with your cheek. Blow until her chest rises, giving the first five breaths quickly.*

4 *Turn your head to watch the casualty's chest fall. Replace your mouth and blow again, this time at the normal breathing rate of 15 per minute.*

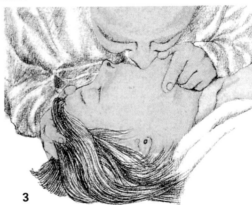

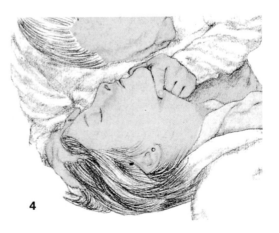

EXTERNAL CARDIAC COMPRESSION

If a casualty's heart stops beating, expired air resuscitation will be useless by itself. External cardiac compression squeezes the casualty's heart between the breastbone and the spine, acting as a hand pump which forces the blood around the circulatory system.

Compression should preferably be applied by a person trained in first aid — and only after he or she has established conclusively that the casualty's heartbeat has stopped. Do this by feeling the pulse of one of the carotid arteries in the casualty's neck. The pulse on the wrist is not a reliable enough indication. See the box 'How to

take a victim's pulse' on page 355.

Instructions for applying compression on an adult are given in this box. Treat children and babies as follows: Use one hand only in the middle of the breastbone to compress the chest of a child under ten. The depth of compression should be approximately 20-30 mm, and the speed a little greater than for an adult — 80-100 times a minute, depending on the child's age. Give cardiac compression to a baby using two fingers only. The depth of compression must be no more than 10-20 mm, and the speed 100 times a minute.

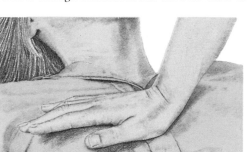

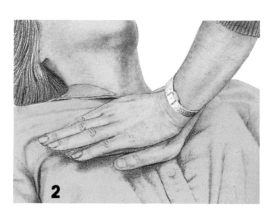

1 If, after five inflations of the lungs, you cannot detect a carotid pulse and if breathing is absent, locate the middle point of the casualty's breastbone. Place the heel of one hand on the casualty's chest, immediately below the middle point of the breastbone. Keep your thumb and fingers raised, so that they do not press on the ribs.

2 Keep the heel of your hand in place and put your other hand over it. Press down 40-50 mm with thumbs and fingers raised. Let the chest rise again without removing your hands.

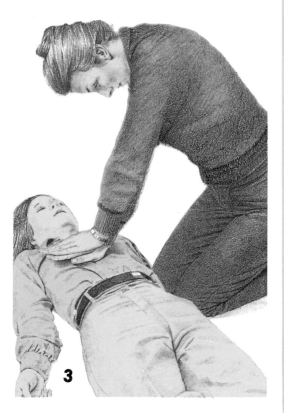

3 With the elbows straight and shoulders over the breastbone, give 15 compressions at normal pulse rate, then inflate the lungs twice by expired air (see opposite page). Repeat the sequence four times a minute. Check the carotid pulse (page 355) after one minute, then every two minutes.

THE LATERAL RECOVERY POSITION

All unconscious casualties who are breathing properly should be turned as quickly as possible on to their side and arranged in the lateral recovery, or unconscious, position. This position prevents the casualty's tongue from falling into the back of his throat and choking him, and allows fluid — such as blood or vomit — to drain from his mouth.

Handle any serious wounds gently while arranging the casualty. But remember that the unconscious position is a life-saver, and in most cases involving a breathing, unconscious victim it has priority over other treatment.

An unconscious person loses body heat very quickly, and in order to maintain a normal temperature he should be covered.

With the casualty on his back, place the far arm at right angles to the body and bend the near leg at the knee. Place the near arm across the chest. Then with one hand under his near shoulder and the other under his hip, turn him gently away from you onto his side. Place the upper arm across the lower arm at the elbow. Tilt his head back, supporting his jaw, and make sure that his face is turned slightly downwards.

Severe wounding

*People can survive massive injuries — even from shark and crocodile attacks —
if clear-headed companions stop blood loss.*

Witnesses of a shark or crocodile attack are horrified, outraged and sometimes panic-stricken. Often their senses reel at the sight of appalling injuries. None of these reactions can help the victim, who is bound to die unless someone has the presence of mind to stem the loss of blood at once.

First aid must start in the water, even as the victim is being brought to safety. Fingers should be pressed hard into or just above points where blood is spurting. Rescuers need have little fear that the attacking animal will turn its attention to them — such behaviour by sharks and crocodiles is virtually unknown.

Once ashore, the victim should be carried only far enough to be clear of the water — NOT rushed to a house or to hospital, even if transport is available. Summon medical help as soon as possible, but not at the expense of giving attention to the control of bleeding.

Gently lay the victim down with the head lowermost and wounded parts raised if possible. Pack any kind of cloth — towels are best but shirts or dresses will do — where blood is pumping and press hard on it. Do not remove soaked material but keep packing more dry cloth on top.

If heavy bleeding continues, seek the arterial pressure point nearest the wound and try

A pressure point (shown as a white dot above) is a spot where an artery can be felt passing close to the skin. It can be identified by its pulsating rhythm. If blood is spurting from a wound, press firmly on the artery nearest to the wound to staunch the flow.

STAUNCHING HEAVY BLEEDING

To reduce the flow of blood to the wound, lie the casualty on her back with feet raised and supported. Lift the injured part of the body and press hard on the wound with any kind of cloth, or your bare hands. This will compress the blood vessels and reduce the bleeding. If bleeding persists pack more dressing over the wound (without removing the soaked material), seek the arterial pressure point nearest the wound (see right) and press it to reduce the flow.

HOW TO MOVE AN INJURED PERSON

It is always best to leave an injured person where they are unless there is further danger. When you must move them, get someone else to help if possible. Take great care with the neck and spine of a casualty.

1 *Roll a blanket lengthways for half its width, then place the rolled edge alongside the victim. Turn the victim carefully towards you, then slide the blanket flush up against her back.*

2 *Turn the victim onto the blanket, unroll it behind her, then gently ease her onto her back.*

3 *Carefully drag the blanket, with the victim headfirst, keeping her back straight.*

pressing it firmly. Otherwise make a tourniquet with a constrictive bandage, a belt or a strip of rubber — but not string or shoelaces that cut into the flesh. Bind the tourniquet tightly over the nearest long bone above the wound — not over a joint.

The victim of an attack by a shark or crocodile may be part-drowned, and if the injuries are massive is very likely to go into shock. Be prepared to clear the airway if he or she should lose consciousness, and to use resuscitation techniques (page 356) if breathing or circulation fail. But avoid any unnecessary movement of the victim.

When bleeding is under control, cover the victim lightly. Do not attempt to remove or loosen clothing unless it is interfering with breathing. A tight-fitting garment — especially a wetsuit — may help retain damaged tissue and organs. If the victim is conscious keep him still. Do your utmost to be reassuring, but give him nothing to eat or drink.

If medical help cannot be obtained, or some further emergency forces the removal of the victim, it is best if possible to fashion a stretcher or to roll him gently onto a blanket and drag it Pressures created by trying to carry a victim bodily are likely to cause renewed bleeding.

All of this advice should be followed in the event of wounding by other animals if there is a heavy, spurting loss of blood. Staunching that bleeding and maintaining breathing take precedence over getting the victim to a doctor and any other consideration — including hygiene. In the case of lesser bites and cuts, where bacterial infection is the chief concern, priority should be given to cleaning the wound and keeping it clean until a doctor is reached.☐

Two kinds of shock

Shock can cause emotional as well as physical collapse.

Shock in medical terms is a state of collapse from failure of blood circulation. It can lead to unconsciousness and death. Shock is a likely result of massive injuries and can also be brought on by heart attacks, poisonings, dehydration and extreme allergic reactions. The early signs are dizziness and blurred vision with a pale, clammy skin, fast and shallow breathing and a feeble, racing pulse. Advice given on these pages for tending victims of severe woundings includes treatment for shock. The key point is keeping the head low, to promote the blood supply to the brain.

Anyone mauled by a shark or crocodile also suffers deep emotional shock. Terror from the speed and savagery of the attack is intensified if the victim has been held under water. It makes first aid more difficult and may increase the chance of a state of physical shock setting in. So do everything you can to calm the victim, assuring him that the danger has passed and the injury is not grave. Try not to let him see the extent of his wounds. And keep away onlookers who are unable to help — especially those who shout advice or exclaim at the severity of the attack. ☐

IMPROVISING A STRETCHER

Two people can move an injured person a long distance by making an improvised stretcher out of two or more coats with their sleeves turned inside out and a pair of poles, such as broomsticks or straight branches, pushed through the sleeves. Any flat, solid board such as a door (including a car door) can also be used in an emergency.

To get the person onto the stretcher, one of the stretcher bearers should roll the casualty gently onto her uninjured side. While he is doing that, the other bearer should put the open stretcher flat against the casualty's back. Then the stretcher, with the casualty on it, can be rolled back gently onto the ground.

If the casualty is unconscious, put the open stretcher against her front so that she is carried in the recovery position (page 357).

1 *Take two or three coats or jackets and turn the sleeves inside out. Pass a strong pole or branch through one of the sleeves of each jacket, then a second pole through the other sleeves. Zip or button up the jackets. If possible, get an uninjured person to lie on the stretcher first, and lift it to make sure that it can take the weight.*

2 *Roll the casualty onto her side while your helper pushes the open stretcher into place and lifts it on edge so that it is flat against the casualty's back. Then lower both stretcher and casualty to the ground before lifting the stretcher. If the casualty is unconscious, put her onto the stretcher face down.*

Snakebites

A simple technique buys ample time for victims to receive medical attention and the right antivenom — if it's needed.

Modern first aid methods, proved in the 1970s, give victims of even the most dangerous snakes an excellent chance of survival. The spread of venom is delayed — for many hours if need be, with little discomfort — until the effects of the bite can be medically countered. Deliberately cutting the flesh and tying painful tourniquets are things of the past.

Unless you are an expert in identifying species, any bite from a snake as thick as your finger should be regarded as potentially dangerous — especially if the victim is a child. If it turns out later that little or no venom was injected, the precaution of pressure bandaging and immobilisation, illustrated here, will have done no harm. Even if a bite goes unnoticed until symptoms of illness develop, the technique is still beneficial.

It works by shutting down tiny blood vessels and lymph channels near the skin surface. The main blood circulation is not impeded. The method is particularly successful in Australia because the venoms of our snakes do no significant damage at the site of the bite. And one venom component is thought to lose potency while it is trapped.

To receive the full benefit of the technique, the victim should stay as still as possible from the moment that he or she is bitten. Then even if there is a delay in finding bandaging material and fashioning a splint, little harm will be done.

Only a doctor should remove the binding

Though the dressing must be tight — like the bandage on a sprained ankle — make sure the patient is comfortable. Once applied, the binding must not come off or be loosened except under medical direction. Even at a hospital, doctors usually will not remove it until they have completed tests and the correct antivenom and life-supporting drugs are on hand.

Movement should be avoided even after the bandages and splint are on. A vehicle or a stretcher should be brought to the victim if possible. But if there is no alternative to carrying him, it must be done with the least disturbance to the bitten area. Only as a last resort should someone bitten in the foot or leg try to walk, even with assistance.

Nearly all snakebites are on the feet and hands or not far up the limbs, making the pressure/immobilisation technique easy if materials are available. It can also be used on the trunk provided that the binding does not

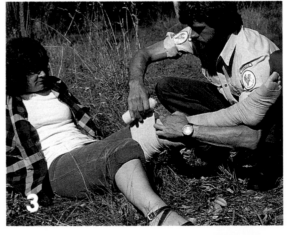

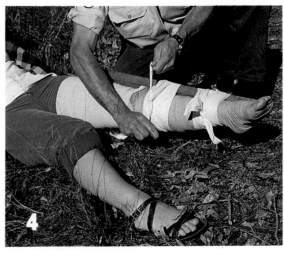

1 You will need a broad, firm, flexible bandage; crepe is ideal. If you do not have any bandages, tear a towel or some clothing into strips. Roll trouser leg up if necessary. Place one end of the bandage over the bite area, covering the fang marks.

2 Wind the bandage round and round the leg (or arm), making it as firm as you would if bandaging a sprained ankle. Do not try to move the victim or take off any of her clothing.

3 Take the bandage as far up the leg as possible — but leave some bandage over for attaching a splint. Reassure the victim and ask her to lie still to prevent the venom from spreading.

4 Find a flat piece of timber or some other rigid object. Place it beside the leg and, starting at the ankle, wrap bandage around both the leg and the splint. The splint should extend from the foot to above the knee.

Note: *If the bite is on the hand or lower arm, bind the arm up to the elbow with bandages. Attach a splint, then place the arm in a cloth sling to keep it still.*

5 *Keep on winding bandages around until the splint is firmly strapped to the length of leg, helping to hold it immobile. If bandages and splint have been put on correctly they should be reasonably comfortable. Do not take either off until antivenom is at hand.*

interfere with breathing. But it cannot be applied safely on the neck, or effectively in the region of the buttocks, groin and genital organs. To be bitten there, and for the bite to be one of the 15 per cent or so that are seriously venomous, would be extraordinarily unlucky.

If a bite victim is alone
Anyone bitten by an apparently dangerous snake while alone and unable to summon help should attempt pressure binding and immobilisation to whatever degree is practical. It will have a partial effect in slowing the spread of venom. In the case of a foot or leg bite, the victim will fare best if he or she can improvise a crutch that allows walking without pressure on the splinted leg.

People venturing into unfrequented areas of the outback without a radio transmitter run worse risks than snakebite. But if you are in a travelling party it is worth remembering that the correct first aid and complete stillness could keep a bite victim out of danger for six hours or more while someone else goes for help. Better still, make sure that all members of the party know and follow the rules for avoiding snakebite (pages 123-124).□

Other venomous bites and stings

Many creatures besides snakes deliver venom when they bite or sting us. First aid techniques depend on the creature responsible.

The pressure/immobilisation method of snakebite first aid, explained on these pages, works just as well for the bites of sea snakes. It also helps to delay the effects of bites by blue-ringed octopuses and stings by cone shells.

Most importantly, it is a life-saving protection from the deadly venom of funnelweb spiders. Anyone in eastern Australia who is bitten by a big black spider and is not sure of its identity should follow the pressure/immobilisation procedure as a precaution and seek medical attention — taking along the spider if possible.

A bite by the redback spider, however, is intolerably painful if the venom is trapped at the site. This venom is slow-acting and not as potent as a funnelweb's, so provided that medical treatment can be obtained within an hour or two there is no risk in leaving the bite unbound. It should simply be washed and lightly covered.

Local pain from spider bites and many insect stings can be relieved by gently cooling the area with a mixture of ice and water in a plastic bag. *But NEVER place ice directly on the skin — you can damage the tissue with an effect like frostbite.*

Specific first aid measures to deal with paralysis tick bites and the stings of ants, bees, wasps, jellyfishes and stonefishes are highlighted in the relevant main entries.

Venoms from the spines of other stinging fishes are mainly local in their effects. Immersion in hot — but not scalding — water usually relieves the pain. It stimulates local blood circulation and hastens the dispersal of venom, and it may also create a chemical change that inactivates some venom components. If anyone suffers extreme pain from a fish stinging, companions should prepare for the possibility of collapse from shock — a dangerous situation in the sea or on a small boat.□

LET MEDICAL AUTHORITIES KNOW YOU'RE ON THE WAY
If you have to drive a long distance to a hospital or a doctor's surgery with someone who has been dangerously bitten or stung, take the first chance you get to telephone ahead. An ambulance or a police escort may be sent to meet you. And precious time will be saved in getting ready the testing equipment and antivenoms likely to be needed.

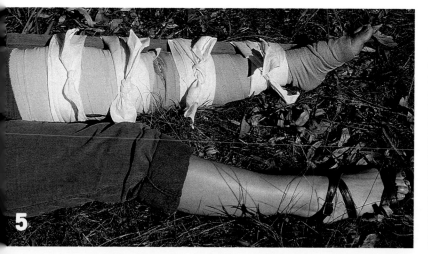

Index of scientific names

General index

Page numbers in red type indicate that first aid advice, or advice on how to avoid trouble, is given on those pages.

Acknowledgments

The publishers, editors and author of *Australia's Dangerous Creatures* are grateful for advice and assistance from the staff of many organisations. These include commonwealth and state government health and primary industry departments and wildlife conservation services, the Commonwealth Serum Laboratories, the Commonwealth Scientific and Industrial Research Organisation, the Commonwealth Institute of Health, the Commonwealth Department of Aviation, the Australian National Botanic Gardens, the Australian Museum, the Queensland Museum, the Australian Institute of Marine Science, the School of Public Health and Tropical Medicine, the Macquarie University school of chemistry, the University of Queensland zoology department, the James Cook University department of marine biology, the Australian National University Curtin school of medical research, the Queensland Institute of Medical Research, the Walter and Eliza Hall Institute of Medical Research, the NSW Police community relations department, the Royal North Shore Hospital, Sydney, the Royal Children's Hospital, Brisbane, the Surf Life Saving Association of Australia, and the Royal Society for the Prevention of Cruelty to Animals of NSW.

Reference sources Special credit is due to the publishers of and contributors to the *Medical Journal of Australia.* Invaluable information was also obtained from learned contributions to *Australian Natural History, Wildlife, Geo* and *Habitat* magazines. Much incidental material was drawn from daily newspapers. We are grateful to all of the journalists and publishers concerned, and to the library filing services of News Ltd and John Fairfax and Sons. Indebtedness is also acknowledged for information gained from the following books: *A Colour Atlas of Clinical Parasitology,* Tomio Yamaguchi, ed. (Wolfe Medical Publications); *A Comparative Study of Funnelweb Spider Venoms,* D.D. Sheumack, B.A. Baldo, P.R. Carroll, F. Hampson, A. Skorulis and M.E. Howden, in *Comparative Biochemistry and Physiology* Vol. 78C (Pergamon Press); *A Guide to Medical Entomology,* M.W. Service (Macmillan); *A Handbook for the Identification of Insects of Medical Importance,* John Smart, Karl Jordan and R.J. Whittick (The British Museum); *Animal Parasites: Their Life Cycles and Ecology,* O. Wilford Olsen (University Park Press); *Animals Parasitic in Man,* Geoffrey Lapage (Pelican); *Arthropods and Human Skin,* J. O'D. Alxander (Springer-Verlag); *Arthropods of Medical Importance,* J.H. Grundy (Noble Books); *A Treasury of Australian Wildlife,* D.F. McMichael (Taplinger); *Animal Toxins and Man,* John Pearn, ed. (Queensland Health Department); *Australian Animal Toxins,* Struan K. Sutherland (Oxford University Press); *Australian Encyclopaedia,* John Shaw, ed. (Collins); *Australian Harmful Arachnids and Their Allies,* R.V. Southcott (self-published); *Australian Insect Wonders,* Harry Frauca (Rigby); *Australian National Course on Mosquito Vector Control,* Richard C. Russell, ed. (Shell Public Health); *Australian Sea Fishes North of 30°S,* Neville Coleman (Doubleday); *Australian Sea Fishes South of 30°S,* Neville Coleman (Doubleday); *Australian Sharks,* G.P. Whitley (Lloyd O'Neil); *Australian Ticks,* F.H.S. Roberts (CSIRO); *Australia's Venomous Wildlife,* John Stackhouse (Hamlyn); *Canberra, History and Legends,* John Gale (Library of Australian History); *Dangerous Australians* (Bay Books); *Dangerous to Man,* Roger Caras (Penguin); *Dangerous Snakes of Australia and New Guinea,* Eric Worrell (Angus & Robertson); *Encounters with Australian Animals,* Harry Frauca (Heinemann); *Entomology in Human and Animal Health,* Robert F. Harwood and Maurice T. James (Macmillan); *Family Guide to Dangerous Animals and Plants of Australia,* Struan K. Sutherland (Rigby); *Fishes of the Great Barrier Reef,* Tom C. Marshall (Angus & Robertson); *Foundations of Parasitology,* Gerald D. Schmidt and Larry S. Roberts (Mosby); *Guide to Fishes,* E.M. Grant (Queensland Government); *Guide to the Australian Coast* (Reader's Digest); *Guinness Book of Animal Facts and Feats,* Gerald Wood (Guinness Superlatives); *Injuries to Man from Marine Invertebrates in the Australian Region,* Sir John Cleland and R.V. Southcott (Commonwealth Government of Australia); *Insects and Other Arthropods of Medical Importance,* Kenneth G.V. Smith, ed. (The British Museum); *Kingdom of the Octopus,* Frank W. Lane (Jarrold); *Library of Essential Knowledge* (Reader's Digest); *Life on Earth,* David Attenborough (Reader's Digest/Collins/BBC); *Living with Animals,* David Fleay (Sport Shelf); *Looking at Animals,* David Fleay (Boolarong Publications); *Man is the Prey,* James Frederick Clarke (Stein & Day); *Marine Animal Injuries to Man,* Carl Edmonds (Wedneil); *Mosquitoes of Public Health Importance and Their Control,* H.D. Pratt, R.F. Darsie and K.S. Littig (Centre for Disease Control, U.S. Public Health Service); *Mosquitos,* J.D. Gillett (Weidenfeld & Nicholson); *Motoring Guide to Australia* (Reader's Digest); *Parasites of Man in Nuigini,* W.H. Ewers and W.T. Jeffrey (Jacaranda); *Plague Sydney, 1900,* Max Kelly (Doak Press); *Poisonous Plants of Australia,* S.L. Everist (Angus & Robertson); *Recorded Shark Attacks in Australian Waters,* Jack Green (self-published); *Reptiles and Amphibians of Australia,* H.G. Cogger (Reed); *Sea Turtles,* Robert Bustard (Collins); *Shark Attack in South African Waters,* Tim Wallet (Struik); *Sharks — Silent Hunters of the Deep* (Reader's Digest); *Spiders,* Barbara York Main (Collins); *Stay Alive, A Handbook on Survival,* Maurice Dunlevy (AGPS); *Talking of Animals,* David Fleay (Jacaranda); *The Australian Encyclopaedia* (Grolier Society of Australia); *The Australian Fisherman,* Harold Vaughan (Lansdowne Press); *The Australian Museum Complete Book of Australian Mammals,* Ronald Strahan, ed. (Angus & Robertson); *The Insects of Australia,* CSIRO Division of Entomology (Melbourne University Press); *The Marine Stinger Book,* John Williamson, ed. (Surf Life Saving Association of Australia, Queensland); *The Mosquito,* L.K.H. Goma (Hutchinson); *The Paradoxical Platypus: Hobnobbing with Duckbills,* David Fleay (Jacaranda Wiley); *The Pharmacological Basis of Therapeutics,* Alfred G. Gilman, Louis S. Goodman, Theodore Rall and Seriz Murad (Collier Macmillan); *Trawled Fishes of Southern Indonesia and Northwestern Australia,* Thomas Goerfelt-Tarp and Patricia Kailola (Australian Development of Assistance Bureau/Directorate-General of Fisheries, Indonesia/German Agency for Technical Cooperation); *Venomous Australian Animals Dangerous to Man,* J. Ros Garnet, ed. (CSL); *Venomous Creatures of Australia,* Struan K. Sutherland (Oxford University Press); *Walter and Eliza Hall Institute, 1915-1965,* Frank Macfarlane Burnet (Melbourne University Press); *Wild Australia* (Reader's Digest); *Wild Food in Australia,* A.B. and J.W. Cribb (Collins); *Wild Life of Australia and New Guinea,* Charles Barrett (Heinemann); *Wild Medicine in Australia,* A.B. and J.W. Cribb (Collins).

Photographs Position of photographs on the page: t = top, c = centre, b = bottom, l = left, r = right. **Front cover** Gunther Schmida. **Back cover** tl, Keith Gillett; cl, David P. Maitland/Auscape; bl, Ron & Valerie Taylor; bc, Murray Fagg; br, Densey Clyne/Mantis Wildlife Films. Page **1** David P. Maitland. **2-3** David P. Maitland. **4-5** Jean-Paul Ferrero/Auscape. **8-9** M.P. Kahl/Auscape. **10** Jean-Paul Ferrero/Auscape. **11** t, Jean-Paul Ferrero/Auscape; bl, Graham Robertson/Auscape; br, Esther Beaton/Auscape. **12** t, Mrs Joan Schmidt; b, The Photo Library. **13** Graham Robertson/Auscape. **14-15** Frith Foto/A.N.T. Photo Library. **16** t, Hans & Judy Beste/Auscape; b, Jean-Paul Ferrero/Auscape. **17** t, Mitchell Library; b, Mortlock Library of South Australiana. **18** and **18-19** Jean-Paul Ferrero/Auscape. **19** bl, Hans & Judy Beste/Auscape; br, Esther Beaton/Auscape. **20** t and b, Jean-Paul Ferrero/Auscape. **21** t, Jean-Paul Ferrero/Auscape; b, Hans & Judy Beste/Auscape. **22** Don Stephens & Associates. **23** t, Graham Robertson/Auscape; c, K. Hateley/A.N.T. Photo Library; b, M. Gillam/Auscape. **24** Gary Steer/Auscape. **25** t, Hans & Judy Beste/Auscape; bl, Gunther Deichmann/Auscape; br, Jean-Paul Ferrero/Auscape. **26** t, Jiri Lochman/Auscape; b, Hans & Judy Beste/Auscape. **27** l, Hans & Judy Beste/Auscape; r, J.C. Wombey/Auscape. **28** t, Graham Robertson/Auscape; b, Frith Foto/A.N.T. Photo Library. **29** t, Denis & Theresa O'Byrne/A.N.T. Photo Library; b, Paul Mathews/John Fairfax & Sons Ltd. **30** Jean-Paul Ferrero/Auscape. **31** t, Tom & Pam Gardner/A.N.T. Photo Library; b, Bay Picture Library. **32** Bay Picture Library. **33** Greg Fyfe/A.N.T. Photo Library. **34** Denis & Theresa O'Byrne/A.N.T. Photo Library. **35** t, C.A. Henley/Auscape; b, I.R. McCann/A.N.T. Photo Library. **36** both courtesy of W.A. Flick & Co. Pty Ltd (Flick Pest Control). **37** Hans & Judy Beste/Auscape. **38** Jean-Paul Ferrero/Auscape. **39** t, Jean-Paul Ferrero/Auscape; b, Gary Steer/Auscape. **40** Jean-Paul Ferrero/Auscape. **40-41, 41** and **42** Gary Steer/Auscape. **42-43** Jean-Paul Ferrero/Auscape. **43** Colin Townsend/Sydney Freelance Photo Agency. **44** tl, C.A. Henley/Auscape; tr, Ken Griffiths/A.N.T. Photo Library; b, Hedley Matthews. **45** t, C.A. Henley/Auscape; c, David P. Maitland/Auscape; b, I.R. McCann/A.N.T. Photo Library. **46** t, Jean-Paul Ferrero/Auscape; b, Jack Green. **47** t, Paul Mathews/John Fairfax & Sons Pty Ltd; b, courtesy of Karen Hatchett. **48** t, Gordon McInnes/A.N.T. Photo Library; b, Hans & Judy Beste/Auscape. **49** t, Graham Robertson/Auscape; b, Jean-Paul Ferrero/Auscape. **50** t, J. & P. Olsen/Auscape; b, Jack & Lindsay Cupper/Auscape. **51** t, Jean-Paul Ferrero/Auscape; b, courtesy of the Commonwealth Department of Aviation. **52** t, Ralph & Daphne Keller/A.N.T. Photo Library; b, Gunther Deichmann/Auscape. **53** t, Gunther Deichmann/Auscape; c, Jean-Paul Ferrero/Auscape; b, E.A. Pratt/A.N.T. Photo Library. **54** t, Brian Chudleigh/A.N.T. Photo Library; b, M. Cermak/A.N.T. Photo Library. **55** l, Gunther Deichmann/Auscape; r, D.B. Carter/A.N.T. Photo Library. **56** t, J.M. La Roque/Auscape; b, Jean-Paul Ferrero/Auscape. **57** t, Darran Leal; c and b, Michael Jensen/Auscape. **58** t, Jack Green; b, Hans & Judy Beste/Auscape. **59** Northern Territory News. **60** t, Peter Krauss/A.N.T. Photo Library; b, Frith Foto/A.N.T. Photo Library. **61** tl and tr, Gunther Deichmann/Auscape; b, R. Jenkins/A.N.T. Photo Library. **62** tc and tr, Jean-Paul Ferrero/Auscape; c, Gunther Schmida/A.N.T. Photo Library; b, John Cann/A.N.T. Photo Library. **63** t, Sea Life Centre, Bicheno; c, R.H. Kuiter/A.N.T. Photo Library; b, Jean-Paul Ferrero/Auscape. **64** Bay Picture Library. **65** Jarrolds Publishers. **66** t, Mary Evans Picture Library; b, R.H. Kuiter/A.N.T. Photo Library. **67** t and c, Ron & Valerie Taylor/A.N.T. Photo Library; bl and br, R.H. Kuiter/A.N.T. Photo Library. **68** Ron & Valerie Taylor. **69** t, H.G. de Couet/Auscape; b, Ron & Valerie Taylor/A.N.T. Photo Library. **70** t, Kevin Deacon/Auscape; c, Ron & Valerie Taylor/A.N.T. Photo Library; b, Kevin Deacon/Auscape. **71** t, Gunther Schmida/A.N.T. Photo Library; tc, R.H. Kuiter/A.N.T. Photo Library; c, Kathie Atkinson; b, R.H. Kuiter/A.N.T. Photo Library. **72** t, Ron & Valerie Taylor/A.N.T. Photo Library. **72-73** Fenton Walsh/A.N.T. Photo Library. **73** t and br, R.H. Kuiter/A.N.T. Photo Library. **74** Ron & Valerie Taylor/A.N.T. Photo Library. **75** t and c, Captain Peter Bristow; b, Ron & Valerie Taylor/A.N.T. Photo Library. **76** t, R.H. Kuiter/A.N.T. Photo Library; c, Fenton Walsh/A.N.T. Photo Library. **77** t, M. Mallis/A.N.T. Photo Library; b, Hans & Judy Beste/Auscape. **78** and **78-79** Francois Gohier/Auscape. **79** br, Ken W. Fink/Ardea. **80** David Doubilet. **81** Bill Wood. **82** Ron & Valerie Taylor. **83** t, Ron & Valerie Taylor; b, Capricorn Press. **84** c, N.R. Kemp; b, Ron & Valerie Taylor. **85, 86, 88, 89** and **90-91** all Ron & Valerie Taylor. **91** inset, Ron & Valerie Taylor; br, Reader's Digest. **92** t, Ron & Valerie Taylor; bl and br, Australasian Marine Photographic Index. **93** t, Bill Wood; b, Ron & Valerie Taylor. **94** t, Bay Picture Library; b, Howard Hall. **95** t,

News Ltd; b, Ron & Valerie Taylor. **96** and **97** Ron & Valerie Taylor. **98-99** Jean-Paul Ferrero/Auscape. **100** tl and tr, John Fairfax & Sons Ltd; b, courtesy of the NSW Police Community Relations Bureau. **101** and **102** News Ltd. **103** Paul Riley/Topics. **104** b, The Photo Library. **104-105** Jean-Paul Ferrero/Auscape. **105** b, Yann Arthus Bertrand/Auscape. **106** t and br, Bendigo Advertiser. **106-107** Jean-Paul Ferrero/ Auscape. **108-109** Gunther Schmida. **112** all Brett Gregory. **115** Trevor Robbins. **116** t, J. Weigel/A.N.T. Photo Library; b, The Photo Library. **117** Bay Picture Library. **118** tl and cl, courtesy of the Walter & Eliza Hall Institute of Medical Research; tr, courtesy of Dr Saul Weiner; bl, courtesy of the Commonwealth Serum Laboratories. **119** J. Weigel/A.N.T. Photo Library. **120-121** John Cann/A.N.T. Photo Library. **122** t, A.Y. Pepper/National Photographic Index of Australian Wildlife; c and b, J.C. Wombey/Auscape. **123** tl, C. Pollitt/A.N.T. Photo Library; tr, L. Naylor/ National Photographic Index of Australian Wildlife; cl, H.G. Cogger; cr, J.C. Wombey/Auscape; both b, courtesy of J.H. Pearn, Royal Children's Hospital, Brisbane. **124** courtesy of J.H. Pearn, Royal Children's Hospital, Brisbane. **125** l, Esther Beaton/Auscape; r, P. Roach/National Photographic Index of Australian Wildlife. **126** H.G. Cogger. **128** Jean-Paul Ferrero/ Auscape. **129** t, Gunther Schmida; b, J.C. Wombey/ Auscape. **130** H.G. Cogger. **131** and **132-133** J.C. Wombey/Auscape. **133** Jean-Paul Ferrero/Auscape. **134** J.C. Wombey/Auscape. **135** tl and tr, B. Miller; b, Gunther Schmida. **136, 137, 138, 139** and **140** all Gunther Schmida. **141** Mike W. Gillam/Auscape. **142** Gunther Schmida. **143, 144, 145** and **146** all J.C. Wombey/Auscape. **147** t, Ross Bennett/Auscape; b, T.D. Schwaner. **148** J.C. Wombey/Auscape. **149** t, Hans & Judy Beste/Auscape; b, Glen Threlfo/Auscape. **150** Gunther Schmida. **151** t and b, J.C. Wombey/ Auscape. **152** S. Swanson/National Photographic Index of Australian Wildlife. **153** t, Mike W. Gillam/ Auscape; b, J.C. Wombey/Auscape. **154** tl, Gunther Schmida; tc, J.A. Sorley/National Photographic Index of Australian Wildlife; bc, Gunther Deichmann/ Auscape; bl, Jean-Paul Ferrero/Auscape; br, J.C. Wombey/Auscape. **154-155** Gunther Schmida. **155** tl, Hans & Judy Beste/Auscape; tr, Gunther Schmida; bl, H.G. Cogger; br, J.C. Wombey/Auscape. **156, 157** and **158** all H.G. Cogger. **159** tl and b, H.G. Cogger; tr, H.G. de Couet/Auscape. **160** t, Jean-Paul Ferrero/ Auscape; b, Andrew Dennis/A.N.T. Photo Library. **161** t, Densey Clyne/Mantis Wildlife Films; c, Andrew Dennis/A.N.T. Photo Library; b, G.A. Wood/A.N.T. Photo Library. **162** t, Cyril Webster/A.N.T. Photo Library; b, Densey Clyne/Mantis Wildlife Films. **163** Jean-Paul Ferrero/Auscape. **164** and **165** all Kathie Atkinson. **166** t, Kathie Atkinson; b, Densey Clyne/ Mantis Wildlife Films. **166-167** David P. Maitland/ Auscape. **167** t, David P. Maitland/Auscape; c, Densey Clyne/Mantis Wildlife Films. **169** t, Kathie Atkinson; b, M.R. Gray. **170** Adrienne Reid, courtesy of Donald Crombie. **171** t, Kathie Atkinson; b, courtesy of G.M. Wheatley. **172** both Densey Clyne/Mantis Wildlife Films. **173** c, Jean-Paul Ferrero/Auscape; b, Kathie Atkinson. **174** t, Esther Beaton/Auscape; c, Densey Clyne/Mantis Wildlife Films; bl and br, Australian Museum. **175** t, Otto Rogge/A.N.T. Photo Library; b, Densey Clyne/Mantis Wildlife Films. **176** Esther Beaton/Auscape. **177** C.A. Henley/Auscape. **178** t, C.A. Henley/Auscape; b, courtesy of Struan Sutherland. **179** both D.M. Starr. **180** tr, R.V. Southcott, bl, Densey Clyne/Mantis Wildlife Films. **181** t, J.C. Wombey/Auscape; bl, Jean-Paul Ferrero/ Auscape; cr, M.R. Gray; br, Barbara York Main. **182** tl and tr, Densey Clyne/Mantis Wildlife Films; bl, R.V. Southcott. **183** tl and cl, R.V. Southcott; tr, D.B. Hirst; cr, Jean-Paul Ferrero/Auscape; bl and br, Densey Clyne/Mantis Wildlife Films. **184** Densey Clyne/ Mantis Wildlife Films. **185** t and c, Densey Clyne/ Mantis Wildlife Films; b, R.V. Southcott. **186** t and br, R.V. Southcott; bl, courtesy of CSIRO Division of Entomology. **187** t and c, C.A. Henley/Auscape; bl and br, Hans & Judy Beste/Auscape. **188** c, Densey Clyne/Mantis Wildlife Films; b, Jean-Paul Ferrero/ Auscape. **189** Esther Beaton/Auscape. **190** C.A. Henley/Auscape. **191** t, Jean-Paul Ferrero/Auscape; c and b, C.A. Henley/Auscape. **192** Densey Clyne/ Mantis Wildlife Films. **193** tl, r and cl, Anne & Jacques Six/Auscape; b, Densey Clyne/Mantis Wildlife Films. **194** c, Esther Beaton/Auscape; b, courtesy of Department of Agriculture & Rural Affairs, Victoria. **194-195** Anne & Jacques Six/Auscape. **195** David P. Maitland/Auscape. **196** courtesy of N.S.W. Department of Agriculture. **197** David Underhill. **198** c, Kathie Atkinson; b, Otto Rogge/A.N.T. Photo Library. **199** t, Brian Chudleigh/A.N.T. Photo Library; b, courtesy of N.S.W. Department of Agriculture. **200** both Kathie Atkinson. **200-201** Otto Rogge. **202** inset, Otto Rogge/A.N.T. Photo Library; b, courtesy of the Keith Turnbull Research Institute. **202-203** Otto Rogge/A.N.T. Photo Library. **203** c, C.B. & D.W. Frith/ A.N.T. Photo Library; b, courtesy of N.S.W. Department of Agriculture. **204** Gunther Schmida/ A.N.T. Photo Library; b, Jim Frazier/Mantis Wildlife Films; inset, Kathie Atkinson/A.N.T. Photo Library. **205** all Densey Clyne/Mantis Wildlife Films. **206** Hans & Judy Beste/Auscape. **207** and **208** Jean-Paul Ferrero/Auscape. **209** Australian Picture Library. **210** t, Kathie Atkinson; inset, Mark Wellard/A.N.T. Photo Library. **211** Kathie Atkinson. **212-213** Australasian Marine Photographic Index. **214** Ron & Valerie Taylor/A.N.T. Photo Library. **215** both H.G. de Couet/ Auscape. **216** tl, Ron & Valerie Taylor/A.N.T. Photo Library; tr, Australasian Marine Photographic Index; cl, Pat Manly; cr, R.H. Kuiter/A.N.T. Photo Library. **217** Pat Manly. **218** t, Ron & Valerie Taylor/A.N.T. Photo Library; c, Bill Wood; b, R.H. Kuiter/A.N.T. Photo Library. **219** l, R.H. Kuiter/A.N.T. Photo Library; r, R.V. Southcott. **220** all Kathie Atkinson. **221** H.G. de Couet/Auscape. **222** l, Ron & Valerie Taylor; r, Ben Cropp/Auscape. **223** t, R.H. Kuiter/A.N.T. Photo Library; c and b, Ron & Valerie Taylor. **224** t, R.H. Kuiter/A.N.T. Photo Library; b, H.G. de Couet/ Auscape. **224-225** Kathie Atkinson/A.N.T. Photo Library. **225** tl, R.H. Kuiter/A.N.T. Photo Library; c, A.A. Bicskos/Auscape; bl, Ron & Valerie Taylor; br, Australasian Marine Photographic Index. **226** t, Bill Wood; c and bl, Ron & Valerie Taylor/A.N.T. Photo Library; br, both Ron & Valerie Taylor. **226-227** Gunther Schmida/A.N.T. Photo Library. **227** t, Ron & Valerie Taylor/A.N.T. Photo Library; c, Esther Beaton/ Auscape; b, all three R.H. Kuiter/A.N.T. Photo Library. **228** t, Pat Manly; bl, Gunther Schmida/A.N.T. Photo Library; br, Ron & Valerie Taylor/A.N.T. Photo Library. **229** tl, Australasian Marine Photographic Index; tr, cr and br, Pat Manly; c and bl, R.H. Kuiter/A.N.T. Photo Library; cl, Ron & Valerie Taylor/A.N.T. Photo Library. **230** c, Keith Gillett; b, Robert Hartwick/Ben Cropp Productions. **231** t, Keith Gillett; b, Kathie Atkinson. **232** Ben Cropp. **233** t, Keith Gillett; b, Australian Picture Library. **234** t, Z. Florion/Ben Cropp Productions; c and b, Robert Hartwick. **235** four top, Townsville General Hospital/Ben Cropp Productions; b, Lynn Cropp. **236** tc and tr, R.V. Southcott; c, courtesy of CSIRO Division of Forest Research. **237** tl and tr, Peter Fenner; b, Ben Cropp. **238** Robert Hartwick. **239** tl, Kathie Atkinson; tr and bl, Australasian Marine Photographic Index; br, Ben Cropp. **240** Kathie Atkinson. **241** t, Keith Gillett; c, Australasian Marine Photographic Index; b, Isobel Bennett. **242-243** Kathie Atkinson. **244-245** Glen Millott. **246** Bill Wood. **247** t, Keith Gillett; b, Glen Millott. **248, 249** and **250** all Bill Wood. **251** all Australasian Marine Photographic Index. **253** courtesy of Barbara McGrath, State Pollution Control Commission of NSW. **254** and **255** all Kevin Williams. **256** The Photo Library. **257** Hans & Judy Beste/Auscape. **258** Francois Gohier/Auscape. **259** Bill Wood. **260** t, Ben Cropp/Auscape; bl, Esther Beaton/ Auscape; br, The Photo Library. **261** Jean-Paul Ferrero/Auscape. **262** t, Mawson Institute for Antarctic Research; b, Colin Monteath. **266** The Photo Library. **267** both courtesy of Jeff Brown, State Pollution Control Commission. **268** Peter Hunt. **270** t, The Photo Library; b, courtesy of the Australian Trade Commission. **271** both courtesy of NSW Department of Agriculture. **272-273** Mitchell Library, Sydney. **276** Graeme Chapman. **277** David Moore. **278** t, Kathie Atkinson; b, Ralph & Daphne Keller/A.N.T. Photo Library. **279** all courtesy of Tessa Guilfoyle, School of Public Health & Tropical Medicine. **281** t, courtesy of B.H. Kay, Queensland Institute of Medical Research; c, both courtesy of Peter Whelan, Northern Territory Department of Health; b, courtesy of Austra Violet Pty Ltd. **282** t and bl, courtesy of Peter Whelan, Northern Territory Department of Health; br, courtesy of Ian Fanning, Queensland Institute of Medical Research. **286** courtesy of the School of Public Health & Tropical Medicine. **287** David P. Maitland/Auscape. **289** D.E. Moorhouse, University of Queensland. **291** Belinda Penglis, University of Queensland. **293** t, I.R. McCann/A.N.T. Photo Library; b, B.F. Stone. **295** t, S. Fiske; b, B.F. Stone. **296** The Photo Library. **297** E. Mainke/A.N.T. Photo Library. **300** tl, Jean-Paul Ferrero/Auscape; tr and cr, Kathie Atkinson. **301** both courtesy of the School of Public Health & Tropical Medicine. **302** Stock Photos. **303** t, courtesy of the School of Public Health & Tropical Medicine; b, courtesy of Reg Hannay, Department of Agriculture & Rural Affairs, Victoria. **304** t, courtesy of Chris McCaughan, Department of Agriculture & Rural Affairs, Victoria; b, Gwynne Roberts/The Sunday Times. **305** t, courtesy of Chris McCaughan, Department of Agriculture & Rural Affairs, Victoria; b, Denis Waugh/The Sunday Times. **306** t, Philip Hauser; b, courtesy of NSW Department of Agriculture. **307** all courtesy of Russell Stewart, Animal Research Institute, Queensland. **309** Liverpool School of Tropical Medicine. **310** courtesy of the School of Public Health & Tropical Medicine. **312** Robin Morrison/Reader's Digest. **313** John Fairfax & Sons Ltd. **314** Jean-Paul Ferrero/Auscape. **316** The Photo Library. **317** t, The Photo Library; b, Peter Solness. **318** t, Graeme Chapman; b, Southern Media Services. **319** Australian Picture Library. **321** t, Peter Solness, b, Graham Robertson/Auscape. **322** courtesy of NSW Department of Agriculture. **323** R. Herd/National Photographic Index of Australian Wildlife. **324** G. Anderson/National Photographic Index of Australian Wildlife. **325** The Photo Library. **326** t, Jean-Paul Ferrero/Auscape; b, The Photo Library. **328** courtesy of Rentokil Pty Ltd. **329** courtesy of Paul B. Hughes. **330** both courtesy of CSIRO Division of Entomology. **331** t, Eric Hansen; b, courtesy of CSIRO Division of Entomology. **332** t and br, Otto Rogge/A.N.T. Photo Library; bl, David P. Maitland/Auscape. **333** t, courtesy of Paul B. Hughes; b, Kathie Atkinson. **334-335** Jean-Paul Ferrero/Auscape. **336** t, R.V. Southcott, cr, Michael Viard/Auscape. **337** t, The Photo Library; b, courtesy of A.R.H. Martin. **338** inset, Esther Beaton/Auscape; Michel Viard/Auscape. **339** t, Michel Viard/Auscape; c, Colin Totterdell; b, J. Aberdeen. **340** t and cr, Stirling Macoboy; cl, A.C. Henley/Auscape; ccl, Jean-Paul Ferrero/Auscape; ccr, Michel Viard/Auscape; bl and bc, R.V. Southcott. **340-341** R.V. Southcott. **341** tl, cc, cr and br, Stirling Macoboy; tr, cl and bc, Michel Viard/Auscape; bl, R.V. Southcott. **342** tl and b, R.V. Southcott; cl, Stirling Macoboy. **342-343** R.V. Southcott. **343** tl, tr and cl, R.V. Southcott; cr and bl, Stirling Macoboy; br, Michel Viard/Auscape. **344** t and b, Stirling Macoboy; c, Mirror Newspapers, Queensland. **345** Stirling Macoboy. **346** t and c, Stirling Macoboy; b, Michel Viard/Auscape. **347** tc and cr, Stirling Macoboy; tr, t lower r, cl, c lower l and bl, R.V. Southcott; bc and br, Keith Williams. **348** tl, C.A. Henley/Auscape; tc and tr, Michel Viard/Auscape; ct, Stirling Macoboy; cb, bl and br, R.V. Southcott. **349** tl, c and bl inset, R.V. Southcott; tc and tr, Michel Viard/Auscape; bl and br, Stirling Macoboy. **350** t, bl and br, Stirling Macoboy; c, Reader's Digest. **351** tl and tr, Keith Williams; cl, Tony Rodd; cr, Densey Clyne/Mantis Wildlife Films; b, C.A. Henley/Auscape. **352** tl, R.V. Southcott; bl, Keith Williams; br, Murray Fagg. **353** both C.A. Henley/Auscape. **354-355** Barry Davies. **358** and **359** all Esther Beaton/Auscape. **360** and **361** all Jean-Paul Ferrero/Auscape.

Front cover: copperhead *Austrelaps superbus*

Back cover, clockwise from bottom right: Adelaide trapdoor spider *Blakistonia aurea;* gympie bush *Dendrocnide moroides;* white pointer shark *Carcharodon carcharias;* bulldog ant *Myrmecia gulosa;* chironex box jellyfish *Chironex fleckeri.*

Endpapers: European wasp *Vespula germanica* (Otto Rogge).

Typesetting by Keyset Phototype, Sydney
Reproduction by Bright Arts, Hong Kong
Printed and bound in 1987 by
Dai Nippon Printing Company (HK) Limited,
Hong Kong